CHEMISTRY OF NEW MATERIALS

CHEMISTRY OF NEW MATERIALS

Editor

Ingomar Schwenke

scitus
academics

Chemistry of New Materials
Edited by **Ingomar Schwenke**

This Book is Printed on Acid-Free Paper. ♾

ISBN: 9781681178882
Library of Congress Control Number: 2017948915

SCITUS Academics LLC,
616, Corporate Way, Suite 2, 4766,
Valley Cottage, NY 10989

www.scitusacademics.com

Printed in United States of America.

Preface

The fundamental theory behind materials science includes concerning the microstructure of a material to its macromolecular physical and chemical properties. The development of new materials to address pressing future needs is one of the biggest challenges of the 21st Century. These new materials will play a key role in shaping the future. Among other things, they will have to provide pathways for sustainable resource management and energy supply, mobility, the future viability of the consumer society and new diagnostic and therapeutic procedures in the healthcare sector. A deeper understanding of materials and their chemical make-up, architecture, functionalization, processing and potential applications creates the foundation which the manufacturing and process industry needs to remain competitive. Chemistry plays a predominant role in materials science because chemistry provides information about the structure and composition of materials, as well as the processes to synthesize and use them. By understanding and then altering the microstructure, material scientists adapt the properties to create custom, or even brand new, materials with specific properties for specific uses. It describes how individual substances interact and studies their stability and reactivity. To optimize material properties, researchers need an in-depth understanding of material structure and composition including how additives work, along with much more. Experience in chemistry is also needed in order to understand how to optimize the functions and quality of materials production, processing and applications, and it makes an important contribution to materials science all along the value-added chain.

'Chemistry of New Materials' outlines the contribution made by chemistry to materials science in meeting a variety of needs. It summarizes the development potential and research needs over the next ten years. It focuses on the use of chemistry for the design and synthesis of materials with interesting or potentially useful physical characteristics, such as magnetic, optical, structural or catalytic properties. It also involves the characterization, processing and molecular-level understanding of these substances. The role of chemistry goes beyond the synthesis of new materials, materials development and optimization for specific applications. It also makes a major contribution to material re-use and the substitution of critical materials.

Table of Contents

CHAPTER 1

Synthesis, Characterization and Hexavalent Chromium Adsorption Characteristics of Aluminum- and Sucrose-Incorporated Tobermorite

Zhiguang Zhao [1,2], Jiangxiong Wei [1,2,*], Fangxian Li [1,2,*], Xiaoling Qu [3], Liang Shi [1,2], Haidong Zhang [1,2] and Qijun Yu [1,2]

[1]*School of Materials Science and Engineering, South China University of Technology, Guangzhou 510640, China*

[2]*Guangdong Low Carbon Technologies Engineering Centre for Building Materials, South China University of Technology, Guangzhou 510640, China*

[3]*School of Civil Engineering, Shaoguan University, Shaoguan 512005, China*

ABSTRACT

Tobermorites were synthesized from the lime-quartz slurries with incorporations of aluminum and sucrose under hydrothermal conditions, and then used for adsorption of Cr(VI). The chemical components, and structural and morphological properties of tobermorite were characterized by X-ray diffraction (XRD), thermogravimetric-differential scanning calorimetry (TG-DSC), Fourier transform infrared spectroscopy (FT-IR), nuclear magnetic resonance (NMR), scanning electron microscopy (SEM), X-ray photoelectron spectroscopic (XPS) and N_2 adsorption–desorption measurements. The formation and crystallinity of tobermorite could be largely enhanced by adding 2.3 wt.% aluminum hydroxide or 13.3 wt.% sucrose. Sucrose also played a significantly positive role in increasing the surface area. The adsorption performances for Cr(VI) were tested using a batch method taking into account the effects of pH, the adsorption kinetics, and the adsorption isotherms. The adsorption capacities of the aluminum- and sucrose-incorporated tobermorites reached up to 31.65 mg/g and 28.92 mg/g, respectively. Thus, the synthesized tobermorites showed good adsorption properties for removal of Cr(VI), making this material a promising candidate for efficient bulk wastewater treatment.

Keywords: tobermorite; aluminum; sucrose; adsorption; hexavalent chromium

1. INTRODUCTION

Hydrothermal products of CaO-SiO_2-H_2O system are highly complex and comprised of crystalline to amorphous phases, collectively denoted as C-S-H phases [1]. Among the C-S-H phases, tobermorite ($Ca_5Si_6O_{16}(OH)_2 \cdot 4H_2O$) is technologically important, not only as the main binder in autoclaved building materials, but also due to its essential roles as adsorbents and ion exchangers relating to multifarious application in catalysis and waste management [2,3,4]. Tobermorite is formed below 180 °C undergoing an phase transition path: $CaO + SiO_2 \rightarrow$ C-S-H gel$\rightarrow$ tobermorite, it always appears within the first 60 min, and the growth is completed within 7 h [5,6]. Three types of tobermorite exist, namely 14 Å, 11 Å and 9 Å tobermorite, with their names relating to the d-spacing of (002) in X-ray diffraction patterns depending on the numbers of water molecule per formula unit. The 11 Å tobermorite has layered structure with a central layer of calcium octahedral accompanied by two infinite silicate chains on each side [5]. The composite layers, one calcium and two silicate layers, are bound together by an interlayer containing Ca^{2+}, H^+, Na^+, and H_2O [5,7].

Recently, the research interest in aluminum-incorporated tobermorite becomes hot again due to the utilization of more aluminum bearing raw materials (e.g., zeolite [4], newsprint recycling residue [8] and fly ash [9]) and optimization of its ion exchange performance [3]. It is reported that Al substitutes for Si in the Q^2 or Q^3 sites (symbol Q represents one SiO_4tetrahedron and the superscript indicates the number of other Q units to which it is bonded) [5]. Black et al. [10] deduced that charge imbalance existed in the tobermorite basing on the observation of no change in Ca/(Si + Al) ratio upon Al substitution. Therefore, charge balance has to achieve via a coupled substitution: $\square + OH^- \rightarrow Ca^{2+} + O^{2-}$, whereby □ represents the so-called "empty" position. The most common form of "□" is Na^+, in addition, other metal ions such as Cd^{2+}, Pb^{2+},Co^{2+}, Ni^{2+} and Cs^+ can also act as charge compensator. This characteristic gives tobermorite superior adsorption performance.

Many studies have been devoted to the exchange properties of synthetic tobermorite. Coleman et al. [7,8,11] reported that Al-substituted 11 Å tobermorite exhibited cation exchange properties and was effective in the removal of Cd^{2+}, Pb^{2+}, and Zn^{2+} from acidified aqueous wastes. An 11 Å tobermorite substituted with Na^+ and Al^{3+} allows ion exchange with alkalis (Cs^+, Rb^+, and K^+) and alkaline earth elements (Ba^{2+}, Sr^{2+}, and Mg^{2+}) [12,13]. To the best of our knowledge, few studies have been conducted to study the adsorption of Cr(VI) by tobermorite or Al-substituted tobermorite. It is well known that chromium contamination is a widespread environmental problem arising from chromium residues. In nature, chromium exists mainly in two oxidation states (III and VI). Cr(VI) is a more toxic and soluble specie compared to Cr(III). Among the removal technologies for Cr(VI), adsorption has attracted the most interest because of its cost-effectiveness and simplicity. Although zeolites [14], clay [15,16], fly ash [17], activated carbon [18], magnetite nanoparticles [19], carbon nanotube (CNT) [20], and chitosan [21] are frequently used as general adsorbents of inorganic and organic compounds, alternative adsorbents are needed to be developed to improve the effectiveness. Based on the special structure and excellent performance, tobermorite should have excellent potential in the safe disposal of Cr(VI).

As expected, increases in tobermorite yield and its crystallinity have a positive impact upon the adsorption property of tobermorite [7,8]. It is well known that process kinetics is mainly ruled by the temperature dependent retrograde solubility of portlandite under hydrothermal conditions. Due to a significant increase of the solubility of CaO by sucrose at room temperature, it is possible to effectively increase the concentration of Ca^{2+} ions to generate a higher amount of C-S-H with the addition of sucrose. This study gives attention to modifications of the crystallization process in the CaO-SiO_2-H_2O system by the addition of sucrose. Our aim is to enhance the yield and crystallinity of tobermorite by introducing admixtures and hence to improve the adsorption property of tobermorite. Structures of tobermorite were identified by using XRD, TG-DSC, FT-IR, NMR, SEM, XPS techniques and N_2 adsorption–desorption measurements in detail. The effects of dosage of aluminum and sucrose on the material structures were investigated to optimize synthesis conditions. Furthermore, the synthesized products were applied for removal of Cr(VI) ions from water. The influences of various important parameters, viz., pH, contact time and initial concentration on the removal were investigated. In addition, kinetics and isotherms were examined concerning the adsorption of Cr(VI).

2. EXPERIMENTAL

2.1. Hydrothermal Synthesis

The tobermorite samples were synthesized from slurry of calcium oxide (CaO) and ground quartz powder. Fresh calcium oxide was obtained by calcining calcium carbonate ($CaCO_3$, 99.0%, Fuchen Co., Ltd., Tianjin, China) at 1000 °C for 3 h, immediately after which it was stored under vacuum. The quartz powder with average particle size of around 5 μm (SiO_2 > 99.8%) was dried in an oven at 105 °C for 24 h to remove any water. Aluminum hydroxide and sucrose of analytical purity (Fuchen Co., Ltd., Tianjin, China) were added as admixtures. Deionized water obtained by the standard purification method was used as mix water.

Hydrothermal syntheses were carried out in a Teflon-lined steel autoclave equipped with a power driven stirrer and a heat controller. The molar ratios of initial mixtures were according to the ideal composition of tobermorite ($Ca_5Si_6O_{16}(OH)_2{\cdot}4H_2O$). The molar ratio of Ca/Si for the control sample was designed at 0.83. For the sample with Al, the molar ratio of Ca/(Si + Al) was fixed at 0.83, and the replacement molar ratio of Si by Al was calculated as molar ratio of Al/(Al + Si) which varied from 0.03, 0.06 to 0.09. The addition of sucrose was calculated as molar ratio of sucrose/CaO, marked as Su/Ca, which was selected as 0.05, 0.1 and 0.15. Dry primary mixtures were stirred with deionized water (water/solid ratio of the slurry was 50) for 30 min in advance to guarantee the starting materials uniformly distributed. Then, the slurry was hydrothermally reacted at 180 °C for 8 h, and the synthesis reaction was conducted at a stirring rate of 100 r/min. The resultant products were collected, washed five times with deionized water and dried at 80 °C overnight.

2.2. Characterization Methods

The phase composition of the obtained products were characterized by XRD (X′Pert Pro, PANalytical B.V., Almelo, The Netherlands) analysis with Cu Kα radiation (λ = 1.5418 Å) at a 2θ range of 5°–70° and TG-DSC (STA 449C, Netzsch, Bavaria, Germany, 10 °C/min in N_2 atmosphere) analysis. The chemical functional groups were investigated by FT-IR (VERTEX 70, Bruker, Karlsruhe, Germany) within the wavenumber range of 4000–400 cm^{-1} using KBr as the standard. The morphology of the products was observed by SEM (Nova NanoSEM 430, FEI, Hillsboro, OR, USA) equipped with an energy dispersive X-ray analyzer (EDX). Solid-state NMR was performed to characterize the local atomic structure. NMR spectra for

^{29}Si and ^{27}Al were acquired at 9.4 T using a Bruker Avance 400 spectrometer (Bruker, Karlsruhe, Germany). Spectra were recorded at 79.49 MHz and a spinning rate of 5 kHz with 5 s relaxation delay for ^{29}Si NMR experiments and at 104.26 MHz and a spinning rate of 12 kHz with 1 s relaxation delay for ^{27}Al NMR experiments. Brunauer–Emmett–Teller (BET) specific surface areas and pore size distributions were measured with a surface area and pore size analyzer (Nova Win2, Quantachrome, Boynton Beach, FL, USA). XPS analysis was exerted to identify the surface chemical composition. The XPS spectra (Kratos, Manchester, UK) were recorded on a Kratos AXIS Ultra with a monochromatic Al X-ray source at 75 W and in the constant pass energy mode (160 eV for the survey scan, 40 eV for the high-resolution analysis).

2.3. Adsorption Experiments

Adsorption experiments were conducted using a batch method in stoppered conical flasks at room temperature. All batch experiments were conducted in a mechanical shaker at a speed of 150 rpm. The Cr(VI) solutions were prepared by dissolving $K_2Cr_2O_7$ in distilled water. The initial pH values of the Cr(VI) solutions were adjusted with 0.1 M HCl or NaOH to investigate the effects of pH value (pH = 2–6) on the adsorption ability of the synthetic samples. Solutions with pH value > 6 were not used, to avoid forming the metal hydroxide precipitates. Accurate amounts of the samples (T, T-A and T-S) (0.2 g) were mixed with the Cr(VI) solutions (50 mL, 160 mg/L) at different pH values in 150 mL conical flasks and shaken for 6 h. Then the dispersions were centrifuged to separate solution from solid followed by determination of Cr(VI) concentration in solution. The concentration of Cr(VI) was determined by using a atomic absorption spectrometer (Z-2000, Hitachi, Tokyo, Japan).

Adsorption isotherms were obtained as follows: 0.2 g of samples were in contact with 50 mL of Cr(VI)solutions with varying initial concentrations (20–240 mg/L)at a pH of 4.0 for 6 h, then centrifuged to separate solution from solid followed by determination of Cr(VI) concentration in solution, as described earlier.

The kinetic studies were performed as follows: 0.2 g of samples were equilibrated using an initial concentration of 160 mg/L Cr(VI)at a pH of 4.0. The samples were equilibrated for different time intervals of 5, 15, 30, 60, 90, 120, 180, or 360 min, then the phases were separated, and the Cr(VI) concentration in solution was determined.

3. RESULTS AND DISCUSSION

3.1. Synthesis of Tobermorite

XRD patterns of the synthetic samples are presented in Figure 1. The main diffraction peaks of the control sample are in good agreement with the JCPDS card (19-1364) for 11 Å tobermorite. Minor traces of calcite ($CaCO_3$) with diffraction peak at $2\theta = 29.3°$ are also present and commonly arise from atmospheric carbonation during the preparation of tobermorite. For the samples with aluminum, higher overall intensities at Al/(Al + Si) = 0.03 suggests an increased reaction rate to tobermorite. In particular, highly narrow and sharp peaks imply the high crystallinity of the tobermorite phase. The basal spacing of the (002) diffraction peak from the tobermorite phase is found to be 11.5 Å and is consistent with the previously reported values of Al-substituted 11 Å tobermorite [4]. As the aluminum addition increases, the weaker overall peak intensity of tobermorite alongside additional quartz (JCPDS card No. 70-3755) and hydro garnet (JCPDS card No. 84-1354) peaks indicate a lower reaction degree. Ease of tobermorite crystallization is known to be affected by the precursor C-S-H [2,22]. Aluminum may enhance the formation of precursor C-S-H due to lower binding energy of Al-O than that of Si-O [10] thus prompts the transformation of C-S-H to tobermorite. In the presence of excess aluminum, hydro garnet is significantly formed which retards the formation of tobermorite from the precursor C-S-H due to a lack of available Ca^{2+} in sufficient quantity [23].

For the samples with sucrose, the higher overall peak intensity of tobermorite at Su/Ca = 0.05 indicates an increased reaction rate to tobermorite. At a sucrose concentration of Su/Ca = 0.1 the weaker overall peak intensity of tobermorite alongside additional quartz peaks demonstrate a lower reaction degree. As the sucrose concentration further increases, the diffraction peaks corresponding to tobermorite become broad and ambiguous at Su/Ca = 0.15 suggesting a low crystallinity. This phenomenon may be explained that the concentration of Ca^{2+} is increased by sucrose [24]. As Al-Wakeel [25] pointed out, the concentration of Ca^{2+} has a great impact on the nucleation and growth of C-S-H. At a high level of sucrose concentration, the availability of Ca^{2+} is probably reduced due to excessively complexed calcium by sucrose [26], which retards the hydrothermal reaction.

According to the above, the purer synthesized tobermorites were prepared with aluminum addition of Al/(Al + Si) = 0.03 or at a sucrose concentration of Su/Ca = 0.05. Therefore, the above two samples of tobermorites, denoted by T-A and T-S, respectively, were preferably synthesized to conduct further studies. In addition, the control sample is marked as T to clearly explain the effects of aluminum and sucrose.

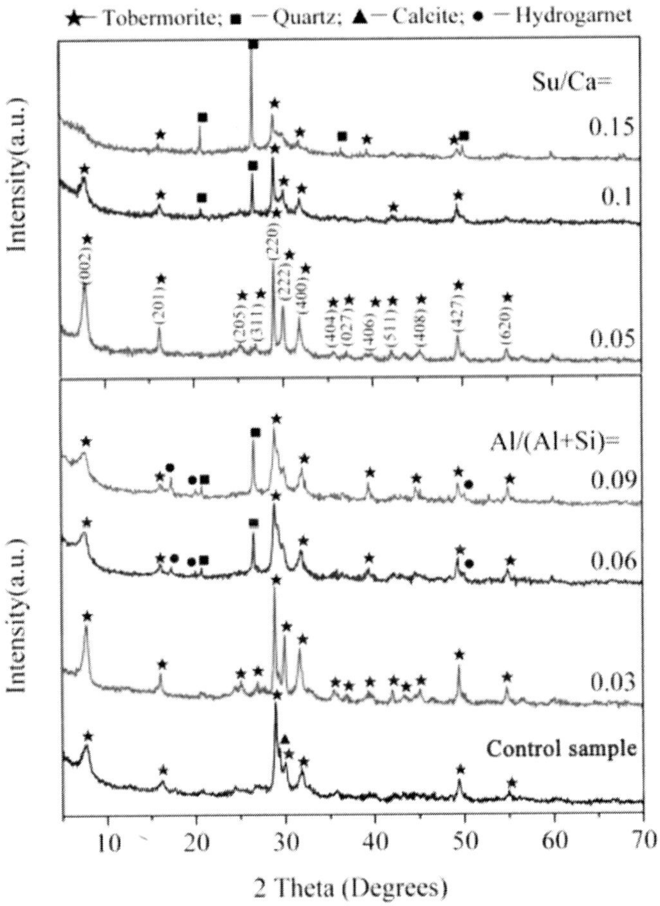

Figure 1. XRD patterns of the synthesized products.

3.2. Effects of Aluminum and Sucrose on Tobermorite

3.2.1. TG-DSC Analysis

TG-DSC curves of the tobermorite samples are shown in Figure 2. Irrespective of the water content a mild endothermic peak occurs between 40 °C and 200 °C which corresponds with the removal of ~10 wt.% of adsorbed water and labile inter-layer water molecules of tobermorite. An exothermic peak around 840 °C is ascribed to recrystallization of tobermorite into wollastonite (β-$CaSiO_3$) [25,27,28]. The weight loss of samples at each stage and the total weight loss are summarized in Table 1. Three stages of weight loss are interpreted as:

a) 40 to 100 °C, weight loss of free water contained in tobermorite;

b) 100 to 400 °C, weight loss of dehydration from the transformation from 11 Å tobermorite to 9 Å tobermorite [28]; and

c) 650 to 720 °C, weight loss of decomposition of calcite [27].

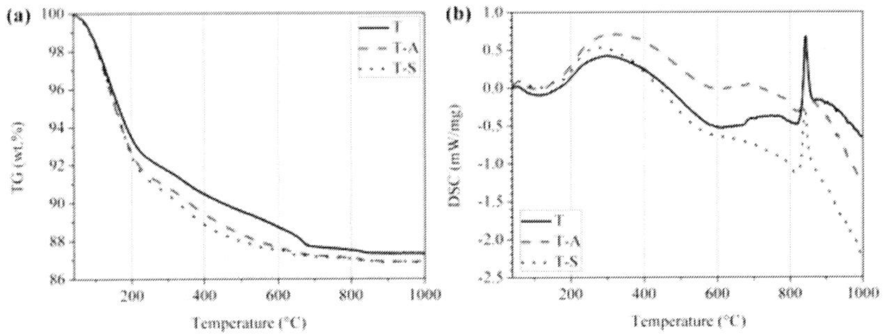

Figure 2. Thermogravimetry (**a**); and differential scanning calorimetry (**b**) curves of the tobermorite samples.

Table 1. Weight change proportion according to Figure 2.

Temperature Range (°C)	Weight Change (%)		
	T	T-A	T-S
40–100	1.48	1.64	1.66
100–400	8.06	8.95	9.48
650–720	0.54	0.13	0.10
Total	12.68	13.12	13.17

It can be observed that the total weight loss of the investigated samples is in the range of 12.7–13.2% and is in the typical range for tobermorite material. Shaw et al. [5] determined a total weight loss of 12.2% for 11 Å tobermorite, which is close to our results and verifies the XRD results. The weight loss in the temperature range of 100–400 °C for the T-A sample and T-S sample is higher than that of the control sample indicates more interlayer water. Thus, there is more Ca^{2+} in the interlayer, which may be favorable for the ion exchange process when tobermorite is directly used as an adsorbent.

3.2.2. FT-IR Analysis

Figure 3 displays that the resulting three FT-IR spectra plots show similar behaviors concerning the wavenumber from 4000 to 400 cm^{-1} indicating that incorporations of aluminum and sucrose have little effect on the change of silicate chain structure. The FT-IR spectra display a main narrow bands around 976 cm^{-1}, typical of the Si-O stretching vibrations generated by Q^2 silicon sites [2,27,29]. Band at 1202 cm^{-1} is due to Si-O stretching vibrations in Q^3 silicon sites specific for tobermorite [27]. A shoulder at 910 cm^{-1} for T-A sample may be assigned to O-H bending vibrations in Al-OH-Al bonds (octahedral aluminum) [2]. The band at 672 cm^{-1} is due to Si-O-Si stretching vibrations in SiO_4 tetrahedra. It should be noted that the higher band intensity of Q^3 silicon sites alongside the weaker band intensity of SiO_4 tetrahedra for T-A sample and T-S sample compared to the cases for the control sample indicate an increased polymerization degree of silicate chains by incorporating aluminum and sucrose. The bands in the range of 533–451 cm^{-1} are assigned to O-Si-O deformation and bending modes of SiO_4 tetrahedra [2,27,29]. The band at 1635 cm^{-1} is due to H-O-H bending vibrations and the band centered at 3450 cm^{-1} is associated with O-H stretching vibrations [2,25,27,29]. In the region of 1475 to 1415 cm^{-1}, the spectra show a very broad band corresponding to asymmetric stretching of CO_3^{2-}.

3.2.3. NMR Analysis

By means of ^{29}Si-NMR investigations, it is possible to detect whether the incorporations of aluminum and sucrose affect the connectivity of the Si-centers of the synthesized tobermorite samples by identifying different Q sites correlating with the linkage of the SiO_4 tetrahedra. There are five peaks (Figure 4a) in the ^{29}Si NMR spectra of the control sample and the chemical shifts are −98.2 ppm, −88.2 ppm, −85.4 ppm, −82.5 ppm and −79.8 ppm, respectively. The signal at −79.8 ppm is attributed to Q^1 silicon sites (isolated dimers) referred to the previous studies [3,28]. Maeshima et al. [28] presented ^{29}Si-NMR data for naturally occurring 11 Å tobermorite with signals for the Q^2 sites (chain groups) located at ~−80.5 to −85.3 ppm and signals for Q^3 sites (chain branching) at ~−91.9 to −96.1 ppm. Our Q^2 signals locate at −88.2 ppm, −85.4 ppm and −82.5 ppm, respectively, and with slight deviations in comparison to results found in the literature. The signal at −98.2 ppm can be assigned to Q^3 sites.

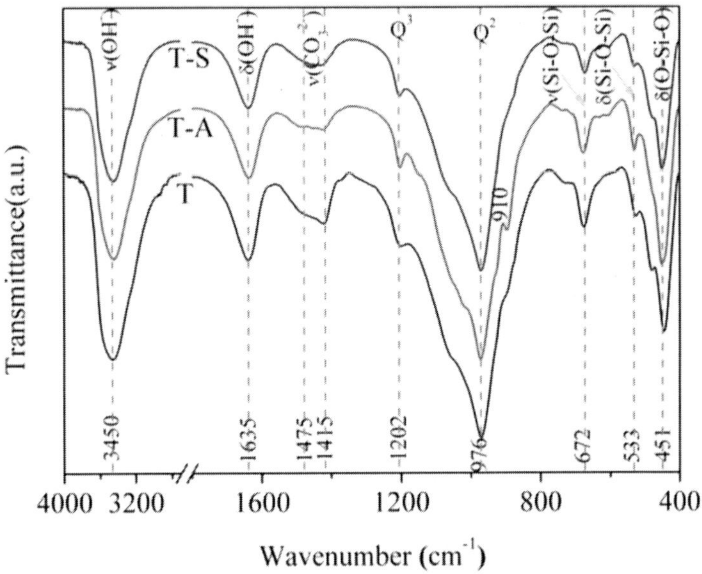

Figure 3. FT-IR spectra of the tobermorite samples.

As shown in Figure 4b,c, Q^1 signals disappear in the spectra of T-A sample and T-S sample. In both cases, only Q^2 and Q^3signals can be observed, which is in good agreement with the FT-IR results and also correlates with the high crystallinity of the samples observed via XRD measurements and the associated high order of the structure which leads to the formation of Q^3 sites. After being incorporated with aluminum, resonances slightly deviate to higher frequency for Q^2(1Al) and Q^3(1Al) sites relative to Q^2 and Q^3 sites, respectively. By comparison, Coleman [11] identified that Q^2 and Q^2(1Al) sites located at −86.5 and −83.5 ppm, while Q^3 and Q^3(1Al) sites corresponded with signals at −98.4 and −93.5 ppm, respectively. Gabrovŝek et al. [30] assigned signals at −86 and −82.5 ppm to Q^2 and Q^2(1Al) sites, those at −96 and −92 ppm were attributed to $Q^3$and Q^3(1Al) sites. We have measured two signals for Q^3 sites (Figure 4b) at −98.5 ppm and −93.9 ppm, respectively, in the spectra of T-A sample. The signals at −88.2 ppm, −86.3 ppm and −84.9 ppm can be assigned to Q^2 sites according the results of Gabrovŝek et al. [30], in which Q^2 sites for Al-substituted tobermorite located at −85 to −89 ppm. Our Q^3(1Al) and Q^2(1Al) sites are located at −91.0 ppm and −82.6 ppm, respectively. The shift into higher frequency is typical for Al-substituted tobermorite and is due to the formation of neighboring atoms [3]. In the spectra of the T-S sample (Figure 4c), there are five peaks, chemical shifts of which are −98.11 ppm, −94.51 ppm, −87.95 ppm, −85.46 ppm and −82.82 ppm, respectively. The signals at −98.11 ppm and −94.51

ppm are attributed to Q^3 sites [11,28]. The signals at −87.95 ppm, −85.46 ppm and −82.82 ppm are corresponding to Q^2 sites [11,30].

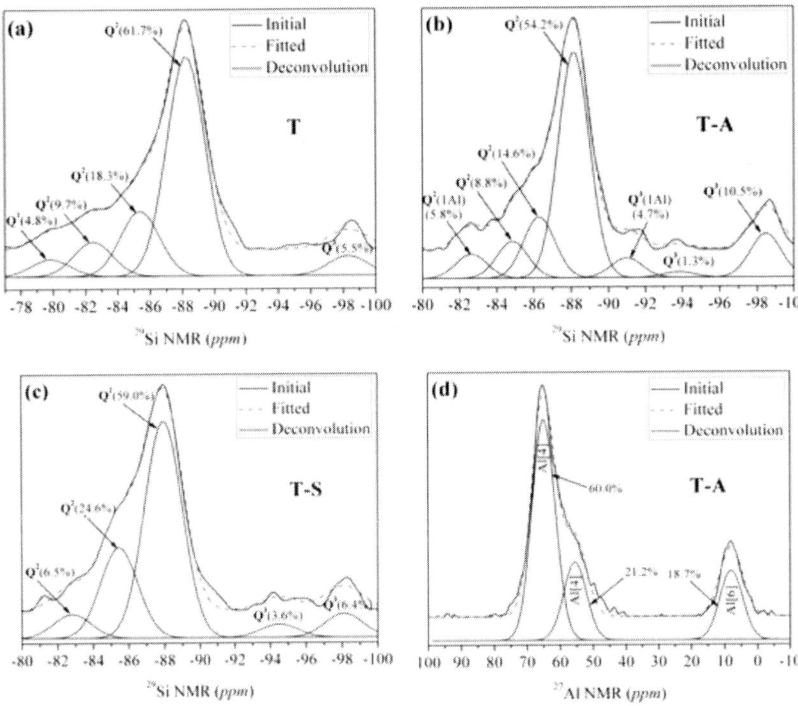

Figure 4. NMR spectra of the tobermorite samples: (**a**) ^{29}Si NMR spectra of the control sample; (**b**) ^{29}Si NMR spectra of the T-A sample; (**c**) ^{29}Si NMR spectra of the T-S sample; (**d**) ^{27}Al NMR spectra of the T-A sample.

It is evident that the proportion of Q^3 and Q^3(1Al) sites is increased by incorporating aluminum and sucrose, indicating a higher cross-linking degree of the silicate chains of tobermorite. Thus, it can be concluded that the tobermorite has a higher polymerization degree with incorporations of aluminum and sucrose. Especially, the occurrence of Q^3(1Al) and Q^2(1Al) sites indicates Al has more opportunity to occupy the bridging position than Si, which may lower the formation energy (ΔG) of tobermorite [10]. The role of aluminum in the aluminosilicate chains of tobermorite structure can be further identified by ^{27}Al NMR. Generally, the presence of aluminum in C-S-H has three forms, i.e., coordination numbers of 4, 5 and 6 [3,9,31]. Al [4] coordination sites with center bands are at ~50–70 ppm, Al [5] ~30–40 ppm, and Al [6] ~0 ppm, respectively. The ^{27}Al NMR spectra for T-A sample, presented in Figure 4d, are predominantly Al [4] with some Al [6]. The two Al [4] resonances at 64.95 and 55.54 ppm correspond well with those reported for Al-substituted 11 Å tobermorite [3,11,32]. The assignments of the two distinct Al [4] have been associated with the isomorphous substitution of Al for Si in two different environments, the 64.95 ppm resonance corresponding to substitution in the middle chain bridging Q^2 and the 55.54 ppm resonance to substitution in the branching Q^3 sites [32]. Thus, Al occupancy of Q^2 bridging and Q^3branching sites indicates that Al links silicate chains together, creating silicate units that are connected by bridging and branching Al tetrahedra. The resonance at 8.04 ppm is attributed to Al[6] probably from some calcium aluminate hydrate phase, which includes $Al(OH)_6^{3-}$ or $O_xAl(OH)_{6-x}^{(3+x)-}$ octahedral [9]. This is consistent with the above FT-IR conclusion about the O-H bending vibrations in Al-OH-Al bonds at 910 cm^{-1}.

3.2.4. SEM Analysis

Micromorphology of the tobermorite samples are displayed in Figure 5. The control sample shows a thin platy morphology (Figure 5a), which is the crystal habit of tobermorite [2], with a size of approximately 1–2 μm. The morphology of tobermorite has almost no change after incorporated with aluminum except for more arranged of foils, as shown in Figure 5b. In the presence of sucrose, some of tobermorite display a lathy morphology with around 1 μm long and 300 nm wide (Figure 5c). It should be noted that in all cases the tobermorite samples present irregular spherical particles with an open framework surface structure (Figure 5d). This porous microstructure may develop from the stacking of tobermorite plates, foils or lathes, and means a large surface area and pore volume to which adsorbates are readily accessible. Figure 6 shows the aluminum-incorporated tobermorite predominantly consists of Ca and Si with trace of Al and K. The Ca/Si molar ratio of this tobermorite is 0.69, which is slightly smaller than the ideal composition of tobermorite (0.8).

3.2.5. BET Analysis

The N_2 adsorption–desorption isotherms and pore size distributions of the tobermorite samples are shown in Figure 7. All the isotherms are similar and can be classified as type IV [33], which indicates that the framework structure of tobermorite does not change by incorporating aluminum and sucrose. The isotherms show a H4-type hysteresis loop, which is typically observed for aggregates of plate-like particles [33,34]. This result is consistent with the SEM observations (Figure 5). Moreover, the pore size distributions of the tobermorite samples confirm the pore size ranges from 3 to 100 nm and is bimodal with trace of small mesopores (~4 nm) as well as dominantly larger ones (~20 nm). The small mesopores and larger ones probably come from the aggregation of primary particles and secondary particles, respectively. The average pore diameters of T, T-A and T-S samples are 10.56 nm, 8.71 nm and 6.56 nm, respectively. Thus. the pore structure of tobermorite is refined by incorporating aluminum and sucrose. The BET specific surface areas of T, T-A and T-S samples are 67.04 m^2/g, 74.39 m^2/g and 89.19 m^2/g, respectively, which are higher than the previous value for the synthetic tobermorite (57 m^2/g) [35]. It should be noted that the incorporations of aluminum and sucrose also play a productive role in increasing the porosity of tobermorite. The larger surface area provides a larger number of adsorption sites, which may be ideally effective for pollutant adsorption when it is directly used as an adsorbent.

Figure 5. SEM images of the tobermorite samples: (**a**) the control sample; (**b**) and (**d**) the T-A sample; (**c**) the T-S sample.

3.3. Cr(VI) Adsorption Characteristics

3.3.1. Effects of Solution pH

The adsorption of Cr(VI) from aqueous solutions is highly dependent on the pH of the solution because the pH affects not only the surface properties of the adsorbents but also the ionic state of chromium during the reaction. In this study, the effects of pH on the adsorption amount of the tobermorite samples are shown in Figure 8. The tendencies of adsorption amount with the increase of pH are quite similar for different tobermorite samples. The adsorption amount increases when pH increases from 2.0 to 4.0, and the maximum adsorption amount of Cr(VI) by T-S sample is found at about pH 4.0 while those of T-A and T samples are at about pH 5.0. Moreover, the adsorption amount of Cr(VI) by all samples slightly decreases as pH increases to 6.0. This result is consistent with the previous results [15,16] and can be explained as follows: different species of Cr(VI) (e.g., $Cr_2O_7^{2-}$, CrO_4^{2-}, and $HCrO_4^-$) coexist at acidic condition. The predominant Cr(VI) species at pH 2.0–4.0 is the monovalent $HCrO_4^-$, which is favorable adsorbed since it has a low adsorption free energy [19]. On the other hand, when the pH value is low, adsorbent (tobermorite samples) static charges are presented in positively charged form, so the adsorption rate of Cr(VI) is relatively high by electrostatic interaction. Apart from the electrostatic interaction effect, complexation of $HCrO_4^-$ with the interlayer or Ca^{2+} structural calcium in the structure of tobermorite may occur during the absorption process. The absorption mechanism will be interpreted in detail in Section 3.3.4. The decrease of adsorption amount at pH 6.0 may be due to: (1) more adsorption sites required for the bivalent $Cr_2O_7^{2-}$and CrO_4^{2-}, which are the predominant species at pH 6.0; (2) competitions from increasing OH^- ions with Cr(VI) anions for the adsorption sites; and (3) more negative charges on the surface of tobermorite [15,16]. In comparison, the adsorption amounts of Cr(VI) by T-S and T-A samples are higher than that of T sample at the same pH value. This may be because the amount of adsorption sites in tobermorite incorporating aluminum and sucrose is substantially more than those in the control sample.

3.3.2. Effects of Contact Time and Adsorption Kinetic Study

Figure 9 displays the adsorption amount of Cr(VI) by tobermorite as a function of time. All the materials show fast adsorption during the first 60 min of the reaction and a much slower subsequent removal followed by an equilibrium state within 3 h. This equilibrium time is close to those of magnetite nanoparticles (2 h) [19], natural clay (2–3 h) [14,15,16] fly ash (3 h) [17] and is much higher than that of activated carbon (24 h) [18]. It should be noted that the adsorption of Cr(VI) onto tobermorite is a combination of physical adsorption and chemical adsorption since the physical adsorption is usually a rapid adsorption process [15,16].

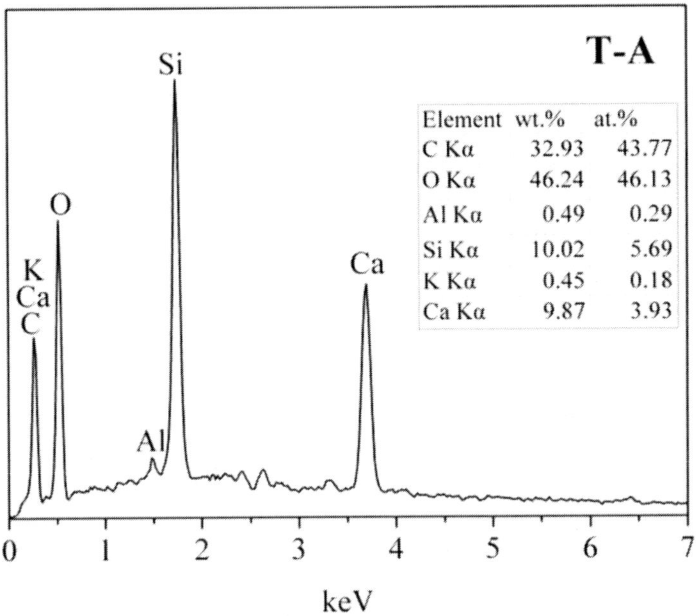

Element	wt.%	at.%
C Kα	32.93	43.77
O Kα	46.24	46.13
Al Kα	0.49	0.29
Si Kα	10.02	5.69
K Kα	0.45	0.18
Ca Kα	9.87	3.93

Figure 6. EDX spectra of the T-A sample in Figure 5d.

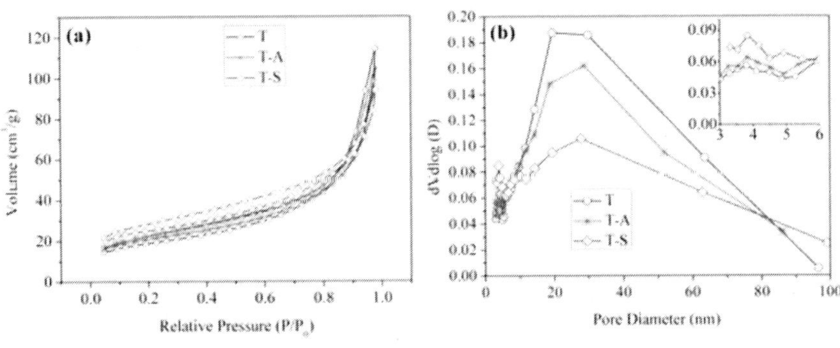

Figure 7. N_2 adsorption–desorption isotherms (**a**); and pore-size distribution curves (**b**) of the tobermorite samples.

To investigate the rate-controlling mechanism of the adsorption process, adsorption kinetic was modeled by the pseudo-first-order and pseudo-second-order models, which can be expressed as follows [15,16]:

$$\text{pseudo-first-order}: \ \ln(q_e - q_t) = \ln q_e - k_1 \cdot t, \quad (1)$$

$$\text{pseudo-second-order}: \frac{t}{q_t} = \frac{1}{k_2 \cdot q_e^2} + \frac{1}{q_e} \cdot t, \quad (2)$$

where q_t (mg/g) and q_e (mg/g) are the adsorption amount at time t (min) and equilibrium, respectively. k_1 (min^{-1}) and k_2(mg/g/min) are the pseudo-first-order and pseudo-second-order rate constants of adsorption, respectively.

As listed in Table 2, the adsorption of Cr(VI) follows the pseudo-second-order model better than the pseudo-first-order model. This indicates that the rate-limiting step may be chemical adsorption involving valency changes through sharing or exchange of electrons between tobermorite and Cr(VI) [14,15,16]. Comparing the q_e values for each sample, they follow the order: q_e T-S > q_e T-A > q_e T. The initial adsorption rate (h, mg/g/min) can be calculated from $h = k_2 \cdot q_e^2$ and follows the order: h T-S > h T-A > h T, suggesting a more rapid adsorption of Cr(VI) onto the T-S sample at a lower initial concentration.

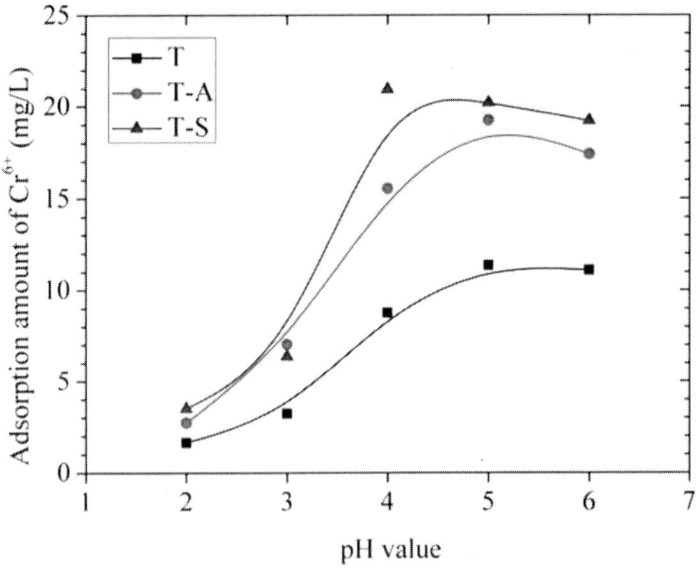

Figure 8. Effects of initial pH on adsorption amount of Cr(VI).

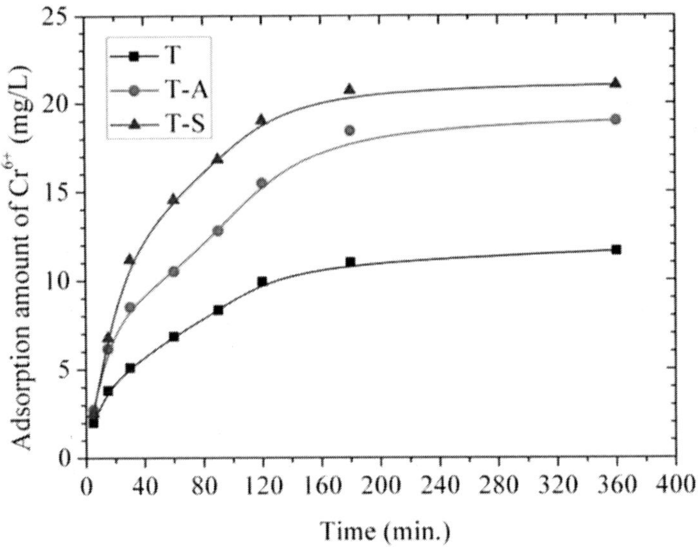

Figure 9. Effects of contact time on adsorption amount of Cr(VI).

Table 2. A comparison of pseudo-first-order and pseudo-second-order kinetic model for the Cr(VI) adsorption by tobermorite.

Sample	Pseudo-First-Order Model			Pseudo-Second-Order Model		
	q_e (mg/g)	k_1 (min^{-1})	R^2	q_e (mg/g)	$k_2 \times 10^3$ (mg/g/min)	R^2
T	10.79	0.01518	0.9851	12.98	1.9040	0.9929
T-A	20.37	0.01757	0.9125	21.48	1.0251	0.9883
T-S	22.57	0.02257	0.9652	23.43	1.2394	0.9967

3.3.3. Effects of the Initial Concentration of Cr(VI) and Adsorption Isotherms

The effects of the initial Cr(VI) concentration on the adsorption amount are shown in Figure 10. It can be seen that the adsorption of Cr(VI) by tobermorite gradually increases as the initial concentration increases until the saturation point is attained at 160 mg/L, therefore a plateau is reached. This may be because available sites for adsorption remain constant for a fixed amount of tobermorite. The experimental data were simulated using the Langmuir and Freundlich models, which can be expressed as follows [15,16]:

$$\text{Langmuir}: \frac{C_e}{q_e} = \left(\frac{1}{k_L \cdot q_{max}}\right) + \left(\frac{C_e}{q_{max}}\right), \tag{3}$$

$$\text{Freundlich}: \ln q_e = \ln k_F + \frac{1}{n} \cdot \ln C_e, \tag{4}$$

where C_e is the equilibrium concentration of adsorbate in solution (mg/L). q_e and q_{max} (mg/g) are the amounts adsorbed at equilibrium and the maximum adsorption capacity for monolayer formation on adsorbent, respectively. k_L (L/mg) is the Langmuir constant related to the maximum adsorption capacity and the energy of adsorption. k_F is the Freundlich constant and n is the heterogeneity factor.

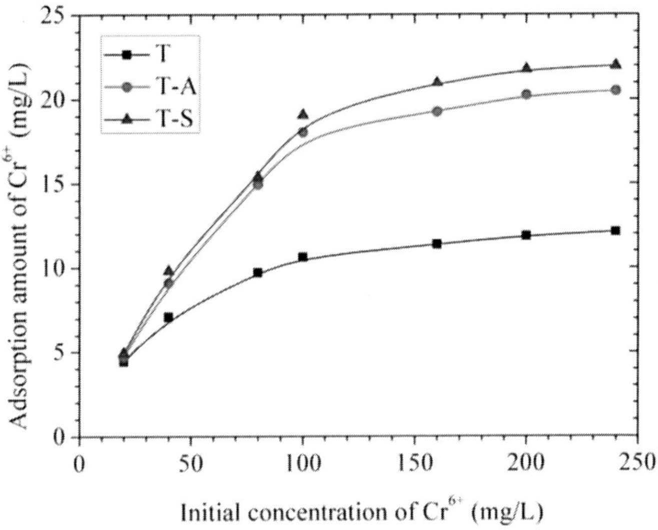

Figure 10. Effects of initial concentration of Cr(VI) on adsorption amount of Cr(VI).

The value of theoretical parameters and constants were calculated from the intercept and slope of the linear equation, respectively, and the results are summarized in Table 3. The Langmuir isotherm correlates better than Freundlich due to the higher R^2 values, showing that the adsorption of Cr(VI) on the whole surface of tobermorite samples is uniform. In other words, the whole surface has identical adsorption activity and therefore the adsorbed Cr(VI) ions do not interact or compete with each other, and they are absorbed in the form of an almost complete monolayer coverage on the surface of tobermorite adsorbent [15,16]. The Langmuir maximum adsorption amount for Cr(VI) is in the order: T-S (31.65 mg/g) > T-A (28.92 mg/g) > T (14.27 mg/g). The incorporations of aluminum and sucrose promote the formation and crystallinity of tobermorite during hydrothermal synthesis. On the other hand, the surface area of tobermorite is significantly increased by sucrose, resulting in more adsorption sites for Cr(VI). More "empty" positions exist needing for charge balance in the Al-substituted tobermorite [10]. All the modifications make the tobermorite samples have enhanced adsorption capacity. In the case of Langmuir isotherm the dimensionless coefficient K_L values lie between 0 and 1 showing favorable adsorption. Moreover, the values of n in the Freundlich isotherm are in the range of 1–10, representing a good adsorption [15,16]. In this work, the exponent of n is 1.67–2.54 ($1 < n < 10$), which reflects that the adsorption of Cr(VI) onto tobermorite is favorable.

Table 3. Langmuir isotherm model for the Cr(VI) adsorption by the tobermorite samples.

Sample	Langmuir Isotherm			Freundlich Isotherm		
	q_m (mg/g)	K_L (L/mg)	R^2	k_F	n	R^2
T	14.27	0.0243	0.9991	1.5362	2.54	0.9293
T-A	28.92	0.0116	0.9768	0.9557	1.70	0.9277
T-S	31.65	0.0109	0.9767	0.9704	1.67	0.9341

Table 4 compares the maximum adsorption capacities of Cr(VI) by tobermorite obtained from the Langmuir model in the present work with those reported in literatures. It can be seen that the adsorption capacities T-A sample and T-S samples are not the highest among the materials in literatures, but are higher than those of zeolite and clay, which have been used as adsorbent for years owing to their high cation exchange ability and high affinity for heavy metals [14,15,16]. The high adsorption ability of the tobermorite adsorbent can be ascribed to several factors such as the mesoporous structure, high specific surface area and intrinsic chemical nature of tobermorite. The adsorption mechanism will be discussed in detail below. In other words, it can be predicted that the aluminum- and sucrose-incorporated tobermorites prove to have promising potential for the treatment of Cr(VI) waste water applicable in industry, agriculture, livelihood and other fields in large scale for its low toxicity and cost and high efficiency in the future.

Table 4. Comparison of adsorption ability of Cr(VI) on various absorbents.

Adsorbents	S_{area} (m^2/g)	q_{max} (mg/g)	Equilibrium Time (h)	Optimum pH	T (°C)	References
T	67.04	14.27	3	4	20	This study
T-A	74.39	28.92	3	4	20	This study
T-S	89.19	31.65	3	3	20	This study
Zeolite-rich tuff	5.29	1.16	2	3	18	[14]
Vermiculite	-	27	3	4	25	[15]
Bentonite	-	24	3	4	25	[15]
Attapulgite	-	15	3	4	25	[15]
Zeolite	-	13	3	4	25	[15]
Natural clay	116	4.5	2	5	20	[16]
Natural clay	140	10.9	2	5	20	[16]
Activated carbons	1462	262	24	2	60	[18]
Magnetite nanoparticles	-	15.3	2	2.5	25	[19]
CNT	700 (SWCNT); 270 (MWCNT)	2.35 (SWCNT); 1.26 (MWCNT)	1	2.5	25	[20]
chitosan	-	215	-	4	25	[21]

3.3.4. Adsorption Mechanism

The adsorption of Cr(VI) by tobermorite is a complex process, to better understand the adsorption mechanism, the concentrations of Ca^{2+} were analyzed. Figure 11 displays a nearly linear relationship of the molar amount of absorbed Cr(VI) and the molar amount of Ca^{2+} in the solution. To better understand the adsorption mechanism, XPS measurement was carried out as shown in Figure 12. It can be seen that the Cr 2p1/2 and Cr 2p3/2 photoelectron peaks are located at 588.07 and 578.34 eV, respectively [36], which are assigned to the characteristic peaks of Cr(VI) and Cr(III), respectively. The results indicate that both Cr(VI) and Cr(III) are present on surface of tobermorite samples during the adsorption process. Cr(VI) may be reduced to Cr(III) in the presence of organic matter [36,37] or at a pH of 2.0–6.0 in an acid environment through following redox reaction:

$$HCrO_4^- + 7H^+ + 3e^- \rightarrow Cr^{3+} + 4H_2O$$

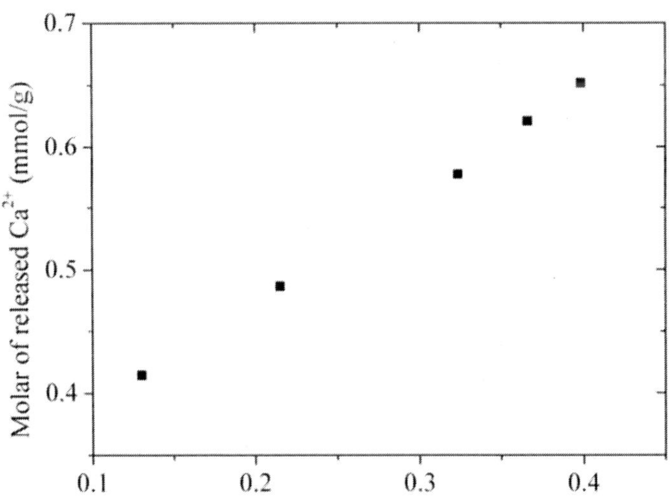

Figure 11. The amounts of Ca^{2+} released from tobermorite at different adsorption amounts of Cr(VI).

Then, the Cr(III) cations are absorbed by cation-exchange reaction with Ca^{2+} in the structure of tobermorite. While, the exchange seem be non-stoichiometric, because the molar amount of Ca^{2+} in the solution after absorption is always higher than the stoichiometrically calculated molar amount of Cr^{3+}. Therefore, the leaching of Ca^{2+} during the reaction process may be due to partial exchange with Cr(VI) and partial hydrolysis of tobermorite in the acidic solution. It is well known that the structure of tobermorite consists of double central layers of CaO octahedra between single silicate chains [5]. The prominent structure feature of tobermorite is the cavity between two adjacent building layers stacked along *c*, the "interlayer space", which may contain water and Ca^{2+} ions. Thus, the exchange reaction probably takes place from the interlayer Ca^{2+} and certainly can occur on the planar surface or edges of tobermorite derived by the non-equilibrium chemical bonds. At the same time, since all oxygen atoms are shared with Si-O chains in the central layer structure of CaO octahedra in tobermorite, the Ca-O interaction is relatively weaker and easy to break attributed to the acidic nature of the initial solution, accordingly, extra Ca^{2+} ions are released in the solutions. As a result, the structure of tobermorite should be destroyed to some extent, then the reaction of Cr(VI) leads to a partial loss of crystallinity of tobermorite as can be deduced from the decrease of intensities of diffraction peaks in the XRD pattern and the infrared spectra as well as from the morphology observed by SEM.

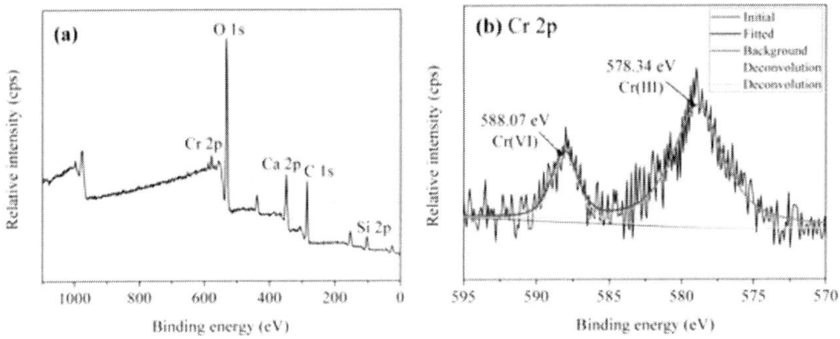

Figure 12. XPS spectra of the tobermorite sample after absorption: (**a**) survey; and (**b**) Cr 2p.

It can be seen from Figure 13a that the relative intensities of diffraction peaks from the T-A sample reacted with Cr(VI) solution (adsorption conditions: initial concentration of 160 mg/L; pH of 4.0; contact time of 6 h) at 7.8° (2θ), 28.9° (2θ), 29.9° (2θ) and 31.8° (2θ) decrease with respect to the unreacted sample (Figure 1). This is also reflected in the infrared spectra (Figure 13b). The band attributed to Q^3 silicon sites disappears indicating a decreased polymerization degree of silicate chains. In addition, The reaction of tobermorite with Cr(VI) leads to precipitation of chromatite (JCPDS card No. 75-0936, $Ca(CrO_4)$) as shown in the corresponding XRD pattern (Figure 13a). Infrared analyses show the presence of a single, characteristic, sharp band at 870 cm^{-1} (Figure 13b), assigned to the stretching vibrations of the chromate [CrO_4] group [38]. The sharp band at 964 cm^{-1} is due to the Si-O stretching vibrations generated by Q^2 silicon sites. The band at 666 cm^{-1} is due to Si-O-Si stretching vibrations in SiO_4 tetrahedra and the band at 454 cm^{-1} is assigned to the bending modes of SiO_4tetrahedra. Abroad band centered at 1460 cm^{-1} corresponds to the asymmetric stretching of CO_3^{2-}. The band at 1650 cm^{-1}is due to H-O-H bending vibrations. In the O-H stretching region, the reacted tobermorite yields two bands centered at 3597 and 3312 cm^{-1}. The first band is associated with vibrations involving less hydrogen-bonded protons in water molecules and the second one is probably due to O-H stretching of more strongly hydrogen-bonded water molecules in the interlayer of tobermorite [2,27,29].

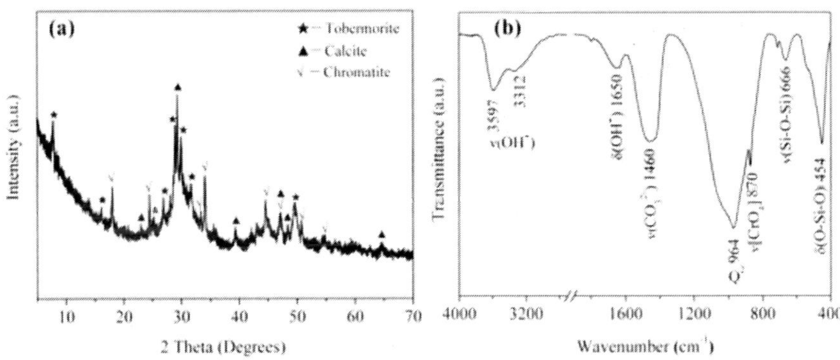

Figure 13. XRD pattern (**a**); and FT-IR spectra (**b**) of the tobermorite reacted with Cr(VI).

SEM image illustrates that the tobermorite reacted with Cr(VI) shows a leaf-like morphology with a size of around 1 μm long and 300 nm wide (Figure 14a). It is noteworthy that this tobermorite seems to be smaller than the unreacted sample (Figure 5) and has a sharp tip. EDX spectra of the reacted tobermorite demonstrate the existence of Ca and Si as well as traces of Cr, K, Mg, and Al (Figure 14b). The detected Si peak is significantly less intense than the Ca peak by more than 50% and also less intense than the unreacted sample (Figure 6). These results imply partial hydrolysis of tobermorite and support the XRD and IR data. Higher intensity of Ca peak may be due to the precipitation of chromatite, which is difficult to distinguish among the tobermorite crystals.

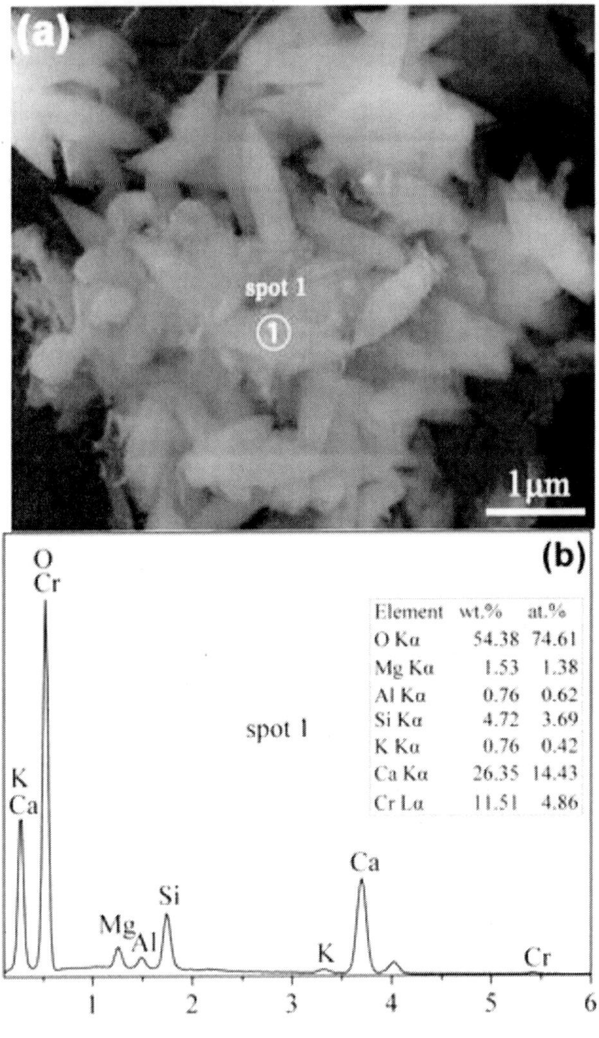

Figure 14. SEM image (**a**); and EDX spectra (**b**) of the tobermorite reacted with Cr(VI).

The above analyses indicate that the adsorption of Cr(VI) onto tobermorite is driven by multiple mechanisms, including physisorption and chemisorption. Undoubtedly, the high surface area arising from the porous structure of tobermorite contributes to the adsorption performance via electrostatic interaction, but the intrinsic nature of tobermorite sometimes is more crucial for the adsorption capacity. The adsorption process of Cr(VI)may arise as a result of the following steps: (1) exchange with Ca^{2+} released from the planar surface or edges of tobermorite crystals, at the same time, partial Cr(VI) may be immobilized though complexation reaction with oxygen linking the building layers; and (2) exchange with the interlayer Ca^{2+} for charge balance and finally with the structural calcium. This adsorption mechanism of Cr(VI) by tobermorite is illustrated in Figure 15. The exchange reaction leads to a loss of crystallinity of tobermorite but not the only reason, because tobermorite inevitably decomposes due to the acidic nature of the initial solution. On the other hand, the reaction of tobermorite with Cr(VI) leads to precipitation of chromatite, which may hinder the exchange process and hence promote the adsorption to achieve equilibrium. The incorporations of aluminum and sucrose increase the crystallinity and yield of tobermorite. It is surprising that the surface area is significantly increased by sucrose, which means more active sites for mass transfer and more broken bonds. Due to Al substitution for Si, more "empty" positions exist needing for charge balance and the (Al)Si-O-Ca linkages become more easy to break because of lower Al-O binding energy [10]. Thus, the aluminum- and sucrose-incorporated tobermorites have superb adsorption capacity.

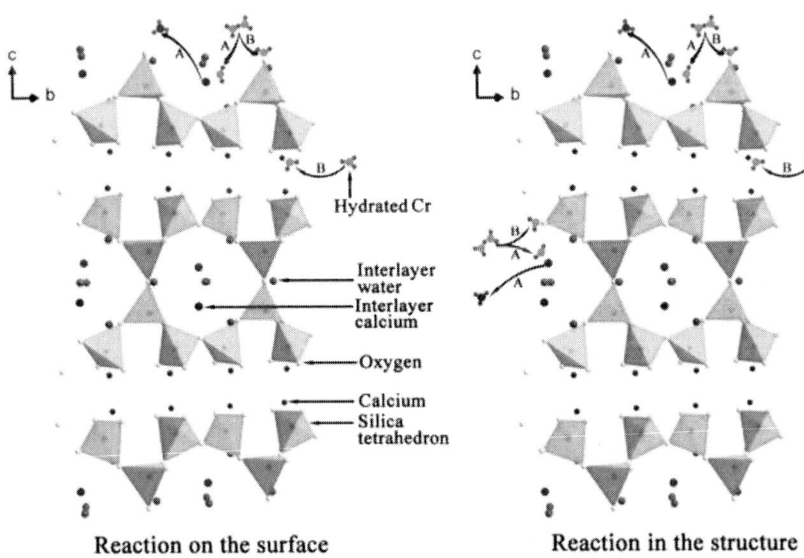

Figure 15. Schematic presentation of the adsorption mechanism of Cr(VI) by tobermorite: (**A**) ion exchange reaction; and (**B**) immobilization effect.

4. CONCLUSIONS

In summary, the formation of tobermorite under hydrothermal conditions was modified by incorporating aluminum and sucrose. It was found that the crystallinity and yield of tobermorite were significantly promoted with aluminum addition of Al/(Al+Si) = 0.03 or at a sucrose concentration of Su/Ca = 0.05. FT-IR and NMR results demonstrate that the polymerization degree of silicate chains was increased under the presence of aluminum and sucrose. SEM observations show tobermorite presented irregular spherical particles with an open framework surface structure to which adsorbates may be readily accessible. In addition, it is surprising that the surface area was significantly increased by sucrose, and the BET specific surface areas of T, T-A and T-S samples were 67.04 m^2/g, 74.39 m^2/g and 89.19 m^2/g, respectively. The adsorption process of Cr(VI) by tobermorite followed the pseudo-second-order model implying a chemical adsorption. Adsorption mechanisms were deduced to be a mixed effect of exchange, immobilization, and precipitation. The adsorption isotherms were well fitted by the Langmuir model which indicates a monolayer adsorption. The maximum adsorption amount for Cr(VI) was in the order: T-S (31.65 mg/g) > T-A (28.92 mg/g) > T (14.27 mg/g) and higher than most of the materials in the literature. Due to superb adsorption capacity, high thermal stability, and low cost for large scale production, this adsorbent has potential for Cr(VI) removal from wastewater.

ACKNOWLEDGMENTS

This study was funded by grand science and technology special project of Guangdong (No. 2013A011401008 and 2013B090600056), key platform and major scientific research project of Guangdong (No. 2016K QNCX151), special fund for building energy conservation of Guangzhou (No. J-2016-03) and the Fundamental Research Funds for the Central Universities. Their financial supports are gratefully acknowledged.

AUTHOR CONTRIBUTIONS

Zhiguang Zhao, Xiaoling Qu, Liang Shi and Haidong Zhang performed the experiments; Zhiguang Zhao and Xiaoling Qu performed the characterizations of the synthesized tobermorite materials; Fangxian Li,

Jiangxiong Wei and Qijun Yu contributed reagents/materials/analysis tools; Zhiguang Zhao wrote the paper.

CONFLICTS OF INTEREST

The authors declare no conflicts of interest.

REFERENCES

1. Hartmann, A.; Buhl, J.-C.H.; Van Breugel, K. Structure and phase investigations on crystallization of 11 Å tobermorite in lime sand pellets. Cem. Concr. Res. 2007, 37, 21–31.
2. Mostafa, N.Y.; Shaltout, A.A.; Omar, H.; Abo-El-Enein, S.A. Hydrothermal synthesis and characterization of aluminium and sulfate substituted 1.1 nm tobermorites. J. Alloy. Compd. 2009, 467, 332–337.
3. Komarneni, S.; Roy, R.; Roy, D.M.; Fyfe, C.A.; Kennedy, G.J.; Bothner-By, A.A.; Dadok, J.; Chesnick, A.S. 27Al and 29Si magic angle spinning nuclear magnetic resonance spectroscopy of Al-substituted tobermorites. J. Mater. Sci. 1985, 20, 4209–4214.
4. Komarneni, S.; Komarneni, J.S.; Newalkar, B.; Stout, S. Microwave-hydrothermal synthesis of Al-substituted tobermorite from zeolites. Mater. Res. Bull. 2002, 37, 1025–1032.
5. Shaw, S.; Clark, S.M.; Henderson, C.M.B. Hydrothermal formation of the calcium silicate hydrates, tobermorite (Ca5Si6O16(OH)2•4H2O) and xonotlite (Ca6Si6O17(OH)2): An in situ synchrotron study. Chem. Geol. 2000, 167, 129–140.
6. Eℓ-Hemaly, S.A.S.; Mitsudab, T.; Taylora, H.F.W. Synthesis of normal and anomalous tobermorites. Cem. Concr. Res.1977, 7, 429–438.
7. Coleman, N.J.; Brassington, D.S. Synthesis of Al-substituted 11 Å tobermorite from newsprint recycling residue: A feasibility study. Mater. Res. Bull. 2003, 38, 485–497.
8. Coleman, N.J. Synthesis, structure and ion exchange properties of 11 Å tobermorites from newsprint recycling residue. Mater. Res. Bull. 2005, 40, 2000–2013.
9. Sun, G.K.; Young, J.F.; Kirkpatrick, R.J. The role of Al in C-S-H: NMR, XRD, and compositional results for precipitated samples. Cem. Concr. Res. 2006, 36, 18–29.
10. Black, L.; Stumm, A.; Garbev, K.; Stemmermann, P.; Hallam, K.R.; Allen, G.C. X-ray photoelectron spectroscopy of aluminium-substituted tobermorite. Cem. Concr. Res. 2005, 35, 51–55.
11. Coleman, N.J. Interactions of Cd(II) with waste-derived 11 Å tobermorites. Sep. Purif. Technol. 2006, 48, 62–70.
12. Tsutsumi, T.; Nishimoto, S.; Kameshima, Y.; Miyake, M. Hydrothermal preparation of tobermorite from blast furnace slag for Cs+ and Sr2+ sorption. J. Hazard. Mater. 2014, 266, 174–181. [PubMed]
13. Mostafa, N.Y.; Kishar, E.A.; Abo-El-Enein, S.A. FTIR study and cation exchange capacity of Fe3+- and Mg2+-substituted calcium silicate hydrates. J. Alloy. Compd. 2009, 473, 538–542.
14. Salgado-Gómez, N.; Macedo-Miranda, M.G.; Olguín, M.T. Chromium VI adsorption from sodium chromate and potassium dichromate aqueous systems by hexadecyltrimethylammonium-modified zeolite-rich tuff. Appl. Clay Sci.2014, 95, 197–204.
15. Thanos, A.G.; Katsou, E.; Malamis, S.; Psarras, K.; Pavlatou, E.A.; Haralambous, K.J. Evaluation of modified mineral performance for chromate sorption from aqueous solutions. Chem. Eng. J. 2012, 211–212, 77–88.

16. Eloussaief, M.; Kallel, N.; Yaacoubi, A.; Benzina, M. Mineralogical identification, spectroscopic characterization, and potential environmental use of natural clay materials on chromate removal from aqueous solutions. Chem. Eng. J.2011, 168, 1024–1031.
17. Adamczuk, A.; Kołodyńska, D. Equilibrium, thermodynamic and kinetic studies on removal of chromium, copper, zinc and arsenic from aqueous solutions onto fly ash coated by chitosan. Chem. Eng. J. 2015, 274, 200–212.
18. Zdemir, E.; Duranolu, D.; Beker, Ü.; Avc, A.Ö. Process optimization for Cr(VI) adsorption onto activated carbons by experimental design. Chem. Eng. J. 2011, 172, 207–218.
19. Yuan, P.; Fan, M.D.; Yang, D.; He, H.P.; Liu, D.; Yuan, A.H.; Zhu, J.X.; Chen, T.H. Montmorillonite-supported magnetite nanoparticles for the removal of hexavalent chromium [Cr(VI)] from aqueous solutions. J. Hazard. Mater.2009, 166, 821–829. [PubMed]
20. Dehghani, M.H.; Taher, M.M.; Bajpai, A.K.; Heibati, B.; Tyagi, I.; Asif, M.; Agarwal, S.; Gupta, V.K. Removal of noxious Cr(VI) ions using single-walled carbon nanotubes and multi-walled carbon nanotubes. Chem. Eng. J. 2015, 279, 344–352.
21. Rojas, G.; Silva, J.; Flores, J.A.; Rodriguez, A.; Ly, M.; Maldonado, H. Adsorption of chromium onto cross-linked chitosan. Sep. Purif. Technol. 2005, 44, 31–36.
22. Pardal, X.; Pochard, I.; Nonat, A. Experimental study of Si-Al substitution in calcium-silicate-hydrate (C-S-H) prepared under equilibrium conditions. Cem. Concr. Res. 2009, 39, 637–643.
23. Mostafa, N.Y. Influence of air-cooled slag on physicochemical properties ofautoclaved aerated concrete. Cem. Concr. Res. 2005, 35, 1349–1357.
24. Cameron, F.K.; Pattern, H.E. The solubility of lime in aqueous solutions of sugar and glycerol. J. Phys. Chem. 1910, 15, 67–72.
25. Al-Wakeel, E.I.; El-Korashy, S.A. Reaction mechanism of the hydrothermally treated CaO-SiO2-Al2O3 and CaO-SiO2-Al2O3-CaSO4 systems. J. Mater. Sci. 1996, 31, 1909–1913.
26. Mackenzie, J.E.; Quin, J.P. CXXVII.—Compounds of hydroxides of the alkaline earths with sugars. J. Chem. Soc. 1929, 951–964.
27. Tränkle, S.; Jahn, D.; Neumann, T.; Nicoleau, L.; Hüsing, N.; Volkmer, D. Conventional and microwave assisted hydrothermal syntheses of 11 Å tobermorite. J. Mater. Chem. A 2013, 35, 10318–10326.
28. Maeshima, T.; Noma, H.; Sakiyama, M.; Mitsuda, T. Natural 1.1 and 1.4 nm tobermorites from Fuka, Okayama, Japan: Chemical analysis, cell dimensions, 29Si NMR and thermal behavior. Cem. Concr. Res. 2003, 33, 1515–1523.
29. Wang, S.P.; Peng, X.Q.; Tang, L.P.; Zeng, L.; Lan, C. Influence of inorganic admixtures on the 11 Å-tobermorite formation prepared from steel slags: XRD and FTIR analysis. Constr. Build. Mater. 2014, 60, 42–47.
30. Gabrovšek, R.; Kurbus, B.; Mueller, D.; Wieker, W. Tobermorite formation in the system CaO, C3S-SiO2-Al2O3-NaOH-H2O under hydrothermal conditions. Cem. Concr. Res. 1993, 23, 321–328.
31. Bae, S.; Taylor, R.; Kilcoyne, D.; Moon, J.; Monteiro, P.J.M. Effects of incorporating high-volume fly ash into tricalcium silicate on the degree of silicate polymerization and aluminum substitution for silicon in calcium silicate hydrate. Materials 2017, 10, 131.
32. Youssef, H.; Ibrahim, D.; Komarneni, S.; Mackenzie, K.J.D. Synthesis of 11 Å Al-substituted tobermorite from trachyte rock by hydrothermal treatment. Ceram. Int. 2010, 36, 203–209.
33. Carja, G.; Nakamura, R.; Aida, T.; Niiyama, H. Texturtal properties of layered double hydroxides: Effect of magnesium substitution by copper or iron. Microporous Mesoporous Mater. 2001, 47, 275–284.
34. Guan, W.; Ji, F.Y.; Chen, Q.K.; Yan, P.; Zhang, Q. Preparation and phosphorus recovery performance of porous Calcium-silicate-hydrate. Ceram. Int. 2013, 39, 1385–1391.
35. Odler, I. The BET-specific surface area of hydrated Portland cement and related materials. Cem. Concr. Res. 2003, 33, 2049–2056.
36. Wu, Q.P.; You, R.R.; Lv, Q.F.; Xu, Y.L.; You, W.J.; Yu, Y. Efficient simultaneous removal of Cu(II) and Cr2O72− from aqueous solution by a renewable amphoteric functionalized mesoporous silica. Chem. Eng. J. 2015, 281, 491–501.

37. Li, T.; Shen, J.F.; Huang, S.T.; Li, N.; Ye, M.X. Hydrothermal carbonization synthesis of a novel montmorillonite supported carbon nanosphere adsorbent for removal of Cr(VI) from waste water. Appl. Clay Sci. 2014, 93–94, 48–55.
38. Doyle, W.P.; Eddy, P. Infra-red of chromates (V) of calcium, strontium and barium. Spectrochim. Acta A 1967, 28, 1903–1907.

CHAPTER 2

Structure and Compressive Properties of Invar-Cenosphere Syntactic Foams

Dung Luong [1,*,†], Dirk Lehmhus [2,3,†], Nikhil Gupta [1,†], Joerg Weise [4] and Mohamed Bayoumi [5]

[1]*Composite Materials and Mechanics Laboratory, Mechanical and Aerospace Engineering Department, Tandon School of Engineering, New York University, Brooklyn, NY 11201, USA*

[2]*ISIS Sensorial Materials Scientific Centre, University of Bremen, Bremen D-28359, Germany*

[3]*MAPEX Center for Materials and Processes, University of Bremen, Bremen D-28359, Germany*

[4]*Fraunhofer Institute for Manufacturing Technology and Advanced Materials (IFAM), Bremen D-28359, Germany*

[5]*Department of Mechanical Engineering, Al-Azhar University, Cairo 11651, Egypt*

ABSTRACT

The present study investigates the mechanical performance of syntactic foams produced by means of the metal powder injection molding process having an Invar (FeNi36) matrix and including cenospheres as hollow particles at weight fractions (wt.%) of 5 and 10, respectively, corresponding to approximately 41.6 and 60.0 vol.% in relation to the metal content and at 0.6 g/cm^3 hollow particle density. The synthesis process results in survival of cenospheres and provides low density syntactic foams. The microstructure of the materials is investigated as well as the mechanical performance under quasi-static and high strain rate compressive loads. The compressive stress-strain curves of syntactic foams reveal a continuous strain hardening behavior in the plastic region, followed by a densification region. The results reveal a strain rate sensitivity in cenosphere-based Invar matrix syntactic foams. Differences in properties between cenosphere- and glass microsphere-based materials are discussed in relation to the findings of microstructural investigations. Cenospheres present a viable choice as filler material in iron-based syntactic foams due to their higher thermal stability compared to glass microspheres.

Keywords: syntactic foams; metal foams; metal matrix composites; high strain rate; strain rate sensitivity; split-Hopkinson pressure bar

1. INTRODUCTION

Syntactic foams (SFs) are foams in which porosity is introduced via hollow particles dispersed within a matrix material. Polymer matrix SFs have found widespread application in the offshore and oil industry, where they serve as buoyancy devices that can withstand much higher hydrostatic loads than other types of foams [1,2]. Such superior mechanical performance is attributed to the considerable strength of hollow particles used not only as lightweight filler, but also in the strengthening phase. Prominent among these filler materials are hollow glass microspheres, fly ash cenospheres, and ceramic and in some cases even metallic hollow spheres [3,4]. Of these, the former two are available in size ranges below a few hundred micrometers, while the ceramic spheres typically come in sizes ranging from micrometers to millimeters.

Metal matrix SFs (MMSFs) also show great application potential, based both on performance and cost, but their adoption in industry has been slow [5,6,7]. Processes for their production are typically either melt-based or powder metallurgical techniques. Melt infiltration of either geometrically confined loose hollow particles or of sintered or otherwise bonded preforms has been demonstrated for several matrix materials: aluminum matrix syntactic foams have, for example, been produced in combination with various fillers by Weise *et al.* (glass microspheres, [8]), Luong *et al.* (cenospheres, [9]) or Szlancsik *et al.* (Fe hollow spheres, [10]) while magnesium matrices have been scrutinized by Gupta *et al.* [11] and Luong *et al.* [12]. Zinc foams have also been synthesized using pressure assisted infiltration [8]. Higher melting matrix materials are typically processed powder metallurgically: there are exceptions in the case of iron and steel matrices [13,14], but titanium, for example, is generally treated using powder metallurgy (PM) techniques due to its even higher melting point and reactivity [15,16].

A qualitative comparison of mechanical properties of many of these materials with conventional types of metal foams is shown in Figure 1, underlining that SFs typically outperform their non-syntactic counterparts in terms of both absolute and density-normalized compressive strength, whereas lowest absolute density is often achieved through other processes [17]. Generally, in return for their high density, the steel matrix SFs have higher strength than MMSFs. In addition, they represent another option for replacing conventional steel foams, which show comparable density, but typically at lower strength.

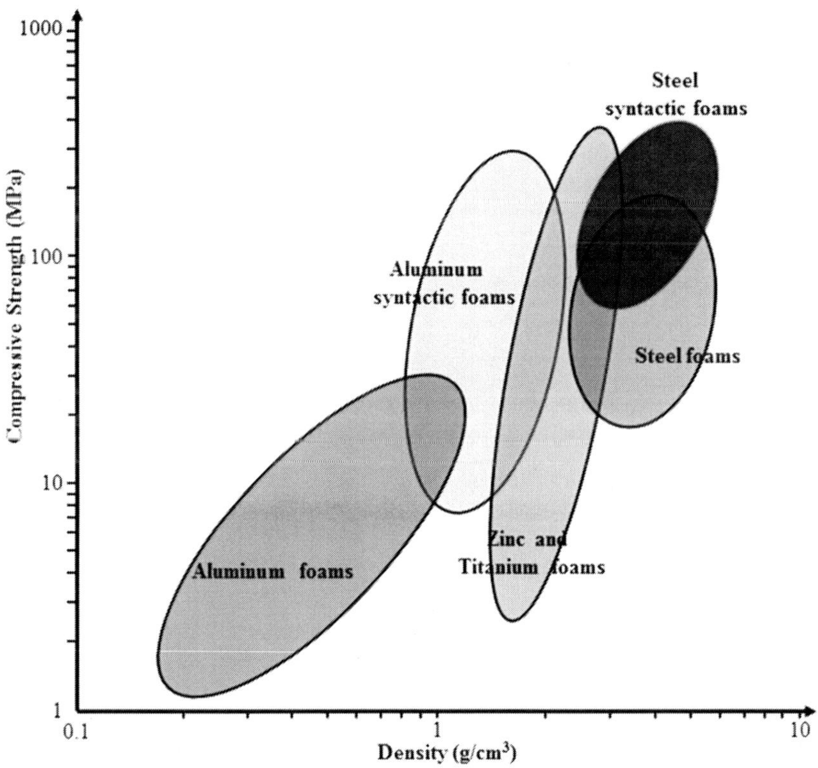

Figure 1. Approximate quantitative comparison of aluminum, titanium, zinc and steel conventional and SFs' (syntactic foams) compressive strength and density, based on a data collection by Weise *et al.* [17].

In this work, two types of SFs filled with 5 and 10 wt.% Fillite alumino-silicate cenospheres, composed, according to supplier specifications, of approximately 55–65 wt.% SiO_2, 27–33 wt.% Al_2O_3 and a balance of Fe_2O_3, are studied [18]. Choice of these materials is based on previous experience from studies dedicated to Invar (FeNi36) matrix SFs [17,19,20,21]. Originally, the metal powder injection molding (MIM) process for production of MMSFs had been successfully developed using Fe99.7 materials as matrix [22,23]. However, observation of the tensile behavior of these foams and notably their low ductility under these conditions has led to the assumption that residual stresses brought in during the cooling step following sintering might account for this behavior, and Invar might thus be a preferable matrix material based on its coefficient of thermal expansion, which is better matched to the glass microspheres used as filler in these studies. Investigation of this effect is still ongoing, without fully conclusive results yet, even though the Invar matrix foams did in fact show the expected higher ductility under tensile load [19]. Despite this positive effect, a microstructural comparison of glass microsphere SFs with matrices of Fe99.7 (sintered at 900 °C), Invar (1000 °C) and 316L (1200 °C) revealed onset of disintegration of glass microspheres at sintering temperatures of 1000 °C, until the full loss of all spheres when sintering at 1200 °C. Figure 2 illustrates this effect on fracture surfaces of the respective materials: for Fe99.7, glass microspheres generally show spherical shape and smooth surfaces (not depicted here, see [20,23]). For Invar, the originally identical microspheres now seem less regular in shape, and satellite glass phases can be observed in between pores within the matrix (Figure 2a) For 316L, the same satellite phases dominate the image, while the inner pore surfaces only occasionally show coverage by glass, which never extends over the entire pore surface area (Figure 2b).

A similar effect is observed when comparing the experimentally measured to the theoretically expected densities (the latter based on rule of mixtures applied to matrix density, filler particle density and volume fraction of filler): the higher the sintering temperature, the greater the level to which the measured density exceeds the theoretical one. This phenomenon is usually understood as reflecting the microsphere's lack of resistance against sinter shrinkage, which gets more pronounced the higher the sintering temperature is chosen [21].

Both effects underline the need for filler of increased thermal stability, and cenospheres comprising higher quality ceramics such as Al_2O_3 and SiO_2 have potential to be useful in this regard [18,21]. Consequently, the present study scrutinizes the structural features of Invar matrix SFs filled with cenospheres rather than glass microspheres. In addition, a link to properties is established through an evaluation of compressive strength across a wide range of strain rates from quasi-static to highly dynamic.

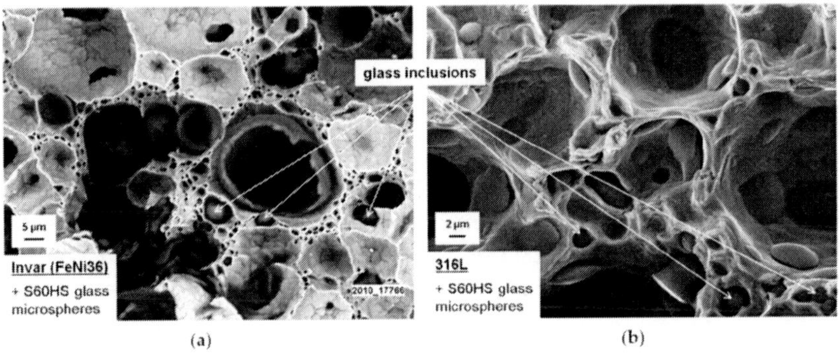

Figure 2. SEM images of fracture surfaces of SFs based on glass microspheres with matrices of Invar (FeNi36, (**a**)) and 316L (**b**)—glass inclusions within the matrix are visible in both cases. S60HS represents S60 hollow glass microspheres manufactured by 3M, St. Paul, MN, USA [24].

2. RESULTS

2.1. Density and Microstructure

The measured and theoretical densities of Invar SFs are listed in Table 1. Determination of densities was done through weighing of machined samples of a well-defined volume. Two types of Invar SF filled with 5 and 10 wt.% cenospheres are denoted as Invar-5CS and Invar-10CS, respectively. For comparison, data from Luong *et al.* [20] on comparable materials containing 3M™ S60HS glass microspheres, for which the average particle density is specified as 0.6 g/cm^3 [25], are added. For these, the theoretical particle density matches the minimum value specified for the cenospheres. The theoretical densities of Invar SFs were calculated based on the density range of cenospheres (0.6–0.85 g/cm^3) as shown in Table 3, assuming that this particle density has been retained in the process and no matrix porosity is present. The real density of both Invar-5CS and 10CS is within the expected range. Had softening of the filler been the case as with the glass microspheres, some deviation would have had to be expected already at the lower filler particle level, an effect which is clearly visible for the glass microspheres included for comparison from the preceding study [20]. Besides, further microscopic observations presented in Figure 3 do not support any such thermal degradation of the cenospheres.

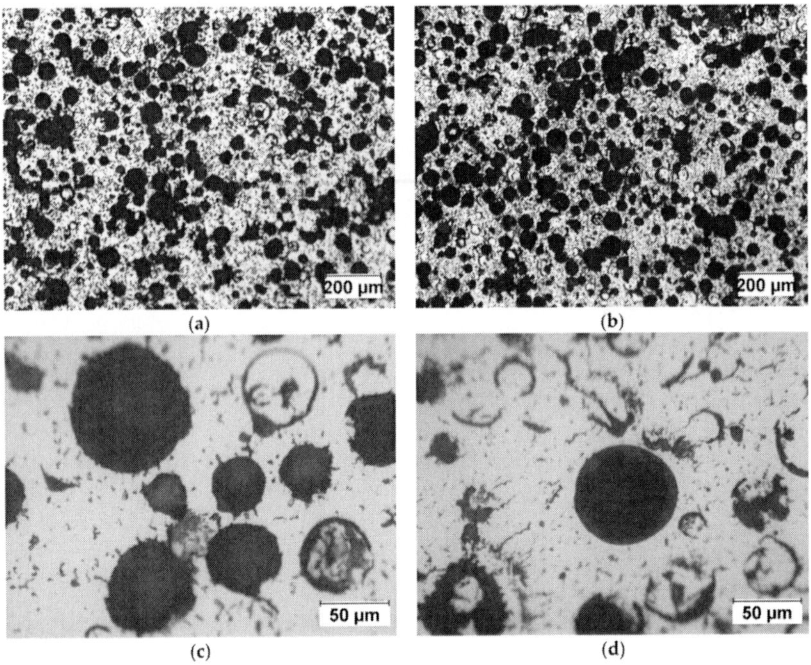

Figure 3. Optical micrographs of (**a**,**c**) Invar-5CS; and (**b**,**d**) Invar-10CS at different magnifications.

Table 1. The measured and theoretical densities of Invar SF with cenospheres (CS) and glass microspheres (GM) as filler.

Sample	Measured Density (g/cm^3)	Theoretical Density (g/cm^3)	
		Min	Max
Invar-5CS	5.09	4.99	5.68
Invar-10CS	4.04	3.60	4.37
Invar-5GM [20]	5.49	4.99	
Invar-10GM [20]	4.21	3.60	

Figure 3 shows the optical microstructure of Invar-5CS and 10CS at different magnifications. Both types of Invar SFs have well-dispersed cenospheres, but the filler distribution is denser for the Invar-10CS. At high magnification in optical micrographs, cenospheres that are broken and filled by the Invar matrix material during the synthesis process are observed. These appear to be more numerous in the structure of Invar-10CS. Some of the filled cenospheres on the exposed surface are due to the deformation of the matrix during polishing before optical microscopy. The density observations in Table 1 show that the fraction of broken cenospheres is small because the experimentally measured density is within the expected range. Figure 3d also shows a seemingly well-bonded interface between the Invar matrix material and the cenospheres. However, no statement about the actual bond strength is possible. Focused Ion Beam (FIB) based investigations on 316L matrices reinforced with glass microspheres in [17] show a clear lack of interface strength, and fracture surfaces of specimens with both 316L and Invar matrices indicate glass microspheres and glass inclusions within the matrix as the main failure initiation sites (see also Figure 2 in this respect). Though differences between glass microspheres and cenospheres in terms of composition suggest the possibility of a deviation in this respect, the current experimental results do not suffice to substantiate this. In order to investigate the effect of the synthesis method on cenospheres, the Invar SFs were immersed in 1.0 vol.% nitric acid for 30 min in order to lay bare the cenospheres originally enclosed in the matrix. The observation of the corrosion face by scanning electron microscope (SEM) equipment would provide the state of cenospheres in the structure of Invar SFs without any physical effects. Figure 4a shows the original cenospheres, while Figure 4b presents the surface of the syntactic foam after having been subjected to the corrosive environment. In Figure 4b, no significant geometrical deformation of the Fillite alumino-silicate cenospheres can be seen in comparison to their original state as shown in Figure 4a. This is explained by the cenosphere's high melting temperature of 1200 °C (Table 3), which is significantly higher than the sintering temperature during the MIM synthesis procedure (1000 °C).

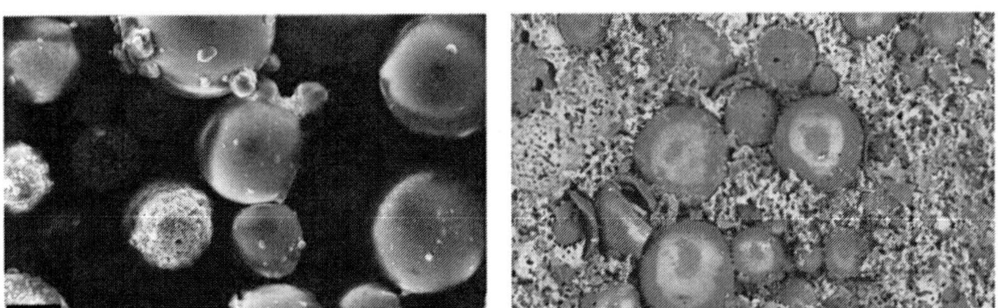

(a) (b)

Figure 4. SEM images of (**a**) Fillite 106 alumino-silicate cenospheres in their virgin state before processing; and (**b**) embedded in the Invar-10CS after feedstock processing, injection molding and sintering.

Table 3. The datasheet of Omya Fillite® FG (106) alumino-silicate cenospheres [18].

Sample	Strain Rate (s^{-1})	Yield Strength (MPa)	Densification Strain (%)	Energy Absorption at 50% Strain (MJ/m^3)	Energy Absorption at Densification Strain (MJ/m^3)
Invar-5CS	10^{-4}	152 ± 6.67	55 ± 2	170 ± 7.2	204 ± 24.6
	10^{-3}	162 ± 3.85	53 ± 3	181 ± 6.8	200 ± 17.9
	10^{-2}	159 ± 11.93	52 ± 3	177 ± 11.2	194 ± 28.4
Invar-10CS	10^{-4}	109 ± 1.26	51 ± 2	113 ± 1.2	118 ± 7.5
	10^{-3}	113 ± 3.60	51 ± 2	115 ± 2.7	119 ± 8.2
	10^{-2}	113 ± 5.58	53 ± 0	123 ± 3.6	136 ± 3.8

2.2. Compressive Properties

The engineering compressive stress-strain behavior of Invar SFs at quasi-static strain rates is plotted in Figure 5. Generally the curve of the Invar SFs can be separated into three regions: elastic, plateau, and densification. A smooth yielding transition is observed between the elastic and plateau regions. The yield strength is therefore specified by the offset method, using 0.2% of plastic strain to define a proof stress-type yield limit and thus separate the elastic and plastic regions. The stress still increases gradually in the plateau region. Since this change is slow but continuous and extends from the plateau into the densification region, there is no significant point separating these two strain ranges. Thus the densification strain has been specified as the intersection of tangent lines formed by the strain range of 10%–30% in the plateau region and another of 60%–80% in the densification region. The yield strength, densification strain and energy absorption at the strain of 50% and densification strain of Invar-5CS and Invar-10CS are listed in Table 2.

In order to study the strain rate sensitivity of compressive properties, Invar SFs were tested at high strain rate loading by using a Split Hopkinson pressure bar (SHPB) set up. The details of the SHPB procedure are presented in Section 4. The strain rate sensitivity of the compressive stress-strain behavior is clearly shown in Figure 6, in which the stress level both for Invar-5CS and Invar-10CS samples shows an increasing trend with strain rate. It should be noted that the end of compression in SHPB testing may not represent specimen densification or failure as the total strain that can be obtained in SHPB for a given strain rate depends on the test instrumentation. It is also noted that the strain rate is recovered from the test results, therefore, behavior of two MMSFs at exactly the same strain rates cannot be compared.

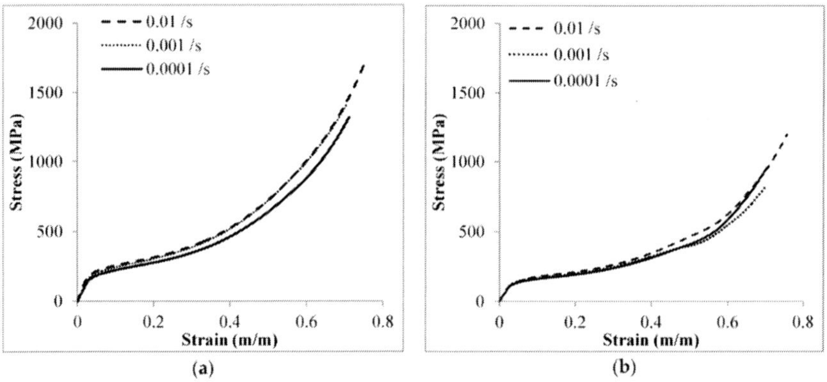

Figure 5. Representative of the compressive stress-strain response of: (**a**) Invar-5CS; and (**b**) Invar-10CS at strain rate of 10^{-2}, 10^{-3} and $10^{-4}\cdot s^{-1}$.

Table 2. The yield strength, densification strain and energy absorption at 50% strain and densification point of Invar SFs under quasi–static loading.

Sample	Strain Rate (s^{-1})	Yield Strength (MPa)	Densification Strain (%)	Energy Absorption at 50% Strain (MJ/m^3)	Energy Absorption at Densification Strain (MJ/m^3)
	10^{-4}	152 ± 6.67	55 ± 2	170 ± 7.2	204 ± 24.6
Invar-5CS	10^{-3}	162 ± 3.85	53 ± 3	181 ± 6.8	200 ± 17.9
	10^{-2}	159 ± 11.93	52 ± 3	177 ± 11.2	194 ± 28.4
	10^{-4}	109 ± 1.26	51 ± 2	113 ± 1.2	118 ± 7.5
Invar-10CS	10^{-3}	113 ± 3.60	51 ± 2	115 ± 2.7	119 ± 8.2
	10^{-2}	113 ± 5.58	53 ± 0	123 ± 3.6	136 ± 3.8

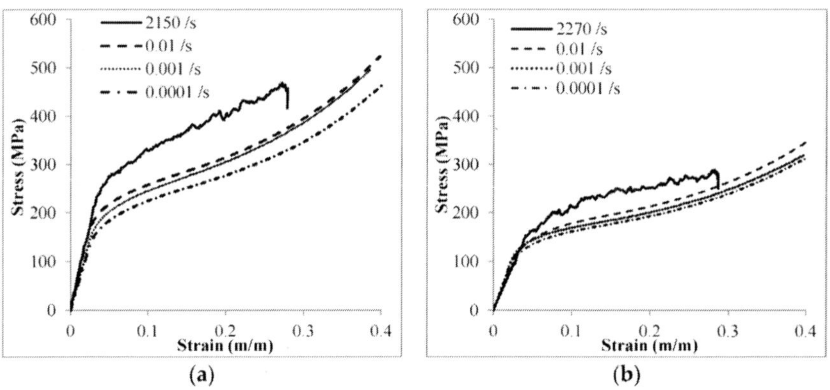

Figure 6. The compressive response of (**a**) Invar-5CS; and (**b**) Invar-10CS at high strain rate loading in comparison to the quasi-static loading.

3. DISCUSSION

Figure 7 shows the engineering compressive stress-strain behavior of the Invar material and its SFs filled with different kinds of hollow particles at a strain rate of $10^{-3} \cdot s^{-1}$. Invar exhibits higher strength in comparison to its SFs at the same strain level. Additionally, the Invar SFs filled with cenospheres have lower strength than the Invar-GM SFs having comparable density [20]. In view of the expected positive influence of the cenospheres on structural features of the foams, the actual realization of which is supported by the density measurements reported in Table 1 as well as the optical micrographs depicted in Figure 3, this may seem surprising. It must however be assumed that the said structural effect is counterbalanced by the generally lower strength of the cenospheres (see Section 4, Table 3, on filler particle characteristics according to supplier specification). This lower strength can be explained by defects in the shell resulting in a generally lower quality of cenospheres, which are an industrial waste product [6]. In general, this observation is another indication of the fact that iron and steel matrix SFs would greatly benefit from higher strength and stiffness, but having at the same time thermally stable microspheres as advanced filler particles. Currently, such alternatives, for which ceramics like SiC or Al_2O_3would be attractive candidates, are not available in the small size ranges matching the requirements of the MIM process, *i.e.*, at diameters roughly below 100 µm. From the available experimental data, the true potential of such optimized iron/steel matrix SFs is hard to estimate, since, in the present case, structural effects and filler particle characteristics are superimposed.

The yielding behavior of SFs is controlled by a complex concurrence of processes, including the matrix material plastic deformation and the hollow particles breakage. Other kinds of SFs such as the aluminum matrix variants reported by [6,26] show a peak stress between the elastic and plateau regions. The cracks that propagate to the matrix material from the broken cell wall of hollow particles were observed, while the stress had a significant drop [26,27]. The cracks lead to the structure collapsing under the compressive loading. The development of cracks in the matrix is due to two reasons: (1) a strong interfacial bonding between matrix and fillers [26]; (2) the insufficient fracture toughness of matrix material to plastically prevent cracks [14]. In contrast, the yielding transition between the elastic and plateau regions of Invar SFs is a smooth curve. It clearly shows that there is no significant collapse in structure of matrix material after the cenopheres fail. Due to the fact that yielding transition is a distinct segment, the yield strength specified by the offset strain of 0.2% should be used to define the end of the elastic region and the beginning of material yielding. The yield strength of Invar SFs from the present work is plotted over density in Figure 8 and contrasted with corresponding values of other steel SFs collected from references [13,14,17,20,21,22,28,29,30,31]. Generally, the yield strength increases with density, thus showing the same trend as reported earlier for other variants of MMSFs in [6,20]. According to Figure 8, Invar SFs filled with cenospheres provide lower yield strength than other steel SFs synthesized via the MIM route, but containing glass microballoons. As discussed above, this observation can be related to the lowered mechanical characteristics of cenospheres (1500–3000 psi or 10.3–20.7 MPa [18], compared to 18,000 psi or 124.1 MPa for S60HS glass microballoons [24]). However, these retain their relevance, as they can provide other options for structural applications requiring high strength over 100 MPa, which only few other MMSFs reach [6].

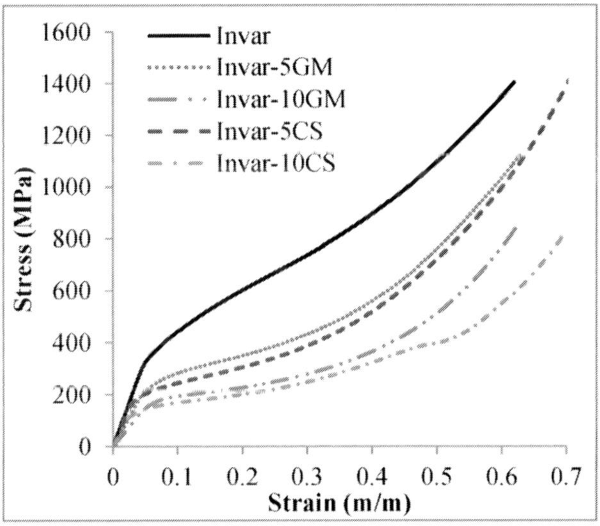

Figure 7. The compressive stress-strain response of Invar steel and its SFs at the strain rate of $10^{-3}\cdot s^{-1}$. The data of Invar and its SFs-filled glass microballoons (GM) is obtained from reference [20].

In order to evaluate the compressive mechanical performance with the contribution of cenosphere strength, the plateau onset strength, which is the stress at the intersection between tangent lines of the elastic and plateau regions, was specified [32]. The plateau onset stress is considered as a point at which plastic deformation of matrix material and filler fracture happen at the same time. Its value for Invar-5CS and Invar-

10CS is 207 and 143 MPa, respectively. In comparison to the yield strength (306 MPa) of the pure Invar matrix material [20], Invar-5CS and Invar-10CS have a 32% and 53% lower strength. This is no surprise result because the Invar matrix material has a solid structure and thus higher density than its SF structure. This point has previously been stressed when contrasting Invar-5GM and Invar-10GM material variants with their cenosphere-based counterparts. The reduced strength of the latter, which at first sight seems to contradict the improvement in structural aspects, is likely the reflection of the lower strength of the filler particles. The plateau onset stress over density of Invar SFs and the matrix yield strength over density are plotted in Figure 9. After eliminating the effect of density on the plateau onset stress value, the results obtained for Invar-5CS materials can be comparable to those associated with a pure Invar matrix. Figure 9 illustrates this effect of the filler on SF strength.

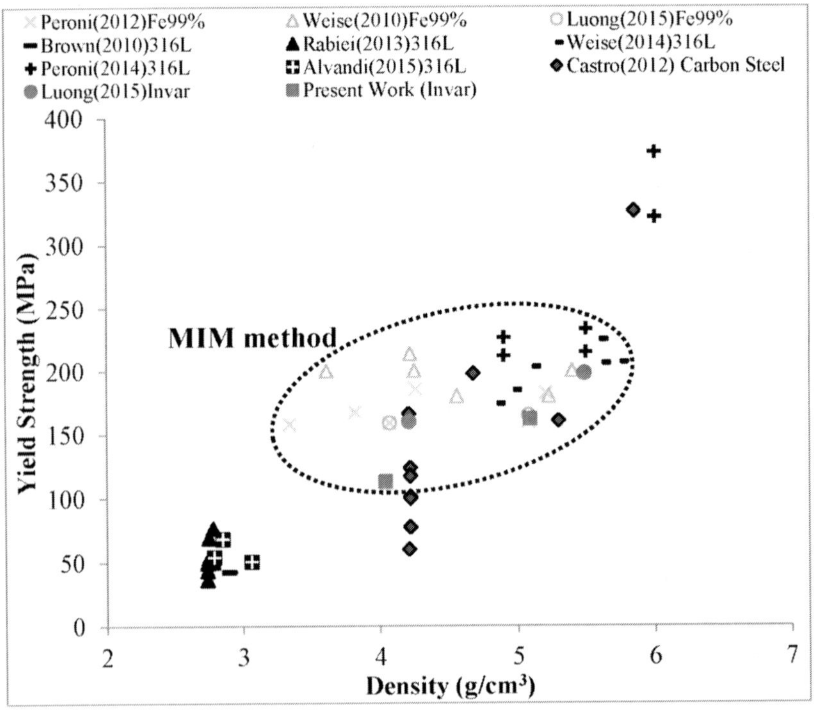

Figure 8. The yield strength of Invar SFs-filled cenospheres in comparison to other steel based matrix SFs. The data is generated and collected from references [13,14,17,20,21,22,28,29,30,31].

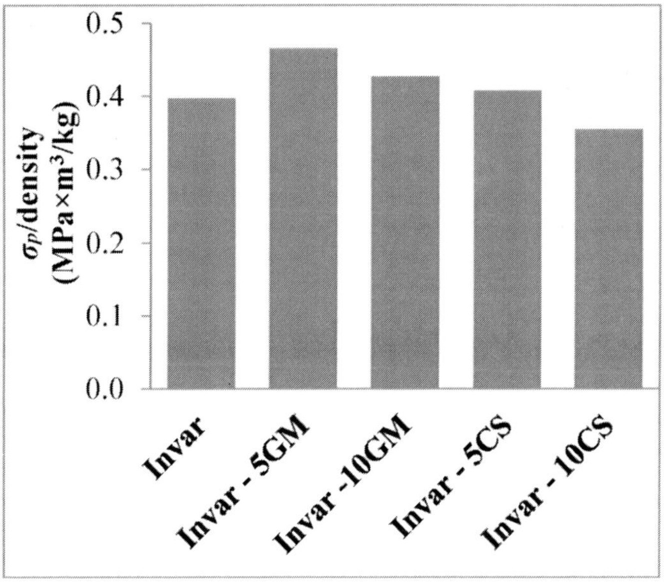

Figure 9. The plateau onset stress over density of Invar SFs in comparison to the yield strength over density of Invar matrix materials. The data of Invar SFs-filled glass microballoons is reported in reference [20].

In the plateau region, stress levels show a gradual increase, impeding the distinction between the former and the densification region. To investigate the state of cenospheres in this region, the Invar-CS specimens were tested at a strain rate of $10^{-3}\cdot s^{-1}$, and deformation was stopped at an early stage within the plateau region (5% strain) and the samples subsequently etched in nitric acid of 1.0 vol.% to reveal the state of the cenospheres at this point. As shown in Figure 10, most cenospheres had already failed at this early stage of the plateau region. The respective crack can be seen on the shell and has the same pattern, developing in the same direction of loading (Figure 11). Possible explanations of this effect include progressive strain hardening as a consequence of density gradients and stress localization in the porous structure. In stochastic foams like those produced in the liquid state via blowing agents, such effects have been alternatively interpreted as consecutive compression of regions of increasing density stacked in the direction of compression [33], as a result of dominating ductile failure of the matrix and ensuing strain hardening [34], or as combinations of both effects [34,35]. In the present case, due to the greater homogeneity of SFs, strain hardening may be considered the dominant effect. It is worth noting in this respect that, contrary to many infiltration-based SFs [8], in MIM-based materials, the filler particles do not form a continuous, brittle network within the matrix, but are dispersed, allowing transfer of part of the ductility of the matrix to the combination of filler and matrix. Alternatively, Peroni *et al.* [23] have suggested a mechanism for development of deformation bands in syntactic foams: the inhomgeneity of the stress distribution might lead to local initiation of failure at peak stress sites. Failure of microspheres at such sites would weaken the respective cross-section and lead to it becoming a preferred site of deformation. Similar effects have been observed computationally in polymer matrix syntactic foams, where fracture of weaker particles causes localized stress concentration, leading to other hollow particles in the vicinity breaking until the stress is sufficiently low [36]. Strain hardening and local densification would lead to strengthening of this initial deformation band and initiation of plastic deformation, first through failure of microspheres in other, now weakest, cross sections [22,23].

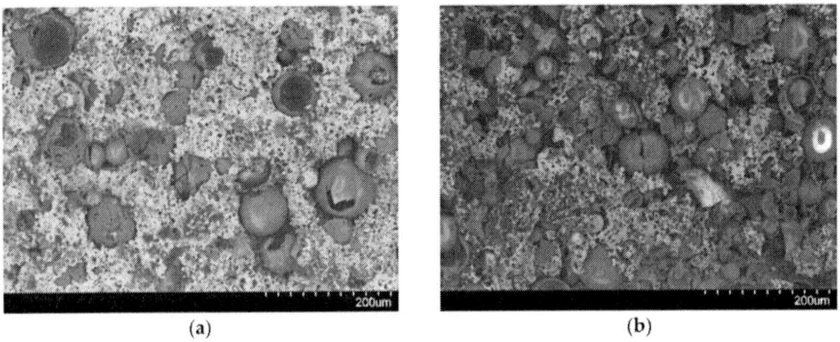

Figure 10. SEM images showing the cenosphere fracture in: (**a**) Invar-5CS; and (**b**) Invar-10CS at a strain of 5%.

Crushing of cenosphere fragments within the pores they originally lined may come in as an additional, secondary explanation. Figure 12a shows the facture face near the border of Invar-10CS specimen after the densification point. The crushed cell walls of cenospheres are easily recognized as darker fragments, while the Invar matrix shows up in a brighter color. The cell walls were debonded from the matrix during the plateau region as shown in Figure 12b. It can be explained by debonding development following the gap at the interface between the matrix and filler that is constituted during the cooling process from the sintering temperature by the difference of their thermal expansion coefficient [28]. Since the SFs are porous structures, the energy absorption capacity of

Invar SFs mainly related to the plateau region is of interest in order to judge the materials' application potential. Figure 13 plots the energy absorption capacity of the presently studied materials at a total strain of 50% in comparison to corresponding results obtained for other types of iron/steel SFs [14,20]. The diagram obtained shows that already, in combination with cenospheres, the material shows characteristics that compare favorably with the related materials considered here. In the lower density range around 4–4.5g/cm^3, the higher carbon content level carbon steel syntactic foams studied by Castro *et al.* [13,14] reach similar energy absorption capacity levels, though at a significantly more brittle overall behavior. Transformation induced plasticity (TRIP) steels also studied by Castro *et al.* [14] promise more characteristics in this respect and almost exactly match MIM-based Invar-CS SFs in terms of energy absorption. In these studies, however, the steel matrices are combined with higher strength ceramic microspheres [14]. As has been pointed out repeatedly above, availability of these materials as micro-scale fillers suitable for the MIM process would be expected to significantly raise the properties of our materials as well. Besides, in view of differences in part sizes (typically larger in the melt-based process suggested by Castro *et al.* [13,14]) and geometrical complexity (typically higher in the powder metallurgical MIM process), these two material synthesis variants must be seen as docile complementary rather than competing technologies.

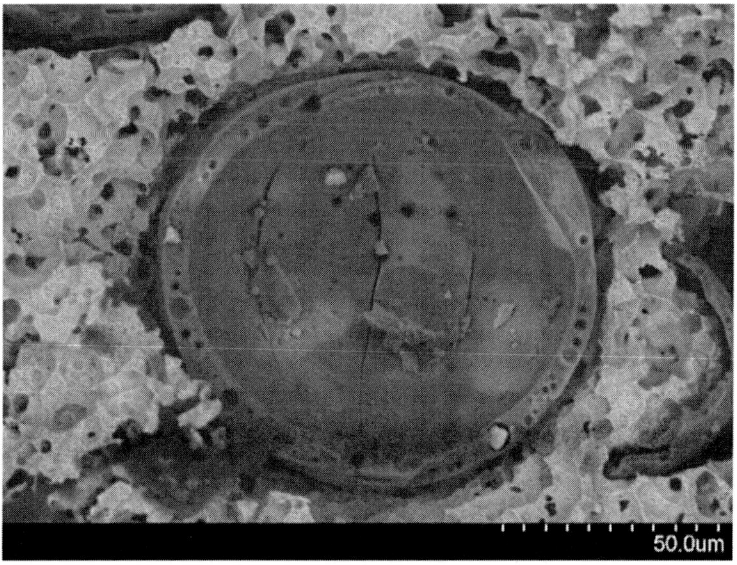

Figure 11. The pattern of crack development on cenospheres under quasi-static loading in the top-bottom direction.

Figure 12. SEM images of fracture face on the Invar-10CS under quasi-static loading at magnification of (**a**) ×400; and (**b**) ×1500.

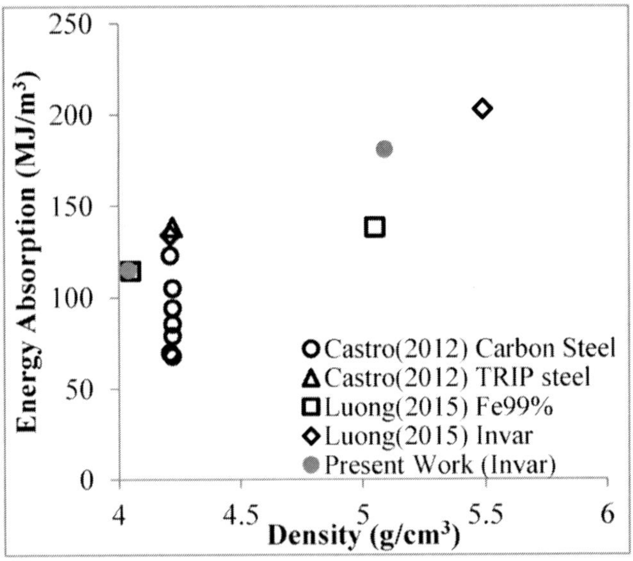

Figure 13. The energy absorption up to the strain of 50% of steel-based matrix SFs.

The strain rate sensitivity of a material at a certain strain level is expressed by [37]:

$$\frac{\sigma_1}{\sigma_2} = \left(\frac{\dot{\varepsilon}_1}{\dot{\varepsilon}_2}\right)^m \quad (1)$$

where σ_1 and σ_2 are stresses at identical strain values corresponding to different strain rates of $\dot{\varepsilon}_1$ and $\dot{\varepsilon}_2$, respectively; the exponent *m* represents the strain rate sensitivity. The stresses at strain from 5% to 25% are plotted over the logarithmic scale of strain rate in Figure 14. The power-law fitting curves are generated for these stresses over strain rate from 10^{-4} to 2500 s^{-1}. The fitting curves are mostly inside the range of stress deviation. The strain rate sensitivity parameter *m* is the slope of the dashed lines as plotted in Figure 14. The parameter *m* is specified as 0.020 and 0.018 for Invar-5CS and Invar-10CS, respectively. Based on this formulation, the stress at an arbitrary strain rate can be predicted if the stress value at certain strain rates and the strain rate relation (*m* parameter) are known. The plateau stresses in the range of 5%–50% strain for Invar-CS SFs are constructed from the one at strain rate of $10^{-4}\cdot s^{-1}$, and the predictions for three individual, deviating strain rates, plotted in Figure 15. The predicted result at each quasi-static strain rate is compared to the average stress-strain curve with the standard deviation (see Figure 15a–d), and the one at high strain rate is directly compared to the experimental curve (see Figure 15e,f). Predicted stresses receive higher error markings at the beginning and end stage of the plateau region, an observation which is associated with the fact that in yielding and densification transition zones, higher complexity of deformation processes makes these less predictable. However, the error between the experimental and prediction results is reported at less than 10%.

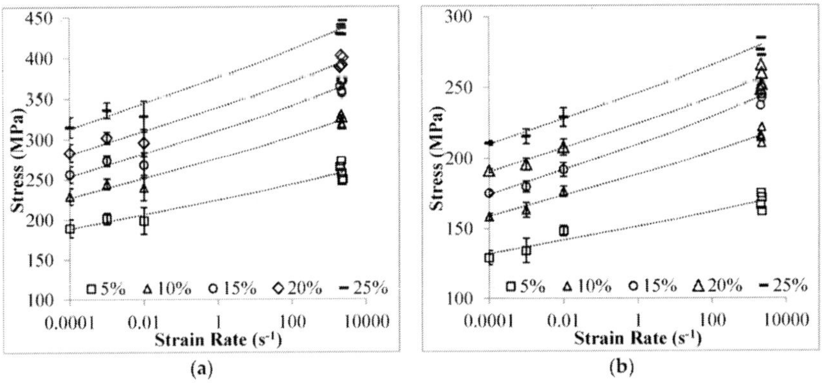

Figure 14. Stresses of (**a**) Invar-5CS; and (**b**) Invar-10CS at the strain from 5% to 25% plotted over strain rate.

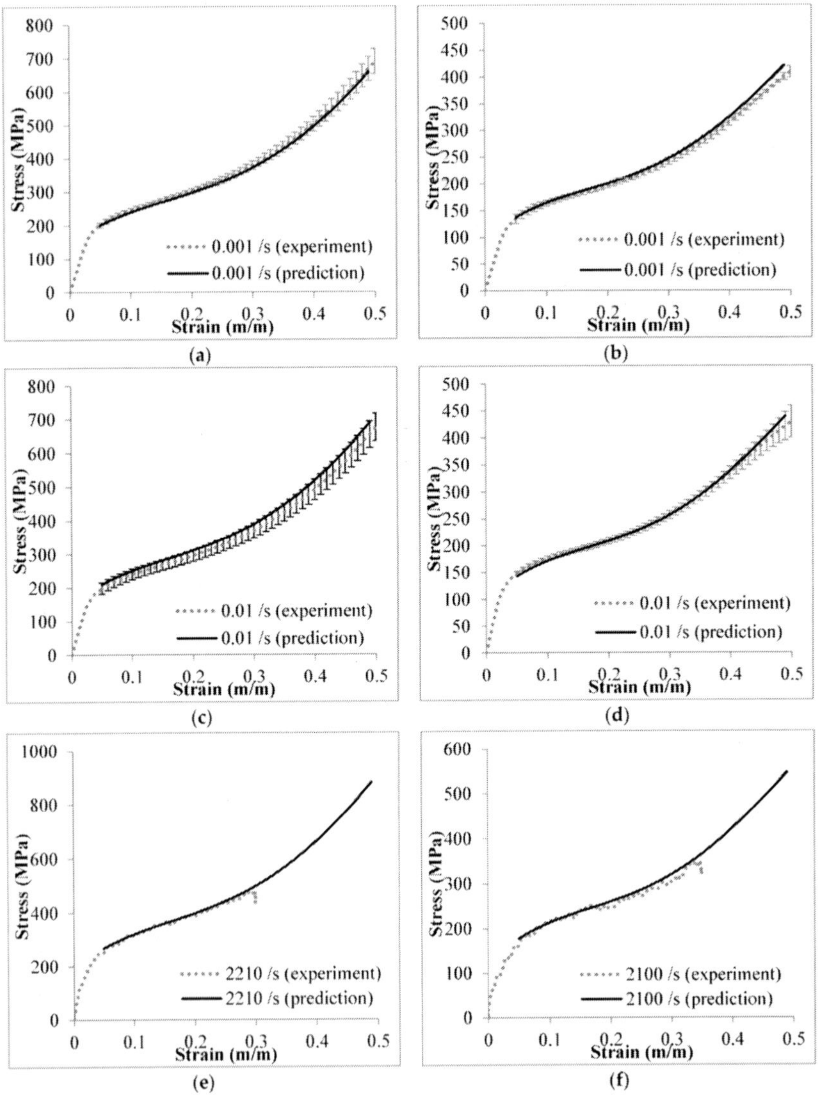

Figure 15. Comparison between the experimental and prediction results of: (**a**,**c**,**e**) Invar-5CS; and (**b**,**d**,**f**) Invar-10CS at different strain rates.

4. MATERIALS AND METHODS

Two types of Invar SFs-filled 5 and 10 wt.% cenospheres were synthesized by the metal powder injection molding method (MIM), which has been described in detail in relation to MMSFs, e.g., by Peroni *et al.* [23] and by Weise *et al.* [19]. The Invar composition of 64 wt% Fe and 36 wt% Ni was achieved by mixing the respective elementary powders. The Omya Fillite® FG (106) Alumino-silicate Cenospheres (Omya GmbH, D-50679 Köln, Germany)were used and salient properties obtained from the datasheet provided by the supplier are listed in Table 3 [18]. The MIM process is depicted in Figure 16. In the first step, the metal powders were mixed with the binder material (polymer-wax blend, 50 vol.% related to the metal-binder mixture without hollow particles) followed by cenospheres to yield the homogeneous feedstock, which is injected into the mold at a pressure of 10 bar. The temperatures of mixing and injection steps are 125 °C and 71–73 °C (mold temperature 39 °C), respectively. The binder was removed from the green part by chemical (Hexane solvent, 24h at 35 °C) and thermal steps (heating up at 0.1 K/min to 500 °C and holding at this temperature for 1 h). Finally the sample was sintered at a temperature of 1000 °C, and this temperature was ramped up at 5 K/min, with a holding time of 1.5 h in a hydrogen atmosphere. Afterwards, sintered parts where slowly cooled to room temperature within the furnace. Previous studies on Invar and 316L matrices have indicated that neither the configuration of the initial powder mixture [19] nor the duration of the final sintering step [17] have a dominant influence on the mechanical properties.

Table 3. The datasheet of Omya Fillite® FG (106) alumino-silicate cenospheres [18].

Properties	Values
bulk density (g/cm^3)	0.350–0.450
particle density (g/cm^3)	0.600–0.850
particle size range (μm)	5.0–106
melting point (°C)	1200–1350
crushing strength (psi)	1500–3000
component elements of shell	Al_2O_3 (27%–33%), SiO_2 (55%–65%), Fe_2O_3 (6% maximum)

In order to study the compressive properties of Invar SFs under the quasi-static and high strain rate loading, cylindrical specimens, with a ratio of length to diameter of 1:1, were machined. The average diameters of Invar-5CS and Invar 10-CS specimens were 4.4 and 4.6 mm, respectively. An Instron 4469 machine (Intron, Norwood, MA, USA) with 50 kN load cell was used to conduct the quasi-static compression tests on 3 specimens of each type with the initial nominal strain rates of 10^{-4}, 10^{-3} and $10^{-2}\cdot s^{-1}$. A lubricant of Dow Corning-Molykote 111 (Ellsworth Adhesives, Germantown, WI, USA) was used to minimize the friction effect between the cylindrical specimens and compression platens.

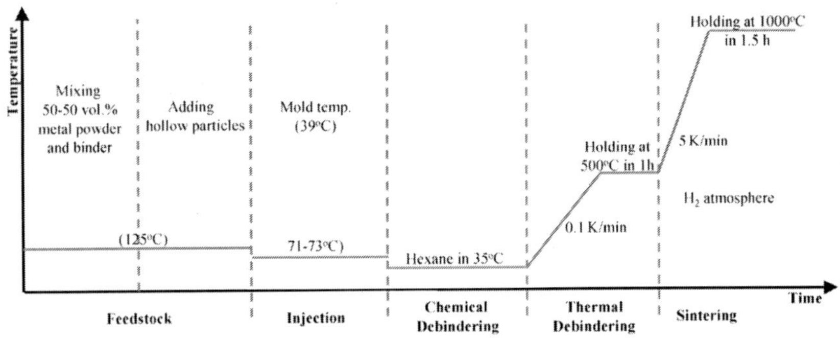

Figure 16. The metal powder injection molding (MIM) process of Invar SFs.

The high strain rate compression tests were conducted by using a split-Hopkinson pressure bar (SHPB) setup. The Invar SF specimen was sandwiched between the incident and transmitter bars. The incident and transmitter bars used in these tests were made of Inconel steel. The stress ((σ);); strain (ε);; and strain rate $(\dot{\varepsilon})$ were calculated by the following equations:

$$\dot{\varepsilon}(t) = \frac{2c_b\varepsilon_r(t)}{l_0} \tag{2}$$

$$\sigma(t) = \frac{AE\varepsilon_t(t)}{A_0} \tag{3}$$

$$\varepsilon(t) = \int_0^{\tau} \dot{\varepsilon}(t)\,d\tau \tag{4}$$

where $\varepsilon_r(t)$ and $\varepsilon_t(t)$ are the reflected and transmitted axial strain pulses, respectively, with respect to time; in addition, *A* and *E* are the cross-section area and Young's modulus of the bar materials, respectively; l_0 and A_0 are the initial length and cross-section area of the specimen, respectively; c_b is the sound wave velocity in the bar; and c_b is a time variable used for integration. The representative strain rate for each stress-strain curve was calculated as the average value of the plateau region of the strain rate-strain graph. The details about the in-house developed SHPB configuration can be found in references [9,12,20,26,38]. The microstructure of Invar SFs before and after the mechanical experiments was observed by using the Hitachi S-3400N scanning electron microscope (SEM, Hitachi, Tokyo, Japan) and a Nikon EPIPHOT 200 microscope fitted with a Nikon DS-Fi1 digital camera (Nikon, Shizuoka, Japan).

5. CONCLUSIONS

In the present work, Invar-cenosphere SFs were investigated in terms of their compressive properties in the strain rate range from 10^{-4} to 2500 s^{-1}. The results show that they are sensitive to strain rate in this range. Their rate-dependent compressive characteristics are experimentally determined by a model that describes stress at given strain levels within the plateau range, allowing the derivation of the respective stress-strain curves for arbitrary rate values within the experimentally covered range. With regard to processing of the materials and the resulting microstructures, this study has confirmed the notion that the higher thermal stability of cenospheres makes them a better choice for powder metallurgy production of Invar matrix SFs than glass microspheres in terms of structural characteristics. Cenospheres clearly survive the synthesis process, while glass microspheres tend to suffer based on their softening point of 600 °C, 400 °C below the sintering temperature employed to produce Invar matrix SFs. At the same time, however, the generally lower mechanical strength of the cenospheres compared to the thermally less stable glass microspheres finds its expression in the mechanical properties of the respective SFs. In this sense, the current study underlines the notion that for optimum performance of metal- and specifically steel-based SFs, higher strength, thermally stable microspheres are mandatory. Candidate materials could be SiC or Al_2O_3. At present, however, these are only available as larger size hollow spheres, which makes them unsuitable for application in the metal powder injection molding or MIM process studied here. For this reason, we strongly advocate the development of micro-scale (less than 100 μm in diameter) ceramic microspheres.

ACKNOWLEDGMENTS

The authors would like to thank Jakub Ostrowski for the sample preparation and gratefully acknowledge support by the German-Egypt Research Fund, Project-No. EGY 08/008. Support from US-Egypt grant through National Science Foundation (IIA-1445686) is acknowledged by Nikhil Gupta.

AUTHOR CONTRIBUTIONS

Dung Luong, Dirk Lehmhus and Nikhil Gupta conceived and designed the experiments; Dung Luong performed the experiments; Dung Luong and Dirk Lehmhus analyzed the data; Joerg Weise, Mohamed Bayoumi and Dirk Lehmhus contributed materials, designed and adapted the manufacturing process and prepared the materials and samples; Dung Luong, Nikhil Gupta and Dirk Lehmhus wrote the paper.

CONFLICTS OF INTEREST

The authors declare no conflict of interest.

REFERENCES

1. Gupta, N.; Zeltmann, S.E.; Shunmugasamy, V.; Pinisetty, D. Applications of polymer matrix syntactic foams. JOM2014, 66, 245–254.
2. Gupta, N.; Pinisetty, D.; Shunmugasamy, V.C. Reinforced Polymer Matrix Syntactic Foams: Effect of Nano and Micro-Scale Reinforcement; Springer International Publishing: Hoboken, NJ, USA, 2013.
3. Friedl, O.; Motz, C.; Peterlik, H.; Puchegger, S.; Reger, N.; Pippan, R. Experimental investigation of mechanical properties of metallic hollow sphere structures. Metall. Mater. Trans. B 2008, 39, 135–146.
4. Fiedler, T.; Öchsner, A. Elastic-plastic properties: Simulation and experiment. In Multifunctional Metallic Hollow Sphere Structures; Öechsner, A., Augustin, C., Eds.; Springer: Berlin/Heidelberg, Germany, 2009; pp. 47–70.
5. Rohatgi, P.K.; Weiss, D.; Gupta, N. Applications of fly ash in synthesizing low-cost MMCs for automotive and other applications. JOM 2006, 58, 71–76.
6. Rohatgi, P.; Gupta, N.; Schultz, B.; Luong, D. The synthesis, compressive properties, and applications of metal matrix syntactic foams. JOM 2011, 63, 36–42.
7. Gupta, N.; Rohatgi, P.K. Metal Matrix Syntactic Foams: Processing, Microstructure, Properties and Applications; DEStech Publications: Lancaster, PA, USA, 2014; pp. 1–370.

8. Weise, J.; Yezerska, O.; Busse, M.; Haesche, M.; Zanetti-Bueckmann, V.; Schmitt, M. Production and properties of micro-porous glas bubble zinc and aluminium composites. Materialwiss. Werkstofftech. 2007, 38, 901–906.
9. Luong, D.; Gupta, N.; Daoud, A.; Rohatgi, P. High strain rate compressive characterization of aluminum alloy/fly ash cenosphere composites. JOM 2011, 63, 53–56.
10. Szlancsik, A.; Katona, B.; Májlinger, K.; Orbulov, I. Compressive behavior and microstructural characteristics of iron hollow sphere filled aluminum matrix syntactic foams. Materials 2015, 8, 7926–7937.
11. Gupta, N.; Luong, D.D.; Cho, K. Magnesium matrix composite foams—Density, mechanical properties, and applications. Metals 2012, 2, 238–252.
12. Luong, D.; Gupta, N.; Rohatgi, P. The high strain rate compressive response of Mg-Al alloy/fly ash cenosphere composites. JOM 2011, 63, 48–52.
13. Castro, G.; Nutt, S.R. Synthesis of syntactic steel foam using mechanical pressure infiltration. Mater. Sci. Eng. A 2012, 535, 274–280.
14. Castro, G.; Nutt, S.R. Synthesis of syntactic steel foam using gravity-fed infiltration. Mater. Sci. Eng. A 2012, 553, 89–95.
15. Xue, X.; Zhao, Y. Ti matrix syntactic foam fabricated by powder metallurgy: Particle breakage and elastic modulus. JOM 2011, 63, 43–47.
16. Mondal, D.P.; Datta Majumder, J.; Jha, N.; Badkul, A.; Das, S.; Patel, A.; Gupta, G. Titanium-cenosphere syntactic foam made through powder metallurgy route. Mater. Des. 2012, 34, 82–89.
17. Weise, J.; Lehmhus, D.; Baumeister, J.; Kun, R.; Bayoumi, M.; Busse, M. Production and properties of 316L stainless steel cellular materials and syntactic foams. Steel Res. Int. 2014, 85, 486–497.
18. Fillite Product Overview—Standard Range. Available online: http://www.omya.com/Documents/Fillite/ProductOverview_StandardRange.pdf (accessed on 25 December 2015).
19. Weise, J.; Salk, N.; Jehring, U.; Baumeister, J.; Lehmhus, D.; Bayoumi, M.A. Influence of powder size on production parameters and properties of syntactic invar foams produced by means of metal powder injection moulding. Adv. Eng. Mater. 2013, 15, 118–122.
20. Luong, D.D.; Shunmugasamy, V.C.; Gupta, N.; Lehmhus, D.; Weise, J.; Baumeister, J. Quasi-static and high strain rates compressive response of iron and Invar matrix syntactic foams. Mater. Des. 2015, 66, 516–531.
21. Peroni, L.; Scapin, M.; Fichera, C.; Lehmhus, D.; Weise, J.; Baumeister, J.; Avalle, M. Investigation of the mechanical behaviour of AISI 316L stainless steel syntactic foams at different strain-rates. Compos. Part B Eng. 2014, 66, 430–442.
22. Peroni, L.; Scapin, M.; Avalle, M.; Weise, J.; Lehmhus, D. Dynamic mechanical behavior of syntactic iron foams with glass microspheres. Mater. Sci. Eng. A 2012, 552, 364–375.
23. Peroni, L.; Scapin, M.; Avalle, M.; Weise, J.; Lehmhus, D.; Baumeister, J.; Busse, M. Syntactic iron foams—On deformation mechanisms and strain-rate dependence of compressive properties. Adv. Eng. Mater. 2012, 14, 909–918.
24. 3M™ Glass Bubbles K Series, S Series and iM Series. 2016. Available online: http://multimedia.3m.com/mws/media/910490/3m-glass-bubbles-k-s-and-im-series.pdf (accessed on 25 December 2015).
25. D'Souza, A.S.; Amos, S.E.; Hendrikson, K. Innovative High Strength Glass Microspheres for Extruded and Injection Molded Plastics. Available online: http://multimedia.3m.com/mws/media/4316900/3mtm-glass-microspheres-for-extruded-inj-molded-plastics.pdf (accessed on 25 December 2015).
26. Luong, D.D.; Strbik, O.M., III; Hammond, V.H.; Gupta, N.; Cho, K. Development of high performance lightweight aluminum alloy/SiC hollow sphere syntactic foams and compressive characterization at quasi-static and high strain rates. J. Alloys Compd. 2013, 550, 412–422.
27. Orbulov, I.N.; Ginsztler, J. Compressive characteristics of metal matrix syntactic foams. Compos. Part A Appl. Sci. Manuf. 2012, 43, 553–561.

28. Weise, J.; Baumeister, J.; Yezerska, O.; Salk, N.; Silva, G.B.D. Syntactic iron foams with integrated microglass bubbles produced by means of metal powder injection moulding. Adv. Eng. Mater. 2010, 12, 604–608.
29. Brown, J.; Vendra, L.; Rabiei, A. Bending properties of Al-steel and steel-steel composite metal foams. Metall. Mater. Trans. A 2010, 41, 2784–2793.
30. Rabiei, A.; Garcia-Avila, M. Effect of various parameters on properties of composite steel foams under variety of loading rates. Mater. Sci. Eng. A 2013, 564, 539–547.
31. Alvandi-Tabrizi, Y.; Whisler, D.A.; Kim, H.; Rabiei, A. High strain rate behavior of composite metal foams. Mater. Sci. Eng. A 2015, 631, 248–257.
32. Lehmhus, D.; Weise, J.; Baumeister, J. Iron matrix syntactic foams. In Metal Matrix Syntactic Foams: Processing, Microstructure, Properties and Applications; Gupta, N., Rohatgi, P.K., Eds.; DEStech Publications: Lancaster, PA, USA, 2014.
33. Avalle, M.; Lehmhus, D.; Peroni, L.; Pleteit, H.; Schmiechen, P.; Belingardi, G.; Busse, M. AlSi7 metallic foams—Aspects of material modelling for crash analysis. Int. J. Crashworth. 2009, 14, 269–285.
34. Lehmhus, D.; Banhart, J.; Rodriguez-Perez, M.A. Adaptation of aluminium foam properties by means of precipitation hardening. Mater. Sci. Tech. 2002, 18, 474–479.
35. Lehmhus, D.; Banhart, J. Properties of heat-treated aluminium foams. Mater. Sci. Eng. A 2003, 349, 98–110.
36. Bardella, L.; Malanca, F.; Ponzo, P.; Panteghini, A.; Porfiri, M. A micromechanical model for quasi-brittle compressive failure of glass-microballoons/thermoset-matrix syntactic foams. J. Eur. Ceram. Soc. 2014, 34, 2605–2616.
37. Vesenjak, M.; Fiedler, T.; Ren, Z.; Öchsner, A. Dynamic behaviour of metallic hollow sphere structures. In Multifunctional Metallic Hollow Sphere Structures; Öechsner, A., Augustin, C., Eds.; Springer: Berlin/Heidelberg, Germany, 2009; pp. 137–158.
38. Gupta, N.; Luong, D.D.; Rohatgi, P.K. A method for intermediate strain rate compression testing and study of compressive failure mechanism of Mg-Al-Zn alloy. J. Appl. Phys. 2011, 109.

CHAPTER 3

Synthesis and Characterization of Poly (D,L-Lactide-co-Glycolide) Copolymer

Cynthia D'Avila Carvalho Erbetta[1], Ricardo José Alves[2], Jarbas Magalhães Resende[3], Roberto Fernando de Souza Freitas[1], Ricardo Geraldo de Sousa[1*]

[1]*Polymer Science and Technology Laboratory, Chemical Engineering Department, School of Engineering, Federal University of Minas Gerais, Belo Horizonte, Brazil;*

[2]*Pharmaceutical Products Department, Pharmacy School, Federal University of Minas Gerais, Belo Horizonte, Brazil;*

[3]*Chemical Department, Exact Sciences Institute, Federal University of Minas Gerais, Belo Horizonte, Brazil.*

ABSTRACT

The copolymer poly(D,L-lactide-co-glycolide) is one of the most interesting polymers for medical applications. This interest is justified by the fact that it is bioreabsorbable, biocompatible and non-toxic, while its degradation kinetics can be modified by the copolymerization ratio of the monomers. In this study, copolymers were synthesised at 175°C by opening the rings of the cyclic dimers of the D,L-lactide and glycolide monomers in the presence of stannous octoate initiator and lauryl alcohol co-initiator. The application of vacuum to the reaction medium, coupled with adequate stirring, is essential for obtaining good results. The following analytical techniques were used to characterise the synthesised copolymers: Differential Scanning Calorimetry (DSC), Thermogravimetry (TG), Nuclear Magnetic Resonance Spectroscopy (NMR) and Fourier Transform Infrared Spectroscopy (FTIR). Both the input monomers and the reaction products were analysed. Important characteristics, such as melting temperature, glass transition temperature, thermal stability, chemical composition and the ratio of the monomers in the synthesised copolymer, were obtained from these analyses. These results helped to infer the absence of residual monomers in the synthesised copolymers.

Keywords: PLGA; Synthesis; Characterization; DSC; TG; NMR; FTIR

1. INTRODUCTION

The most important components of a living cell (proteins, carbohydrates and nucleic acids) are all polymers. Nature uses polymers for construction and as part of the complicated cellular mechanisms [1]. Polymers are a very versatile class of materials and have been changing our daily lives for decades [2] with important applications in the areas of medicine [3,4], agriculture [5,6], and engineering [7].

There has been a rapid growth in the medical use of polymeric materials in many fields including tissue engineering, implant of medical devices and artificial organs, prostheses, ophthalmology, dentistry, and bone repair [8-11].

The polyesters poly(lactic acid) (PLA) and poly(glycolic acid) (PGA) were the two first materials to be successfully used as sutures, and these polymers have extensively been used over the past two decades [12]. These polymers have widespread pharmaceutical use. They belong to a group of polymers that carry hydrolysable groups within their [5], which are susceptible to biodegradation [13]. An interesting alternative to the use of these materials is the copolymer poly(D,L-lactide-co-glycolide) (PLGA), which contains alternate unities of lactic acid and glycolic acid monomers. The structures of PGA, PLA and PLGA are presented in **Figure 1**.

PLA and PLGA are relatively hydrophobic polyesters, unstable in damp conditions and biodegradable to nontoxic by-products (lactic acid, glycolic acid, carbon dioxide, and water).

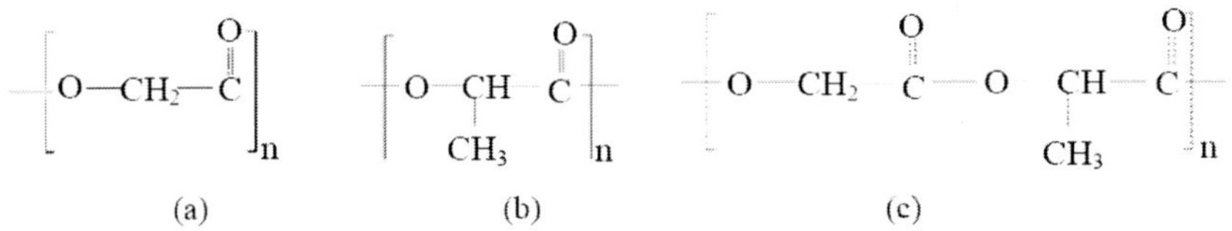

Figure 1. Polymers with hydrolysable chains: (a) poly(glycolic acid); (b) poly(lactic acid) and (c) poly(D,L-lactide-coglycolide) copolymer.

The polymers derived from lactic and glycolic acids have received a great deal of attention in research on alternative biodegradable polymers. Various studies demonstrating their low toxicity, including their approval by the Food and Drug Administration (FDA) for use as drug delivery systems (DDS) can be found in the literature [14,15]. PLGA copolymers have desirable properties, such as a constant biodegradation rate, mechanical resistance, and regular individual chain geometry [10,16,17].

The copolymerization of PLGA can basically be done in two ways: 1) Through a direct polycondensation reaction of lactic acid and glycolic acid, resulting in copolymers of low molecular weight [18,19]; 2) Through an opening polymerization of cyclic dimers of lactic acid (lactide) and glycolic acid (glycolide), resulting in copolymers with high molecular weight and therefore with better mechanical properties [20-23]. The bulk polymerization generally takes from two to six hours for temperatures around 175°C and in the presence of initiator. Lauryl alcohol can be added during the process in order to control the molecular weight [8].

Lactide is more hydrophobic than glycolide, therefore PLGA copolymers rich in lactide are less hydrophilic and absorb less water, leading to a slower degradation of the polymer chains [10]. It is important to stress that the ratio glycolide to lactide and the physical-chemical properties of the corresponding copolymer are not linear. While PGA is highly crystalline, the crystallinity rapidly disappears in copolymers of lactide and glycolide. This morphological change leads to an increase in the rate of hydration and hydrolysis [24]. According to GILDING and REED (1979) [20], the greater reactivity of glycolide compared to lactide means that glycolide is usually found in the final polymer in a greater proportion compared to its portion in the initial mixture of monomers.

The mechanical properties, the capacity for hydrolysis and the degradation velocity of these polymers may be all influenced by the molecular mass and the degree of crystallinity [25,26]. The degradation period of biodegradable polymers is most significantly affected by: the chemical structure and composition of the system; the molecular mass distribution of the polymers; the presence of monomers and oligomers; the size and shape of the system surface; the morphology of the system components and the hydrolysis mechanism [27].

Lactide has asymmetric carbons, which means that the levorotatory (L-PLA), dextrorotatory (D-PLA) and racemic (D,L-PLA) polymeric forms may be obtained. The levorotatory and dextrorotatory forms are semi-crystalline, while the racemic form is amorphous due to the irregularities in the polymer chain [28]. The use of (D,LPLA) is preferred in a DDS system, as the drug can be more homogeneously distributed through the polymer matrix [29]. When polymers are produced for medical applications, two tin salts, tin (II) chloride and tin (II) 2- ethylhexanoate, abbreviated to Stannous Octoate [$Sn(Oct)_2$], may be used as polymerization catalysts [21].

The main objective of this study regards the synthesis of one of the most widely used biopolymers in the medical area, the copolymer poly(D,L-lactide-co-glycolide). New conditions suitable for its synthesis are explored. The obtained biopolymers were characterised by using a variety of analytical techniques in order to identify their chemical composition, as well as the final ratio between monomers within the polymeric structure. Analytical techniques have also been employed in order to investigate the biopolymers thermal behaviour as well as to verify the presence or absence of unreacted monomers in the final reaction mixture.

2. MATERIALS AND METHODS

2.1. PLGA Copolymer Synthesis

PLGA was synthesised by opening the rings of D,L-lactide and glycolide in a mass polymerization. Variables such as the proportion of the stannous octoate initiator (0.02% of the total monomer mass) and the lauryl alcohol co-initiator (0.01% of the total monomer mass) were chosen based on the study conducted by KIREMITÇIGÜMÜSDERELIOGLU and DENIZ (1999) [30]. Other process variable reference values, such as the reaction time (2h) and temperature (175°C) were chosen according to a previous study by BENDIX (1998) [21]. The copolymerization reaction is illustrated in **Figure 2** [30].

Monomer ratios of 70/30 and 50/50 (D,L-lactide/glycolide) have been used in the syntheses. These proportions have been chosen by considering the degradation time of these copolymers [8]. These are the most commonly used ratios in controlled drug delivery systems.

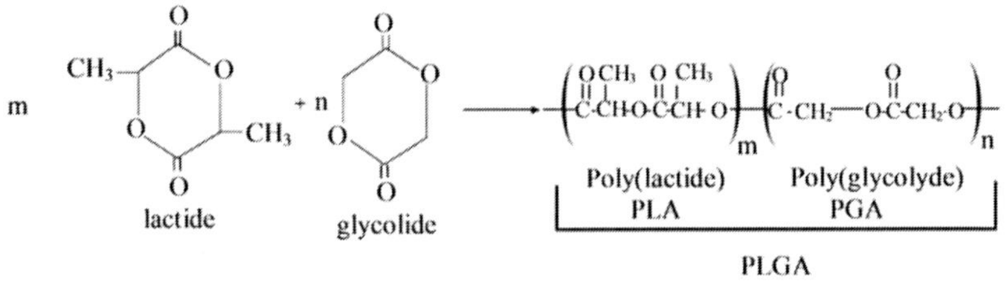

Figure 2. Chemical structure of the dimers and polymers, and the copolymerization reaction.

A total of eight syntheses have been performed, with four of these using the 70/30 ratio and the other four with the 50/50 ratio. The synthesis methodology has been modified in order to optimize the process. In the first two syntheses (syntheses 1 and 2) a round-bottomed flask without a stopper has been used. The reagents were added to the flask, which was immersed in a thermostatic bath (Haake/B12) and stirred during the first 10min of the synthesis. In synthesis 3 a rotary evaporator (Technical/TE-210) was employed. The only difference between this and preceding reaction was the injection of nitrogen into the reaction medium. In synthesis 4, the round-bottomed flask was replaced by vacuometer-type glass flask (Protec), allowing a high vacuum (–60 cmHg) to be applied and controlled. Mechanical stirring was used throughout the synthesis. From this synthesis onwards, nitrogen was no longer injected into the reaction medium. Hence, in the last four syntheses (syntheses 5 to 8), a glass flask with a rubber stopper was used instead of the vacuometer flask and the vacuum was achieved after all of the synthesis reagents had been added. Magnetic stirring was continued until the magnetic bar ceased to move due to the increased viscosity of the solution. **Table 1** summarises the syntheses carried out. They are allocated to groups characterising the 4 different routes (group), as described above.

At the end of the syntheses, the PLGA copolymer was dissolved in anhydrous methylene chloride and precipitated in excess anhydrous methanol. The obtained material was dried at between 60°C and 70°C under vacuum (–60cmHg) in an oven (Vacuoterm/6030A) for a minimum of 48h to remove residual solvents. The copolymer was then ground using a ball mill (Stoneware) at 100 rpm for at least 24 h. The ground PLGA was sieved in a 24- mesh sieve.

2.2. PLGA Copolymer Characterization

2.2.1. Differential Scanning Calorimetry (DSC)

The DSC equipment used, a Shimadzu model DSC-50, was programmed to first heat the samples from room temperature (±20°C) to 120°C at a rate of 10°C·min^{-1}(1st run). All PLGA samples were subjected to this first run to eliminate the material's thermal history. An unsealed aluminium sample vessel was used with nitrogen as the carrier gas at a flow rate of 20 mL·min^{-1}. The mass of the analysed sample varied from 5 to 10 mg. At the end of the first run, the oven was cooled with liquid nitrogen until it reached a temperature of between −20°C and −30°C. The equipment was then programmed for another run up to 300°C, at a heating rate of 10°C·min^{-1} for the monomers and 5°C·min^{-1} for the copolymers. The second-run DSC curve was the reference for determining the glass transition temperature (T_g), phase transition temperature (T_m), and degradation temperature (T_d).

2.2.2. Thermogravimetry (TG)

TG analysis was carried out to measure change in mass with increase in temperature, thermal stability, and maximum degradation temperature for the samples. The test was conducted at a heating rate of 10°C·min^{-1} from room temperature (±20°C) to 400°C in an unsealed platinum sample vessel under nitrogen atmosphere with a flow rate of 20 mL·min^{-1}. The equipment used was a Shimadzu model TGA-50. The mass of the analysed samples varied between 5 and 10mg. This technique allowed us to determine the temperature at which thermal degradation commenced (T_{onset}) and the change in mass as a function of temperature increase. Derived thermogravimetric curves (DrTG) were used to identify the maximum degradation temperature ($T_{deg.max.}$).

2.2.3. Fourier Transform Infrared (FTIR)

The infrared absorption spectra were collected at 20°C from 4000 - 650 cm^{-1} for monomers and copolymers samples. The spectra were recorded on a Perkin Elmer spectrometer operating in the ATR (Attenuated Total Reflectance) mode. 4 scans have been performed for a resolution of 4 cm^{-1}. The FTIR analyses of several samples containing mixtures of PLGA and monomers have also been carried out to investigate the presence of residual monomers in the PLGA copolymers.

Table 1. Syntheses by group.

GROUP	SYNTHESIS	LA/GL (%)	TOTAL MONOMER MASS (g)	LACTIDE MASS (g)	GLYCOLIDE MASS (g)	STANNOUS OCTOATE VOLUME (µL)	LAURYL ALCOHOL VOLUME (µL)
I	1	70/30	17	11.9029	5.1006	2.72	2.06
	2	50/50	17	8.5043	8.5016	2.72	2.06
II	3	70/30	17	11.9050	5.1091	2.72	2.06
III	4	50/50	51	25.5067	25.5107	8.16	6.18
IV	5	50/50	17	8.5078	8.5054	2.72	2.06
	6	50/50	17	8.5053	8.5035	2.72	2.06
	7	70/30	17	11.9104	5.1094	2.72	2.06
	8	70/30	17	11.9049	5.1024	2.72	2.06

2.2.4. Nuclear Magnetic Resonance (NMR)

NMR samples have been prepared by dissolving the copolymers in $CDCl_3$ from Aldrich containing TMS at 0.05%. ^1H and ^{13}C NMR spectra were obtained at, respectively, 400 MHz and 100 MHz. Measurements have been performed at 300 K on a Bruker AVANCE DRX 400 spectrometer equipped with a ^1H-^{13}C 5 mm dual probe. The ^1H NMR spectra were used to determine the ratio of the monomers in the analysed copolymers. In some cases quantitative ^{13}C spectra have been obtained by using inverse-gated decoupling, in order to confirm the results obtained from ^1H NMR spectroscopy. TMS was used as internal reference.

3. RESULTS AND DISCUSSION

3.1. Synthesis of the PLGA Copolymers

During the first two syntheses (1 and 2) it was noted that vapour formed, partially condensed on the wall of the reaction flask, and then crystallised after cooling naturally. In seeking the most favourable copolymerization conditions, rotary vapour equipment was used in synthesis 3. This also enabled nitrogen to be injected, with a vacuum pump providing suction. However, the formation of vapour became more pronounced and intense. Some of the crystals were removed for analysis to identify their composition and, consequently, to define a more appropriate synthesis route.

It is possible to conclude from DSC, TG, and FTIR analyses of the crystals obtained in synthesis 3 that they were produced from the evaporation and crystallisation of the D,L-lactide monomer. Because of this, vacuum was applied in the subsequent syntheses without injection of nitrogen, by using sealed glass flasks.

Glycolide melted during the first minutes of the reaction and occupied the lower part of the reaction flask, while the other monomer, D,L-lactide, needed more time to melt and was suspended in the molten glycolide until its complete melt. Accordingly, efficient stirring was required to avoid preferential polymerization of the glycolide or the formation of a block of copolymer, as favoured by the following conditions: the temperature, the greater reactivity of glycolide compared with D,L-lactide, and the presence of the initiator. Another important synthesis variable is the application of vacuum, since it avoids the presence of moisture—which is extremely harmful to the copolymerization process [8,31,21]—and high pressure in the system due to evaporation of D, L-lactide monomer. With the evaporation of D,L-lactide the pressure inside the reaction flask (closed system) increases, favouring its condensation back to the reaction flask.

It was noted that copolymers with a greater proportion of glycolide are less soluble, requiring a greater volume of solvent (anhydrous methylene chloride) and more time for complete dissolution.

3.2. Characterization of the Monomers and PLGA Copolymers

3.2.1. Differential Scanning Calorimetry (DSC)

Figure 3 shows the DSC curves for the second heating in the thermal study of the analysed monomers. The DSC curves show how heat (mW) varies as a function of temperature. Both curves show two sharp endothermic events.

It can be seen from **Figure 3** that the first thermal event on the DSC curves represents the melting of the monomers (81°C for glycolide and 125°C for D,L-lactide). For D,L-lactide, the second endothermic peak (at 183°C) characterises the evaporation of this monomer, also identified in the monomer characterization conducted by JABBARI and XUEZHONG (2008) [32]. In the case of glycolide, the second peak (at 206°C) relates to the degradation of the monomer, since glycolide does not have a boiling point.

Figures 4 and 5 give DSC curves for the synthesised copolymers, grouped in accordance with the ratio of monomers. It can be seen that the thermal behaviour of the biopolymers is very similar, which means that all of them have a glass transition phase between 31°C and 53°C and degrade above 200°C. From these DSC curves, it is possible to see the thermal transitions related to the glass transition (change in the angle of the DSC curves to the baseline) and the degradation of the polymeric chains (endothermic peak of the DSC curves).

The absence of a melting transition phase event can also be seen from the analysis of these curves. It can be inferred that all of the analysed copolymers are amorphous, which is in agreement with the literature. PLGAs containing less than 85% glycolide are amorphous [33]. Comparing the DSC curves in Figures 4 and 5 with that of **Figure 3**, it can be seen that there is no indication of residual monomers in the synthesised copolymers, since no endothermic event characteristic of theses monomers melting is observed on the respective DSC curves.

The results of the DSC analyses are summarised in **Table 2**. This table gives the glass transition (T_g) and degradation (T_d) temperatures for the PLGA copolymers. The melting (T_m), boiling and/or degradation temperatures are given for the monomers. All of the values were determined by the equipment software. The T_d and T_m values relate to the endothermic peaks of the DSC curves.

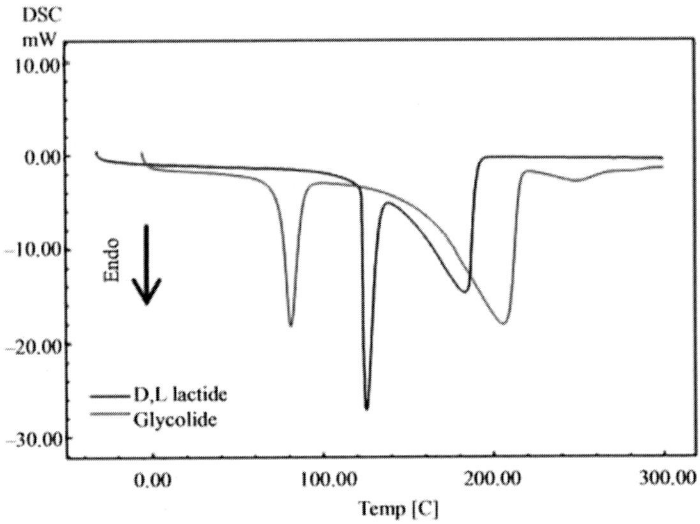

Figure 3. DSC curves of D,L-lactide and glycolide monomers (2nd heating). Heating rate of 10°C·min^{-1} and heating to 300°C.

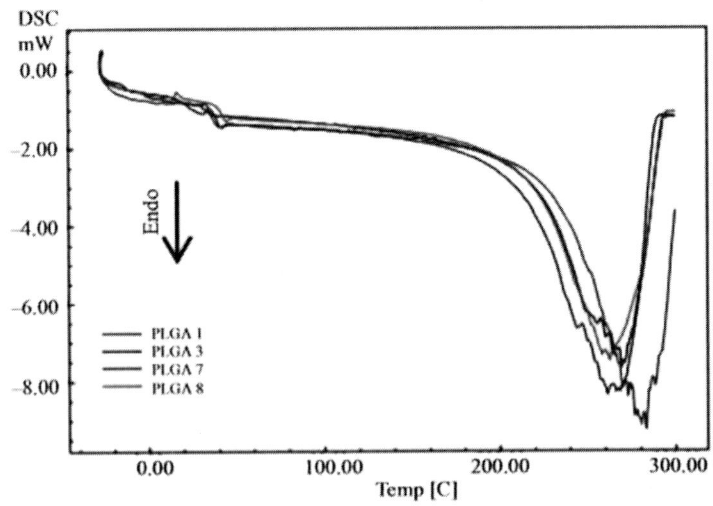

Figure 4. DSC curves of PLGA 70/30 copolymers (syntheses 1, 3, 7, and 8), 2nd heating. Heating rate of 5°C·min^{-1} from –25°C to 300°C.

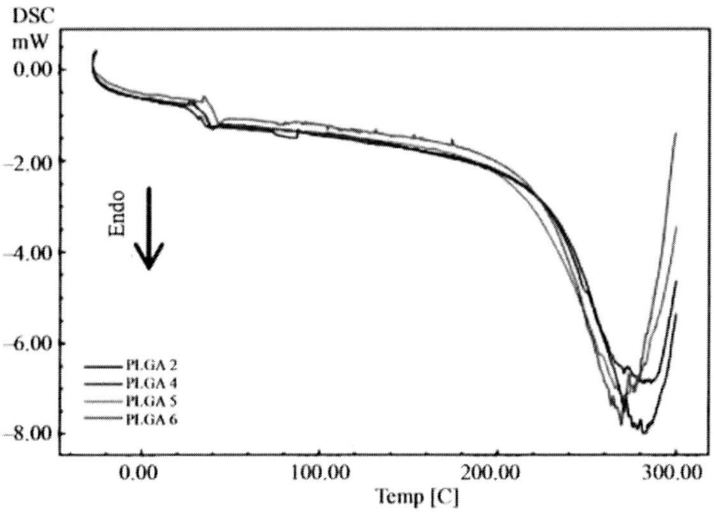

Figure 5. DSC curves of PLGA 50/50 copolymers (syntheses 2, 4, 5, and 6), 2nd heating. Heating rate of 5°C·min^{-1} from –25°C to 300°C.

Table 2. Thermal properties of the monomers and copolymers obtained by DSC.

SAMPLE	T_g (°C)	T_m (°C)	T_d (°C)
D,L lactide	N/A	125	183*
Glycolide	N/A	81	206
PLGA 70/30 synthesis 1	46	N/A	283
PLGA 50/50 synthesis 2	31	N/A	282
PLGA 70/30 synthesis 3	53	N/A	269
PLGA 50/50 synthesis 4	35	N/A	282
PLGA 50/50 synthesis 5	35	N/A	270
PLGA 50/50 synthesis 6	40	N/A	270
PLGA 70/30 synthesis 7	38	N/A	261
PLGA 70/30 synthesis 8	39	N/A	262

*refers to the D,L-lactide monomer evaporation and not its degradation; N/A = Not applicable.

After the synthesis variables were standardised, syntheses 5 and 6 for the PLGA 50/50 and 7 and 8 for PLGA 70/30 had very good repeatability for the polymerization product. The T_g and T_d values for these biopolymers, obtained from duplicate syntheses, are identical or very close. This indicates the same thermal behaviour and, therefore, similar structures and morphologies for the two biopolymers synthesised in different batches.

The degradation temperature of PLGA 70/30 (261.5°C ± 0.5°C; related to the syntheses 7 and 8) is lower than that of PLGA 50/50 (270°C; related to the syntheses 5 and 6). A greater quantity of the D,L-lactide monomer in the PLGA induces a greater branching in the structure due to the presence of the methyl group, making it stereochemically less thermostable. This means that the structure of PLGA 70/30 is more susceptible to thermal degradation than that of PLGA 50/50, thus giving a lower T_d. When the curves obtained in the present study (**Figure 5**) are compared with the DSC curves for the D,L-PLGA copolymer published by SILVA JUNIOR and collaborators (2009) [34] the same behaviour regarding the glass transition and degradation of the biopolymers is observed. The values found for these temperatures are lower than the respective values found in the literature [34], possibly due to the fact that the obtained polymers have lower molecular masses, something that was not determined experimentally. Polymers with lower molecular masses degrade more easily than those with higher molecular masses [35].

3.2.2. Thermogravimetry (TG)

Figures 6 to 8 show the thermogravimetry curves (TG or TGA) and their derivatives obtained for the D,L-lactide and glycolide monomers and for the synthesised copolymers. These curves show the variation of mass (%) and the derivative of the mass loss (DrTG or DrTGA) as a function of temperature (°C). All of the TG curves presented in these figures show a single mass loss event related to the degradation process, except for the D,Llactide monomer, in which this thermal event related to its evaporation is observed.

The results from the TG analyses for the monomers and copolymers are presented in **Table 3**. The T_{onset} and T_{endset} values (respectively, the temperatures at which mass loss begins and ends) were obtained from the TG curves, along with the results for $T_{deg.max.}$ (temperature of maximum mass loss) for the DrTG curves of the respecttive monomers and copolymers.

The existence of only one mass loss stage for the copolymers is confirmed by the single peak observed in the DrTG curves. The copolymers fully degraded (100% mass loss), except PLGA 50/50 obtained from synthesis 4, which was 99%, and for PLGA 50/50 obtained by synthesis 2, which was

97%. This same mass loss value was observed for the degradation of the glycolide monomer. The thermogravimetry results presented above indicate that the degradation onset temperatures (T_{onset}) of the synthesised copolymers are higher than 240°C.

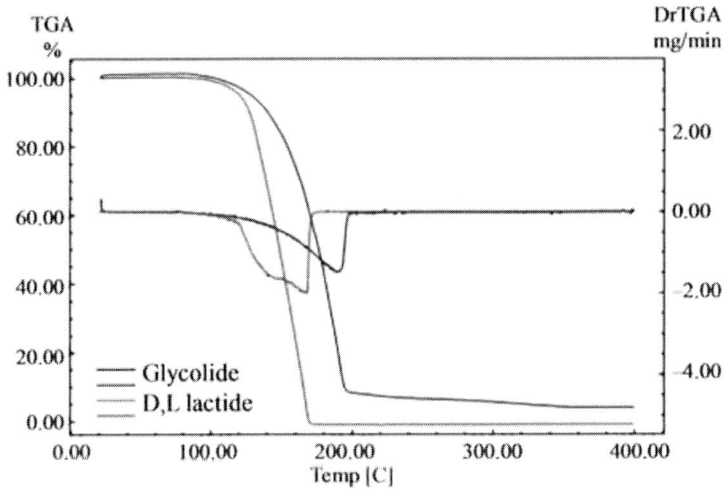

Figure 6. TG and DrTG curves of glycolide and D,L-lactide monomers. Heating rate of 10°C·min^{-1} from room temperature to 400°C.

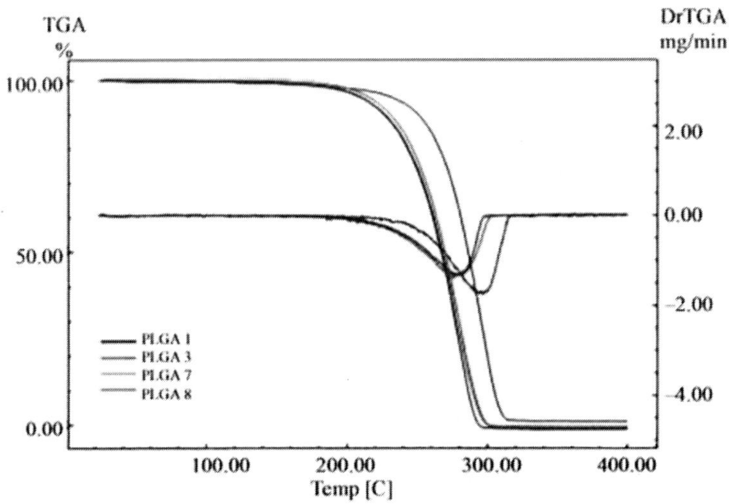

Figure 7. TG and DrTG curves of PLGA 70/30 copolymers (syntheses 1, 3, 7, and 8). Heating rate of 10°C·min^{-1} from room temperature to 400°C.

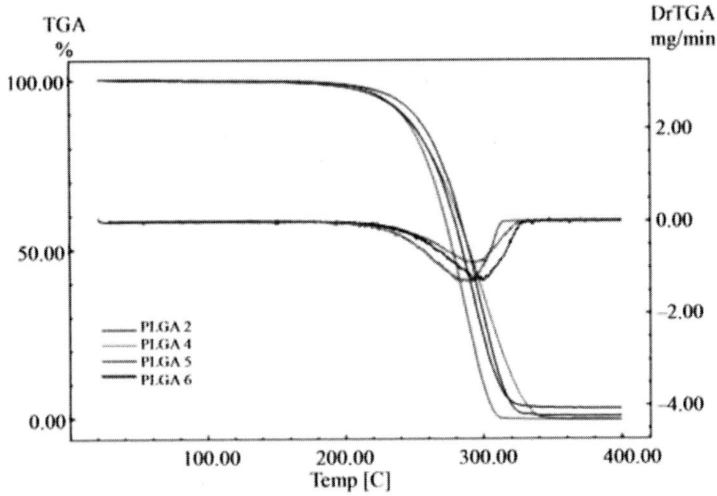

Figure 8. TG and DrTG curves of PLGA 50/50 copolymers (syntheses 2, 4, 5, and 6). Heating rate of $10°C{\cdot}min^{-1}$ from room temperature to 400°C.

The closeness between the respective values of T_{onset} and $T_{deg.max.}$ for PLGA 50/50 obtained from syntheses 5 and 6, and for PLGA 70/30 obtained from syntheses 7 and 8, also indicates the good repeatability of these methodologies. This was achieved through the standardisation of variables carried out during the copolymer synthesis stage. In addition, it can be seen that these temperatures were lower for PLGA 70/30 than for PLGA 50/50, similar to the T_d values observed for these copolymers using DSC. By comparing the $T_{deg.max.}$ values obtained from the DrTG curves of the copolymers with the T_d values determined from the DSC curves (**Table 2**), it can be seen that they are between 261°C and 299°C. The $T_{deg.max.}$ values are, in general, greater than the T_d values for each copolymer, which is as expected since the rate of heating in the TG tests ($10°C{\cdot}min^{-1}$) was greater than that used in the DSC tests ($5°C{\cdot}min^{-1}$, 2nd run).

When compared with the thermogravimetry results published by SILVA JUNIOR and co-authors (2009) [34], the polymers synthesised in this study show similar mass loss behaviour (weight loss near 100% and in a single stage), though with lower thermal degradation temperatures. It is attributed to the different molecular masses of the copolymers, as discussed earlier with regard to the DSC results.

Figure 9 shows the TG curves for all of the copolymers synthesised in this study, together with the curves for the respective monomers. These curves show that there is no residual monomer in the copolymers, since no mass loss events characteristic of theses monomers are observed. The TG curves for the copolymers show no mass loss over the temperature range in which the evaporation and degradation of D,L-lactide and glycolide occur. This observation is in agreement with the DSC curves shown in Figures 4 to 6 (see text for details).

Table 3. Thermal properties of monomers and copolymers obtained by TG.

SAMPLE	MASS LOSS (%)	T_{onset} (°C)	$T_{deg.max.}$ (°C)	T_{endset} (°C)
D,L lactide	100	137	168*	169
Glycolide	97	161	189	195
PLGA 70/30 synthesis 1	100	265	296	307
PLGA 50/50 synthesis 2	97	257	293	312
PLGA 70/30 synthesis 3	100	241	273	295
PLGA 50/50 synthesis 4	99	264	293	314
PLGA 50/50 synthesis 5	100	252	299	326
PLGA 50/50 synthesis 6	100	248	289	305
PLGA 70/30 synthesis 7	100	215	276	290
PLGA 70/30 synthesis 8	100	245	279	295

*relates to evaporation.

3.2.3. Nuclear Magnetic Resonance (NMR)

NMR spectroscopy was used to analyse the composition of the PLGA copolymers. The ratio of D,L-lactide to glycolide was calculated based on the area delimitated by the integral of the D,L-lactide methyl group hydrogens and the area delimitated by the integral of the glycolide methylene group hydrogens [36].

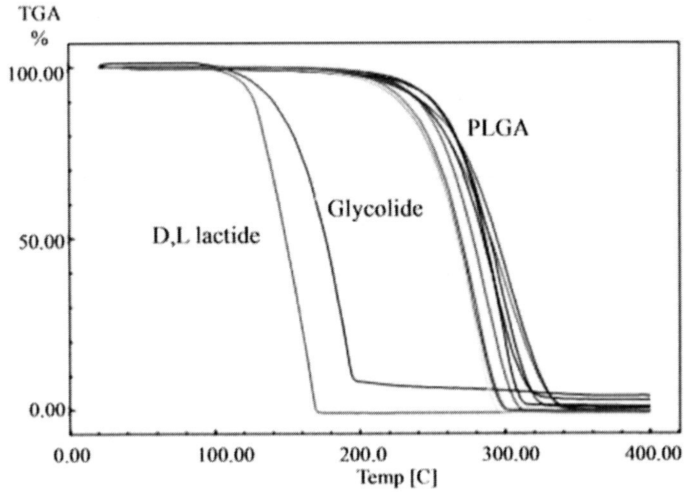

Figure 9. TG curves of D,L-lactide and glycolide monomers and the PLGA copolymers (syntheses 1, 2, 3, 4, 5, 6, 7 and 8). Heating rate of 10°C·min^{-1} from room temperature to 400°C.

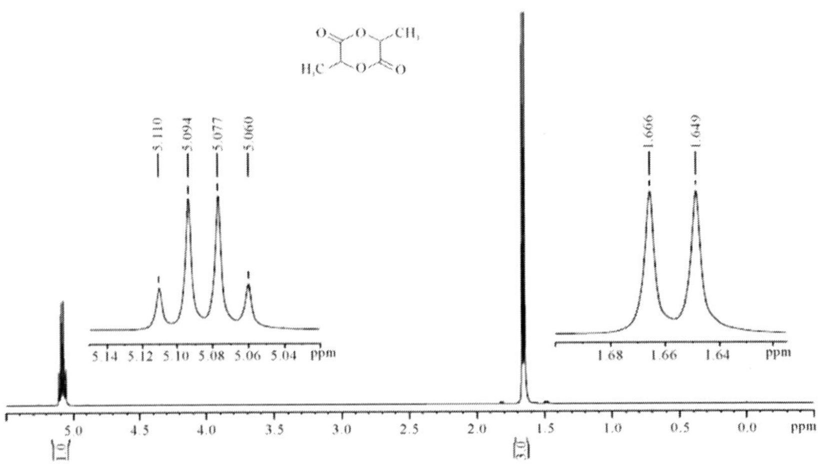

Figure 10. ^{1}H NMR spectrum of D,L-lactide.

Figures 10 and 11 show, respectively, the ^{1}H NMR spectra obtained for D,L-lactide and glycolide monomers. **Table 4** gives the chemical shift values for the CH_3, CH_2, and CH functional groups.

From the results given in **Table 4**, the methyl group (CH_3) was adopted as a reference for the D,L-lactide monomer, with a peak of 1.66ppm. With regard to the glycolide, the reference adopted was the peak observed for the methylene group (CH_2), at 4.99 ppm.

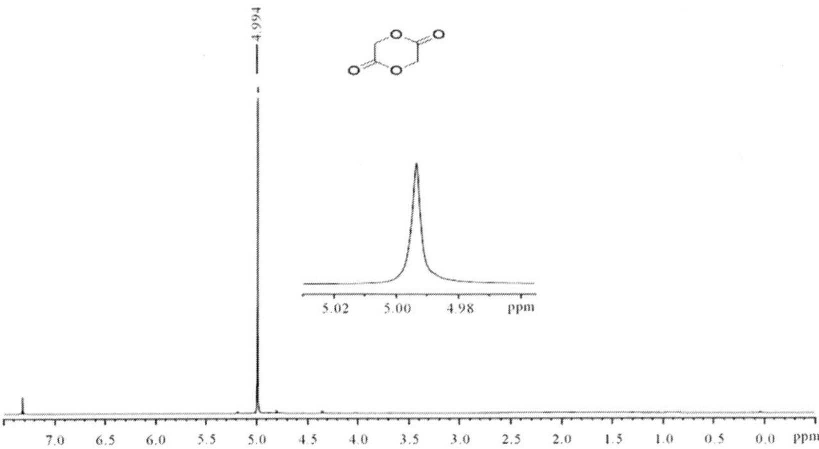

Figure 11. ^{1}H NMR spectrum of glycolide.

Table 4. Chemical shifts obtained from the ^{1}H NMR spectra of D,L-lactide and glycolide monomers.

Functinal Group	CH_3	CH_2	CH
D,L-lactide δ (ppm)	1.66	N/A	5.08
Glycolide δ (ppm)	N/A	4.99	N/A

Figures 12 to 19 show the ^{1}H NMR spectra obtained for the PLGA copolymers. These spectra were used in order to identify the characteristic functional groups and to determine the ratio between the D,L-lactide and glycolide monomers within the PLGA copolymers. This ratio can be evaluated from Equations 1 and 2 [36]:

$$\%\text{D,L-lacticle monomer} = \frac{(CH_3)\text{ integral area}/3}{(CH_3)\text{ integral area}/3 + (CH_2)\text{ integral area}/2} \times 100 \quad (1)$$

$$\% \text{ glycolide monomer} = \frac{(CH_2)\text{ integral area}/2}{(CH_3)\text{ integral area}/3 + (CH_2)\text{ integral area}/2} \times 100 \quad (2)$$

No residual signal of monomers are observed in the NMR spectra of the copolymers, which suggests that non-significant quantities of monomers or no monomers at all are present in the final products.

The ^{1}H NMR spectra of the PLGA copolymers obtained in this study agree with those found in the literature [30]. **Table 5** shows the chemical shift ranges of the CH_3, CH_2 and CH functional group observed in the spectra presented in Figures 12 to 19.

The theoretical and experimental values of the expected and determined monomer ratios of the PLGA samples are presented in **Table 6**.

Amongst the syntheses, two of them (4 and 6) show experimental results for the ratio of monomers identical to the theoretical values. For the other syntheses, the lower proportions of the D,L-lactide were 46% for the PLGA 50/50 and 63% for the PLGA 70/30. These results show strong agreement between the real and theoretical ratios. The difference observed in the majority of the syntheses may be due to the absence of the vacuum (syntheses 1 and 2) or an insufficient vacuum when it was used (syntheses 3, 5, 7, and 8). In addition, it is common to find a greater proportion of glycolide in the PLGA copolymer compared to the initial (theoretical) proportion because its reactivity is greater than the reactivity of the D,L-lactide monomer [20].

The PLGA copolymers obtained from syntheses 3, 7 and 9 were also quantitatively analysed by ^{13}C NMR, for comparison with the results obtained from ^{1}H NMR spectroscopy. **Table 7** shows the results obtained for the composition of PLGA copolymers based on the quantitative ^{13}C NMR data in comparison with those obtained from the ^{1}H NMR spectra, as well as the expected theoretical values.

As can be seen, the ratio between the monomers in the PLGA determined from the ^{13}C NMR spectra is identical to that found using the ^{1}H NMR spectra, except for the synthesis 3 that is about equal. This result confirms the real proportions of the monomers in the copolymers and validates all of the calculations performed with the ^{1}H NMR data.

Table 5. Chemical shifts ranges observed in the ^{1}H NMR spectra of PLGA copolymers.

Functinal Group	CH_3	CH_2	CH
δ (ppm)	1.490 - 1.687	4.675 - 4.909	5.157 - 5.302

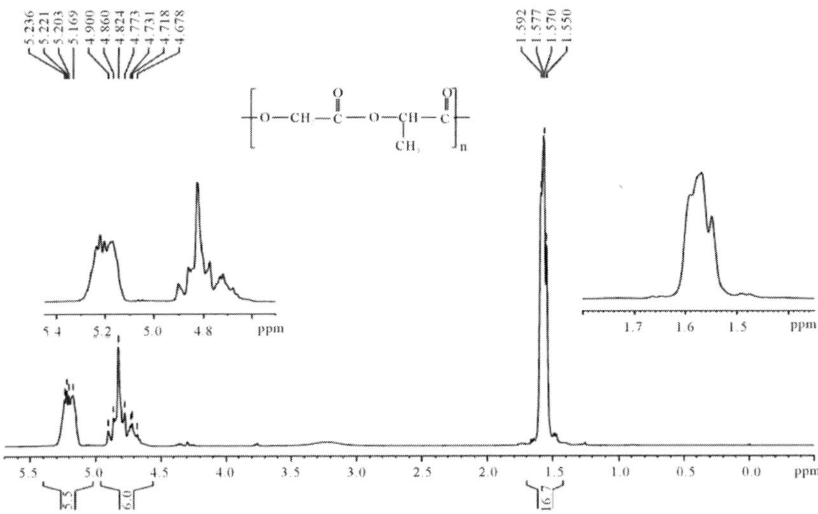

Figure 12. ^{1}H NMR spectrum of PLGA 70/30 (synthesis 1).

3.2.4. Fourier Transform Infrared (FTIR)

The spectra obtained by FTIR for the D,L-lactide and glycolide monomers are presented in Figures 20 and 21, respectively. One can observe, in both the spectra, strong bands in the region between 1760 and 1750 cm^{-1}, due to stretch of the carbonyl groups present in the two monomers. There are also stretching bands due to asymmetric and symmetric C-C(=O)-O vibrations between 1300 and 1150 cm^{-1}. The bands in these regions are useful in the characterization of esters. The 3500 and 3459 cm^{-1} bands in the FTIR spectra for lactide and glycolide (Figures 20 and 21, respectively) are attributed to moisture in the sample (OH group). **Table 8** gives the absorption bands of interest in this study, with their respective attributions.

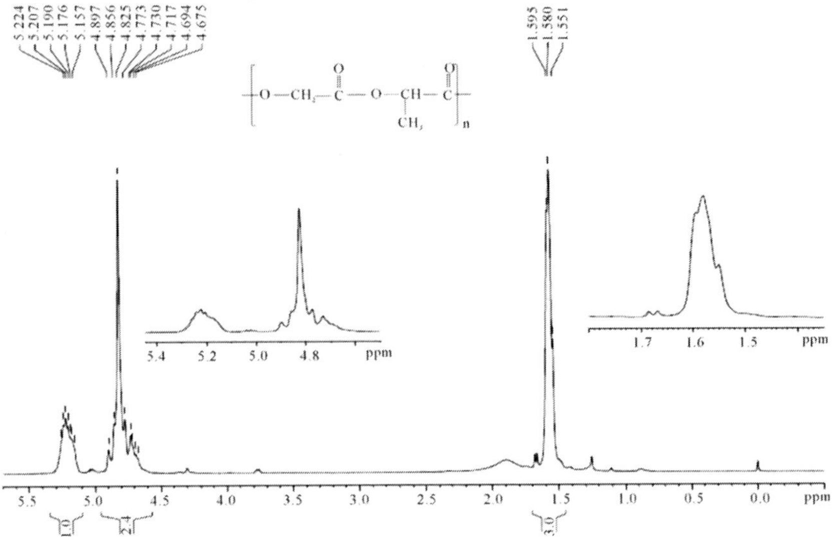

Figure 13. ^{1}H NMR spectrum of PLGA 50/50 (synthesis 2).

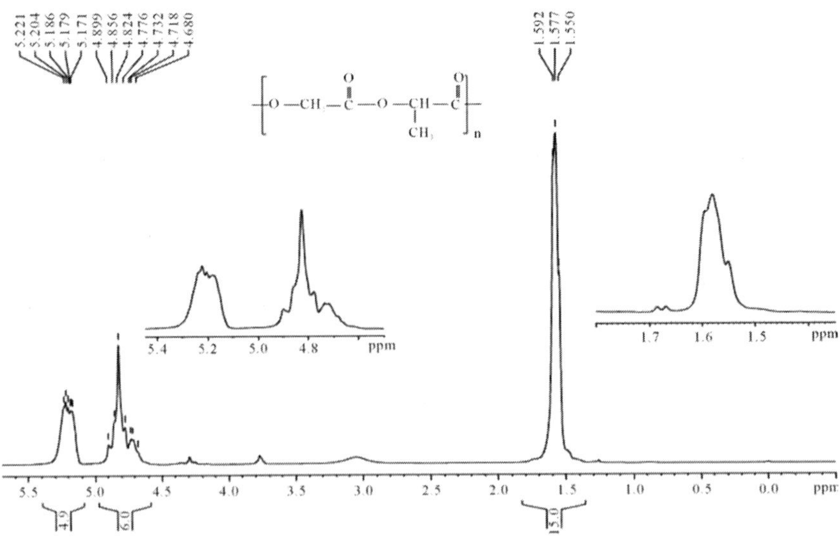

Figure 14. ^{1}H NMR spectrum of PLGA 70/30 (synthesis 3).

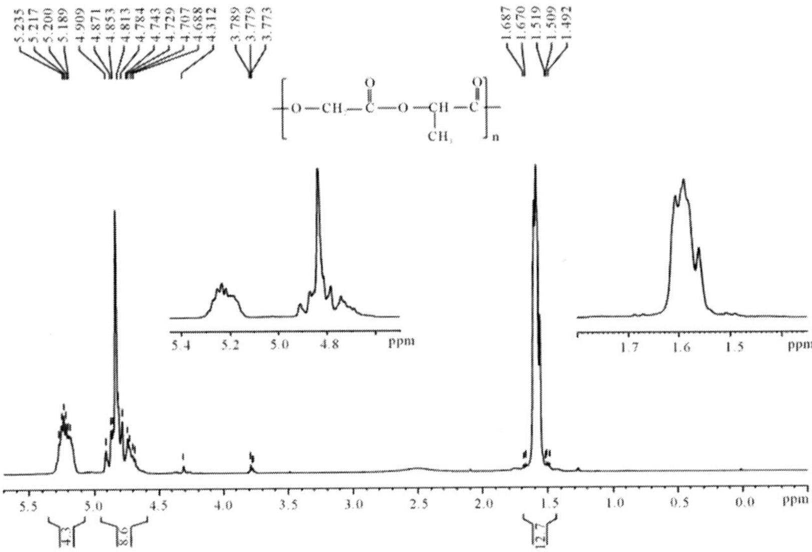

Figure 15. ^{1}H NMR spectrum of PLGA 50/50 (synthesis 4).

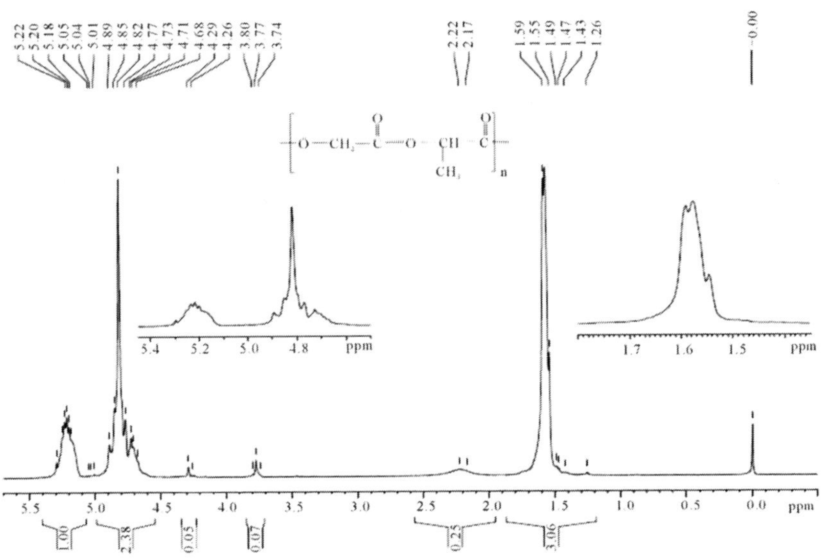

Figure 16. ^{1}H NMR spectrum of PLGA 50/50 (synthesis 5).

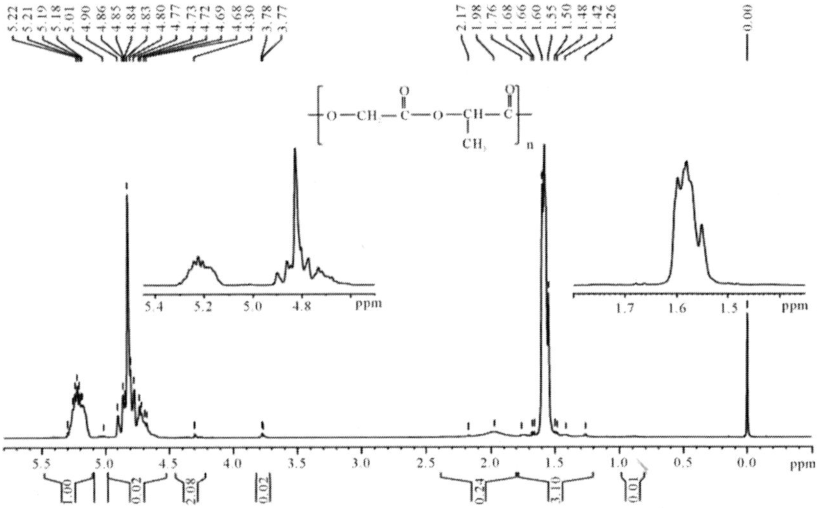

Figure 17. ^{1}H NMR spectrum of PLGA 50/50 (synthesis 6).

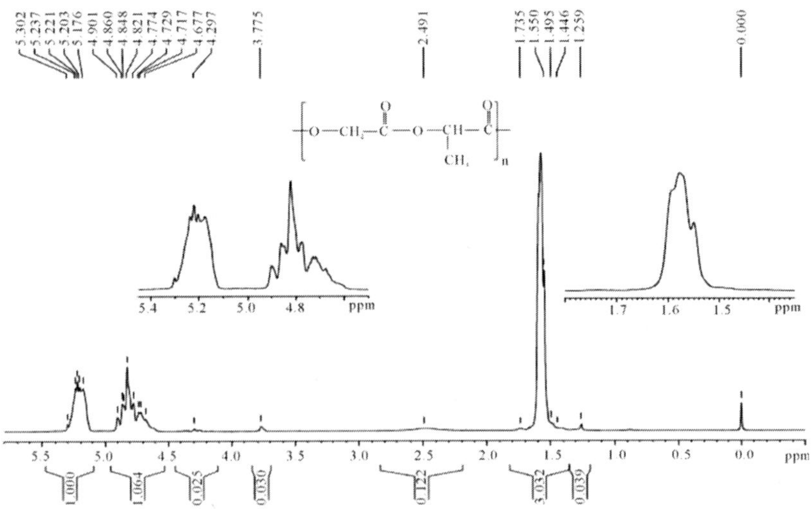

Figure 18. ^{1}H NMR spectrum of PLGA 70/30 (synthesis 7).

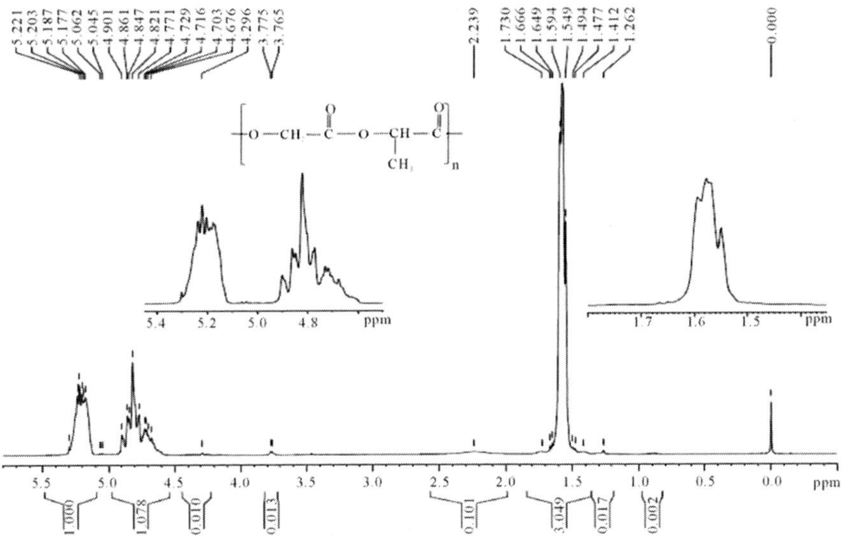

Figure 19. ¹H NMR spectrum of PLGA 70/30 (synthesis 8).

Table 6. Expected and determined composition of D,L-lactide and glycolide monomers present in the PLGA copolymers.

Synthesis	Theoretical D,L-lactide/Glycolide (%)	Experimental D,L-lactide (%)	Experimental Glycolide (%)
1	70/30	65	35
2	50/50	46	54
3	70/30	63	37
4	50/50	50	50
5	50/50	46	54
6	50/50	50	50
7	70/30	66	34
8	70/30	65	35

Table 7. Comparison of the results determined for the real composition of the PLGA copolymers and the theoretical values.

Synthesis	Theoretical D,L-lactide/Glycolide (%)	Experimental D,L-lactide/Glycolide (%)—RMN ¹H	Experimental D,L-lactide/Glycolide (%) RMN—¹³C
3	70/30	63/37	62/38
7	70/30	66/34	66/34
8	70/30	65/35	65/35

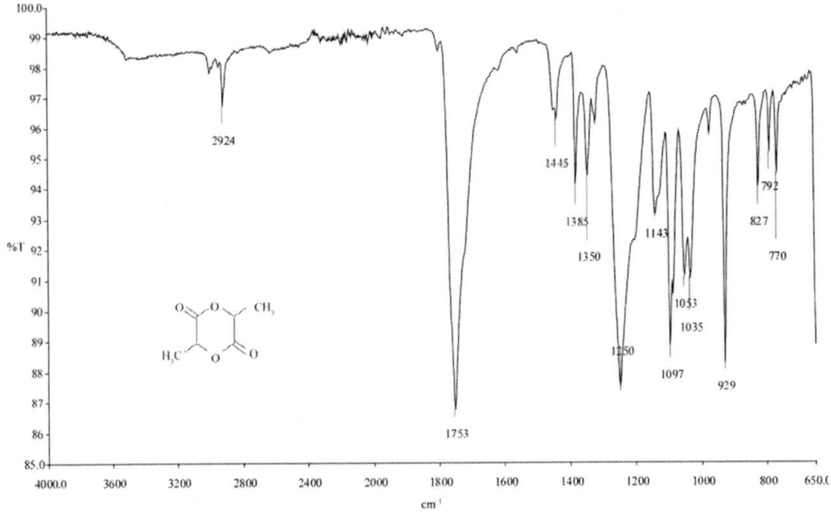

Figure 20. FTIR spectrum for D,L-lactide.

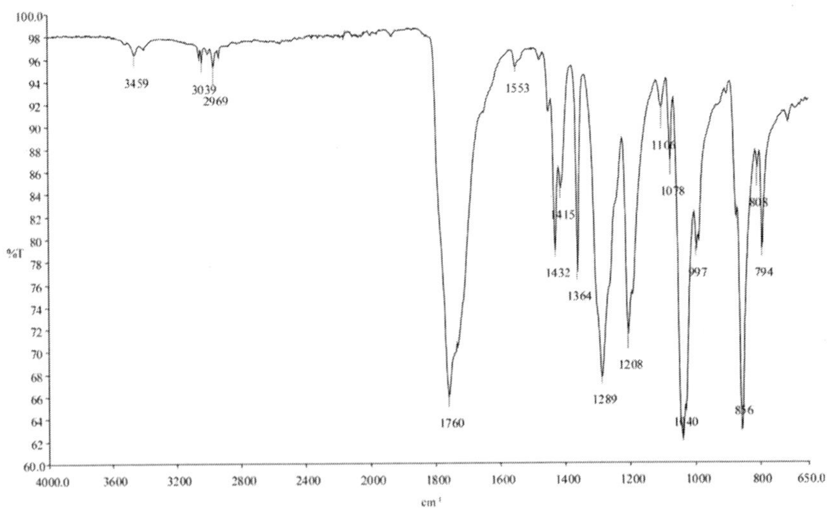

Figure 21. FTIR spectrum for glycolide.

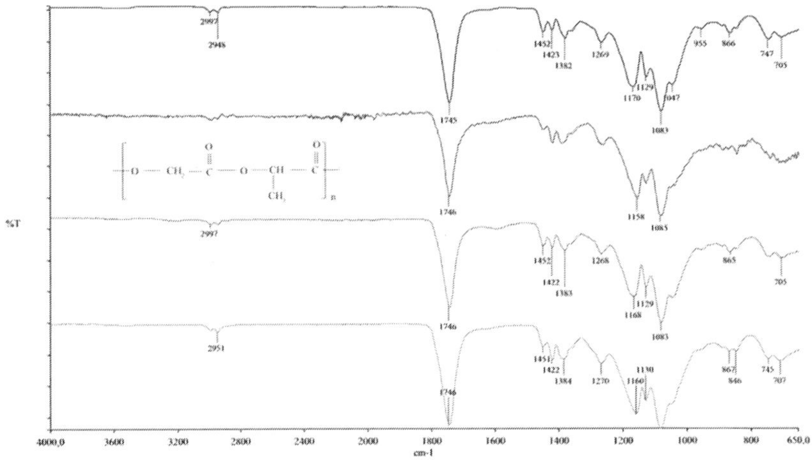

Figure 22. FTIR spectra for PLGA 70/30 (syntheses 1 and 3) and PLGA 50/50 (syntheses 2 and 4).

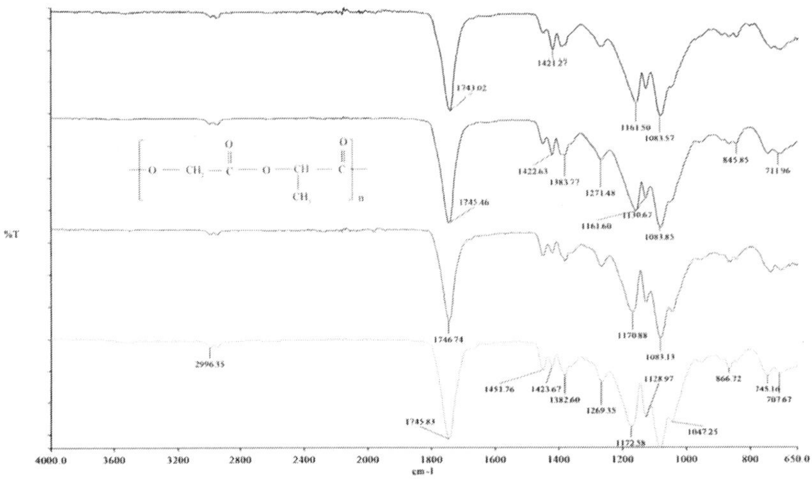

Figure 23. FTIR spectra for PLGA 50/50 (syntheses 5 and 6) and PLGA 70/30 (syntheses 7 and 8).

Figures 22 and 23 give the spectra of the PLGA copolymers synthesised in this work.

The profile of the spectra presented in Figures 22 and 23 are practically identical. All of absorption bands presented in **Table 8** were found in the spectra for the copolymers presented in these figures. These spectra agree with those given in the literature for PLGA copolymers [38]. The absence of absorption bands between 3600 and 3400 cm^{-1} in the spectra presented on Figures 22 and 23, characteristic of the hydroxyl group, indicates that the PLGA copolymers are anhydrous.

Table 8. Assignment of the absorption bands observed in the FTIR spectra of D,L-lactide and glycolide.

Absorption bands (cm^{-1})	Attributions*
3000 - 2700	CH, CH_3 e CH_2 (stretching)
1900 - 1550	C=O (stretching)
1500 - 1250	CH_3 e CH_2 (deformation)
1350 - 1150	CH_2 e CH (wagging vibration)
1300 - 1150	C-O (ester) (stretching)

*Attributions according to COLTHUP *et al.*, 1975 [37].

These FTIR results corroborate the DSC, TG, and NMR results and indicate that the PLGA copolymers have successfully been obtained.

In the attempt to employ FTIR analysis to detect the presence of unreacted monomers, several samples of PLGA 70/30 were prepared with small quantities of the monomers added in the following proportions: 0.5% D,L-lactide, and 0.5 and 3% glycolide. The choice of these proportions was guided by the information that ring opening polymerization produces polymers with about 1 to 3% residual monomers, according to MIDDLETON and TIPTON (1998) [39].

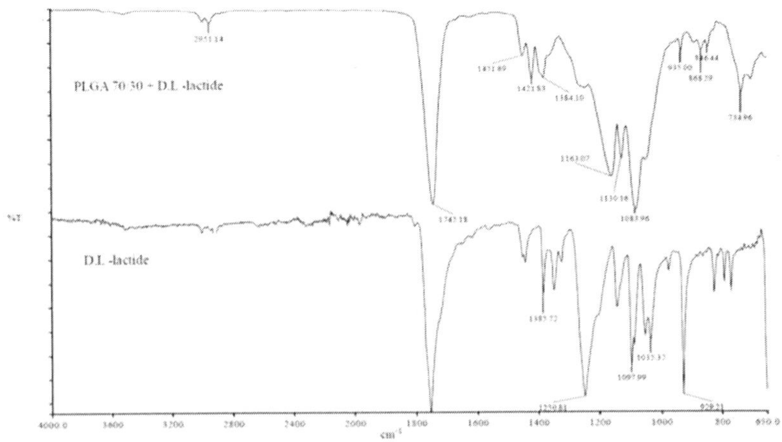

Figure 24. FTIR spectra for the mixture of PLGA 70/30 with 0.5% D,L-lactide and D,L-lactide monomer.

Figure 24 shows the spectra of the PLGA 70/30 copolymer mixed with 0.5% D,L-lactide monomer and pure D,L-lactide monomer. The spectra presented in **Figure 24** shows an absorption band at 935 cm^{-1} that is not observed neither in the spectra presented in Figures 22 and 23. However, a high intensity band in 929cm^{-1} can be seen in the spectrum for D,L-lactide, **Figure 24**, very close to that for the PLGA mixture with 0.5% D,L-lactide. Therefore, it is believed that the D,L-lactide is not present in the obtained PLGA at concentrations above 0.5%.

Regarding the spectra for the mixture of PLGA 70/30 with 0.5 and 3.0% of the glycolide, it was not possible to detect any significant change with the monomer added to enable the same analysis as achieved for the D,L-lactide monomer. All of the absorption peaks that the glycolide monomer presents overlap those of the PLGA copolymer.

4. CONCLUSIONS

With regard to the synthesised copolymers, these have a glass transition temperature between 31°C and 53°C and thermal degradation begins above 240°C. The mass loss occurred in a single step, indicating the total degradation of the polymeric chains.

The NMR analysis allowed the copolymer composition to be determined. The spectra obtained for all of the synthesised copolymers showed chemical shifts in accordance with those anticipated in the literature.

With regard to the FTIR analysis, it was possible to determine all of the reference functional groups (C=O, C-O, CH_3, CH_2, and CH) in the spectra of the synthesised copolymers. The spectra of all of the synthesised copolymers show great similarity and are in agreement with the literature.

To determine the presence of residual monomers in the synthesised copolymers, the results of all of the techniques employed were evaluated together. The DSC technique showed that no endothermic event was detected close to the melting temperatures of the monomers. The thermogravimetry did not present any significant mass loss prior to the onset of copolymer degradation. The NMR analyses did not indicate any significant chemical shift in the spectra for the copolymers that were exactly identical to those in the spectra for the monomers. With regard the FTIR technique, the carbonyl absorption band showed a shift in relation to the monomer bands in all of the copolymer spectra. These results led to the inference that there were no residual monomers in the synthesised copolymers that could be detected by any of the techniques used.

With regard to the special application of FTIR analysis for searching residual monomers, interesting results were obtained when the PLGA mixture was analysed with 0.5% D,L-lactide monomer. These results demonstrate that this technique can detect residual monomers in these copolymers.

The synthesis route employed consisted of the bulk copolymerization of monomers D,L-lactide and glycolide at 175°C, with the initiator stannous octoate and co-initiator lauryl alcohol and application of vacuum in reaction medium. The application of vacuum, coupled with adequate stirring, is essential for obtaining good results. It can be concluded that this new route for synthesising PLGA copolymers is successful and the copolymers can be characterised by FTIR, NMR, DSC, and TG analysis.

ACKNOWLEDGEMENTS

Cynthia D'Avila Carvalho Erbetta, Ricardo José Alves, Jarbas Magalhães Resende, Roberto Fernando de Souza Freitas, Ricardo Geraldo de Sousa Cynthia D. C. Erbetta acknowledges FAPEMIG for supporting grants. Authors gratefully acknowledge the support by FAPEMIG, CAPES and CNPQ.

REFERENCES

1. Y. Galaev and B. Mattiasson, "Smart Polymers and What They Could Do in Biotechnology and Medicine," Trends in Biotechnology, Vol. 17, No. 8, 1999, pp. 335- 340.
2. O. Pillail and R. Panchagnula, "Polymers in Drug Delivery," Current Opinion in Chemical Biology, Vol. 5, No. 4, 2001, pp. 447-451.
3. S. Li, "Hydrolytic Degradation Characteristics of Aliphatic Polyesters Derived from lactic and glycolic acids," Journal of Biomedical Materials Research, Vol. 48, No. 3, 1999, pp. 342-53.
4. S. Hoffman, "Hydrogels for Biomedical Application," Advanced Drug Delivery Reviews, Vol. 54, No. 1, 2002, pp. 3-12.
5. R. Chera and R. Rustgi, "Biodegradable polymers," Progress in Polymer Science, Vol. 23, No. 7, 1998, pp. 1273- 1335.
6. O. Seigo, "Application of superabsorbent polymers in japanese agriculture and greening, Gels Handbook," Academic Press, Waltham, 2001.
7. R. Langer and N. A. Peppas, "Advances in Biomaterials, Drug Delivery, and Bionanotechnology," AlChe Journal, Vol. 49, No. 12, 2003, pp. 2990-3006.
8. M. Chasin and R. Langer, "Biodegradable polymers as Drug Delivery Systems," McGraw-Hill, New York, 1990.
9. C. E. Soma, C. Dubernet, D. Bentolila, S. Benita and P. Couvreur, "Reversion of multidrug resistance by co-encapsulation of doxorubicin and cyclosporin A in polyalkylcyanoacrylate nanoparticles," Biomaterials, Vol. 21, No. 1, 2000, pp. 1-7.
10. R. A. Jain, "The Manufacturing Techniques of Various Drug Loaded Biodegradable Poly(Lactide-Co-Glycolide) (PLGA) Devices," Biomaterials, Vol. 21, No. 23, 2000, pp. 2475-2490.

11. V. Michel, "Polymeric biomaterials: Strategies of the past vs. strategies of the future," Progress in Polymer Science, Vol. 32, No. 8-9, 2007, pp. 755-761.
12. L. Xiaoling and R. J. Haskara, "Design of Controlled Release Drug Delivery Systems," MacGraw-Hill, New York, 2006.
13. N. Angelova and D. Hunkeler, "Rationalizing the design of Polymeric Biomaterials," Trends in Biotechnology, Vol. 17, No. 10, 1999, pp. 409-421.
14. M. M. M. Elnashar, "Review Article: Immobilized Molecules Using Biomaterials and Nanobiotechnology," Journal of Biomaterials and Nanobiotechnology, Vol. 1, No. 1, 2010, pp. 61-77.
15. Q. Soares, L. F. Oliveira, D. Rabelo and A. R. Souza, "Polímeros Biodegradáveis: Novas Perspectivas para as Ciências Farmacêuticas," Revista Eletrônica de Farmácia, Vol. 2, No. 2, 2005, pp. 202-205.
16. M. L. Hans and A. M. Lowman, "Biodegradable Nanoparticles for Drug Delivery and Targeting," Current Opinion in Solid State and Materials Science, Vol. 6, No. 4, 2002, pp. 319-327.
17. Q. Cai, G. Shi, J. Bei and S. Wang, "Enzymatic degradation behavior and mechanism of Poly(Lactide-Co-Glycolide) foams by trypsin," International Journal of Pharmaceutics, Vol. 24, No. 4, 2003, pp. 629-638.
18. J. Lunt, "Large-scale Production, Properties and Commercial Applications of Polylactic Acid Polymers," Polymer Degradation and Stability, Vol. 59, No. 1-3, 1998, pp. 145-152.
19. H. Fukuzaki, M. Yoshida, M. Asano and M. Kumakura, "Synthesis of copoly(D,L-lactic acid) with relatively low molecular weight and in vivo degradation," European Polymer Journal, Vol. 25, No. 10, 1989, pp. 1019- 1026.
20. D. K. Gilding and A. M. Reed, "Biodegradable polymers for use in Surgery Polyglycolid/Poly(Lactic Acid) Homo and Copolymers," Polymer, Vol. 20, No. 12, 1979, pp. 1459-1464.
21. D. Bendix, "Chemical synthesis of polylactide and its copolymers for Medical Applications," Polymer Degradation and Stability, Vol. 59, No. 1-3, 1998, pp. 129-135.
22. P. B. Deasy, M. P. Finan and M. J. Meegan, "Preparation and characterization of Lactic/Glycolic Acid Polymers and copolymers," Journal of Microencapsulation, Vol. 6, No. 3, 1989, pp. 369-378.
23. C. Jéôme and P. Lecomte, "Recent advances in the synthesis of aliphatic polyesters by Ring-Opening Polymerization," Advanced Drug Delivery Reviews, Vol. 60, No. 9, 2008, pp. 1056-1076.
24. B. D. Ratner and A. S. Hoffman, "Biomaterials Science— An Introduction to Materials in Medicine," Elsevier Academic Press, Waltham, 1996.
25. M. J. Blanco-Prieto, E. Fattal, F. Puisieux and P. Couvreur, "The Multiple Emulsion as a common step for the design of polymeric microparticles," In: J. L. Grossiord and M. Seiller, Eds., Multiple emulsions: structure, properties and applications, Éditions de Santé, Paris, 1998, pp. 397-435.
26. D. H Lewis, "Controlled release of bioactive agents from Lactide/Glycolide Polymers," In: M. Chasin and R. Langer, eds., Biodegradable polymers as drug delivery systems, Marcel Dekker, New York, 1990, pp. 1-41.
27. S. L. Fialho, M. G. B. Rego, J. A. Cardillo, R. C. Siqueira, R. Jorge and A. S. Cunha Jr., "Implantes biodegradáveis destinados à administração intra-ocular," Arquivos Brasileiros de Oftalmologia, Vol. 66, No. 6, 2003, pp. 891- 896.
28. Merkli, C. Tabatabay, R. Gurny and J. Heller, "Biodegradable polymers for the controlled release of ocular drugs," Progress in Polymer Science, Vol. 23, No. 3, 1998, pp. 563-580.
29. K. A. Athanasiou, G. G. Niederauer and C. M. Agrawal, "Sterilization, Toxicity, Biocompatibility and Clinical Applications of Polylactic Acid/Polyglycolic Acid Copolymers", Biomaterials, Vol. 17, No. 2, 1996, pp. 93-102.
30. M. Kiremitçi-gümüsderelioglu and G. Deniz, "Synthesis, characterization and in vitro degradation of poly(D,LLactide)/Poly(D,L-Lactide-Co-Glycolide)," Turkish Journal of Chemistry, Vol. 23, No. 2, 1999, pp. 153-161.

31. X. Kaitian, A. Kozluca, E. B. Denkbas and E. Piskin, "Poly(D,L-lactic acid) Homopolymers: Synthesis and Characterization," Turkish Journal of Chemistry, Vol. 20, No. 1, 1996, pp. 43-53.
32. E. Jabbari, E. Xuezhong, "Synthesis and characterization of bioresorbable in situ Crosslinkable Ultra Low Molecular Weight Poly(Lactide) Macromere," Journal of Materials Science: Materials in Medicine, Vol. 19, No. 1, 2008, pp. 311-317.
33. V. Kumar and G. S. Banker, "Chemically-Modified Cellulosic Polymers," Drug Development and Industrial Pharmacy, Vol. 19, No. 1, 1993, pp. 1-31.
34. A. Silva Jr., J. R. Matos, T. P. Formariz, G. Rossanezi, M. V. Scarpa, E. S. do Egito and A. G. Oliveira, "Thermal behavior and stability of biodegradable Spray-Dried Microparticles Containing Triamcinolone," International Journal of Pharmaceutics, Vol. 368, No. 1-2, 2009, pp. 45-55.
35. R. Lindhardt, "Polímeros biodegradáveis para liberação controlada de droga," Springer-Verlag, Now York, 1998.
36. G. Hausberger and P. P. DeLuca, "Characterization of Biodegradable Poly(D,L-Lactide-Co-Glycolide) Polymers and Microparticles," Journal of Pharmaceutical and Biomedical Analysis, Vol. 13, No. 6, 1995, pp. 747-760.
37. N. B. Colthup, L. H. Daly and S. E. Wiberley, "Introduction to infrared and Raman Spectroscopy," 2nd edition, Academic Press, New York, 1975.
38. D. O. Hummel, "Atlas of Polymer and Plastics Analysis," Wiley, New York, 2001.
39. J. C. Middleton and A. J. Tipton, "Synthetic Biodegradable Polymers as Medical Devices," Medical Plastics and Biomaterials Magazine, Vol. 3-4, 1998, pp. 31-38.

CHAPTER 4

Characterization of a Laser Surface-Treated Martensitic Stainless Steel

S. R. Al-Sayed [1,*], A. A. Hussein [2], A. A. Nofal [3], S. I. Hassab Elnaby [1] and H. Elgazzar [3]

[1]*National Institute of Laser Enhanced Sciences (NILES), Cairo University, Giza 12611, Egypt*

[2]*Faculty of Engineering, Cairo University, Giza 12611, Egypt*

[3]*Central Metallurgical Research and Development Institute (CMRDI), Helwan 11731, Egypt*

ABSTRACT

Laser surface treatment was carried out on AISI 416 machinable martensitic stainless steel containing 0.225 wt.% sulfur. Nd:YAG laser with a 2.2-KW continuous wave was used. The aim was to compare the physical and chemical properties achieved by this type of selective surface treatment with those achieved by the conventional treatment. Laser power of different values (700 and 1000 W) with four corresponding different laser scanning speeds (0.5, 1, 2, and 3 m·min^{-1}) was adopted to reach the optimum conditions for impact toughness, wear, and corrosion resistance for laser heat treated (LHT) samples. The 0 °C impact energy of LHT samples indicated higher values compared to the conventionally heat treated (CHT) samples. This was accompanied by the formation of a hard surface layer and a soft interior base metal. Microhardness was studied to determine the variation of hardness values with respect to the depth under the treated surface. The wear resistance at the surface was enhanced considerably. Microstructure examination was characterized using optical and scanning electron microscopes. The corrosion behavior of the LHT samples was also studied and its correlation with the microstructures was determined. The corrosion data was obtained in 3.5% NaCl solution at room temperature by means of a potentiodynamic polarization technique.

Keywords: laser surface treatment; martensitic stainless steel; macrohardeness and microhardeness; impact toughness; wear resistance; potentiodynamic polarization technique

1. INTRODUCTION

The composition and surface microstructure of a component play a decisive role in determining its surface-dependent engineering properties (hardness, wear, fatigue, corrosion and oxidation resistance). Surface engineering processes aim to the control heating and cooling of metallic materials, to alter their physical and mechanical properties without affecting the bulk material. Conventionally, surface treatment processes like flame hardening, induction hardening, carburizing, nitriding, etc. are usually employed in enhancing the wear resistance of Fe-based alloys. These processes suffer from numerous restrictions, i.e., time and energy consumption, complex heat treatment schedule, non-controllable heat affected zones, lack of solid solubility limit, and the necessity of a quenchant [1,2,3].

On the other hand, when a high powered laser beam is used as a source of heat for surface treatment, it will obviate most of the restrictions observed in the conventional surface treatment [4,5]. Martensitic stainless steels are extensively used in engineering applications such as steam and water valves, pumps, screw machining parts, turbines, and many other applications which necessitate high strength and high resistance to wear and corrosion [6,7,8]. Failure of such stainless steel components due to corrosion, fatigue and abrasion is most likely to initiate from the surface. The engineering solution for minimizing or eradicating such surface-initiated failures lies in tailoring the surface composition and/or microstructure of the near surface region of a component without affecting the bulk [9]. In recent years, application of laser surface modification to prolong the service life of the engineering components is one of the most promising techniques for improving the tribological properties of the majority of metals. Among the different types of laser surface treatment processes, laser transformation hardening (LTH) and laser surface melting (LSM) are the simplest, as no additional materials are introduced, and they are particularly effective for processing ferrous alloys with grain refinement and uniformity of the acquired structure. In fact, these processes have been employed for improving the erosion and corrosion resistance of a number of ferrous alloys [9,10,11,12]. Only martensitic grades of stainless steel contain sufficient carbon to be laser hardened. The degree of hardening is strongly dependent on the initial condition of the steel, as well as its chromium content. Laser heating of martensitic stainless steels (MSS) in their annealed delivery condition (a microstructure of ferrite and carbides) causes a significant amount of carbide dissolution. Sufficient carbon is released to form martensite on cooling, although austenite may be retained in regions of particularly high carbon concentration [13]. Ion et al., worked on 12C27 steel. It was observed that laser surface heating resulted in redistribution of carbon to produce hardening by martensitic transformation, but the high alloy content resulted in depression of the martensite finish temperature and, consequently, the retention of austenite at room temperature [14]. Tsay et al. examined the impact toughness of AISI 403 MSS plate and laser-hardened specimens tempered at various temperatures [15]. Tianmin et al. [16] studied the impact of wear behavior of hypo-eutectoid 2Cr13 MSS, and the results of scanning electron microscopy (SEM) analysis and surface profile measurement suggested that the size of the impact wear scars of 2Cr13 specimens decreased with an increase in surface hardness. The effect of processing conditions on the corrosion performance of laser surface-melted of MSS was studied as well, on different grades of MSS [11,17,18,19,20]. The corrosion resistance of all laser surface-melted specimens was significantly improved, as evidenced by a shift from active corrosion to passivity. Steel AISI 416 is "a free machining type of martensitic stainless steel grade, with a minimum sulfur content of 0.15%". Its practical potential is noteworthy and is frequently used in automatic screw-machined components, gears, bolts, valves, pump shafts and turbine blades. The aim of this work is to gain improvement in steel wear resistance and impact toughness as well via LHT, particularly in light of the rather scarce studies reported in this respect [21].

2. MATERIALS AND METHODS

2.1. Materials and Laser Treatment

The martensite steel AISI 416 used in this work has the principal alloying elements (in wt. %) as follows 0.167% C; 12%·Cr; 0.475% Mn; 0.225% S;

0.0232% P; 0.27% Ni; 0.131% Cu, and 0.194% Si. It was received in an annealed condition with a hardness of 155 Hv, in the form of round disks of diameter 150 mm and a thickness of 7.5 mm. Samples measuring 55 mm×10 mm×7.5 mm in size were machined. Preceding the laser treatment, all necessary precautions and safety measures were taken [22]. Precautions with clamping the specimen to circumvent any misalignment with the laser beam scanning were taken as well [23].

A Rofin sinar 2.2 KW continuous wave diode pumped Nd:YAG laser was used to generate a beam which was delivered to the rotating head by a fiber optic. The rotating head moves with variable speeds up to 9 $m \cdot min^{-1}$. The oscillation mode is a multimode with Gaussian energy distribution because the YAG laser has high amplification gain and a refractive index that tends to be non-uniform due to the temperature gradient inside the rod. The laser radiation focused on the sample surface by means of optics consisted of a focusing lens of focal length f = 120 mm, producing a beam spot size of 2 mm on the substrate. The laser treatment was carried out in argon shielding gas with a flux of 10 $L \cdot min^{-1}$ to minimize the oxidation of the hardened surface. Prior to the LHT, the surface of the AISI 416 martensitic stainless steel samples was cleaned with acetone. Preliminary attempts with different values of laser power (P) were carried out, for example 2000 and 1800 W. Unfortunately these trials resulted in the melting of the material surface and changing its dimensions. Therefore, we reduced the range of laser power to the following values to be sure that we are in the hardening range. This range of laser scanning speeds with the laser power was chosen to determine the optimum processing conditions for surface treatment. In the present investigation, two different laser powers were attempted, with four different laser scanning speeds (as listed in Table 1, with the corresponding samples are labeled as LHT P-X-Y, where X: laser power, Y: scanning speed) in order to attain different hardened layer depths. Laser surfacing was achieved by parallel tracks with 25% overlap. Since the laser spot has 2 mm diameter, and in order to achieve a smooth and homogenous track profile, we had to move the laser beam between 2 mm and 1 mm (0% to 50% overlap). We chose to make a 25% overlap to avoid any overheating and any discontinuation in the profile.

For comparison, conventional heat-treated samples (CHT) were subjected to the following heat treatment; samples were austenitized at 980 °C for 15 min with a heating rate of 250 °C/h, and subsequently oil quenched.

Table 1. The laser surface treatment parameters.

Sample No. P-X-Y	Laser Power P (KW)	Laser Scanning Speed v ($m \cdot min^{-1}$)
P-07-0.5	0.7	0.5
P-07-1	0.7	1
P-07-2	0.7	2
P-07-3	0.7	3
P-1-0.5	1	0.5
P-1-1	1	1
P-1-2	1	2
P-1-3	1	3

2.2. Metallographic and Microstructural Analysis

Metallographic samples were prepared using standard procedures of mechanical polishing, and then were etched with acidic ferric chloride solution (25 g of $FeCl_3$, 25 mL of HCl and 100 mL of H_2O) for 10 s. The microstructures of the samples were characterized using the Axiotech 30 optical microscope (Lukas Microscope Service, Inc., Hillview Court, Mundelein, IL, USA), and the hardened layer was examined by the back-scattered electron mode of the QUANTA FEG 250 SEM (FEI, Hillsboro, OR, USA) attached with an energy dispersive X-ray (EDX) microanalyzer. The phases present in the surface layer were determined by X-ray diffraction (XRD) using CuKα radiation.

Vickers hardness values of the as-received (AR), CHT and LHT samples were measured under a load of 30 kg. The measured hardness values were an average of five readings for each condition. Additionally, the distribution of Vickers microhardness along the hardened depth to the base metal (BM) was measured under a load of 200 g for 20 s.

2.3. Tribology Test

The macroscopic wear performances of CHT and LHT samples were evaluated at room temperature by means of a pin-on-ring type tribometer (TNO, The Hague, Netherlands), according to ASM G 132 [24]. In this case, the ring (diameters of 73 mm) of stainless steel with hardness of 63 HRC is rotated on the stationary wear sample, which functions as the pin. The dimensions of the specimens were 7 mm × 7.5 mm × 10 mm. Before each test, the ring was rotated to a defined starting point. The experimental conditions without lubricant used for the wear tests were; normal loads =50 N (0.3 bar) and fixed sliding at speeds =8.4 $m \cdot min^{-1}$ (100 rpm). The wear rate was measured by a weighing method, and each specimen was weighed before and after wear testing by using a sensitive digital balance of accuracy 10^{-4} g. The measured weight loss was an average of three tested samples for each condition.

2.4. Impact Test

The impact test was performed at 0 °C by a Model JB-W500 impact tester (TIME Group Inc., Beijing, China). The impact machine had a maximum impact energy of 500 J, a maximum impact speed of 5.4 $m \cdot s^{-1}$, a minimum impact energy resolution of 0.2 J, and a rising angle of 150°. The test ascertained whether the material was tough or brittle. Un-notched samples measuring 7 mm × 10 mm × 55 mm were used [15]. The result is usually reported as the energy in J required to fracture the test piece. To carry out the test, the standard specimen is supported at its two ends on an anvil, and is struck on the opposite face to the laser treated surface, by the pendulum.

2.5. Potentiodynamic Corrosion Test

The corrosion resistance of the AR, CHT and LHT samples was assessed by means of electrochemical measurements. Linear polarization measurements were performed using an Autolab PGSTAT 302N potentiostat system, which was driven by NOVA 1.10 software (Eco-chemie, Utrecht, Netherlands). The measurements were achieved using a conventional three-electrode setup, where a test sample with approximately 0.2 cm^2 exposed surface area was placed in 3.5% NaCl (pH 7) solution at 23 °C. A constant potential scan rate of 1 $mV \cdot s^{-1}$ was used. All potentials were computed with respect to Ag/AgCl, with a 3 mol/L KCl electrode as a reference electrode. A platinum electrode was used as the counter electrode for current measurement.

3. RESULTS

3.1. Microstructural Analysis

3.1.1. Microstructure of the As-Received Specimen

The microstructure of the annealed AR AISI 416 steel consisted of equiaxed or near equiaxed ferrite grains as shown in Figure 1, with MnS inclusions, carbide particles, and multi-element oxides. These constituents were identified by EDX (Figure 2).

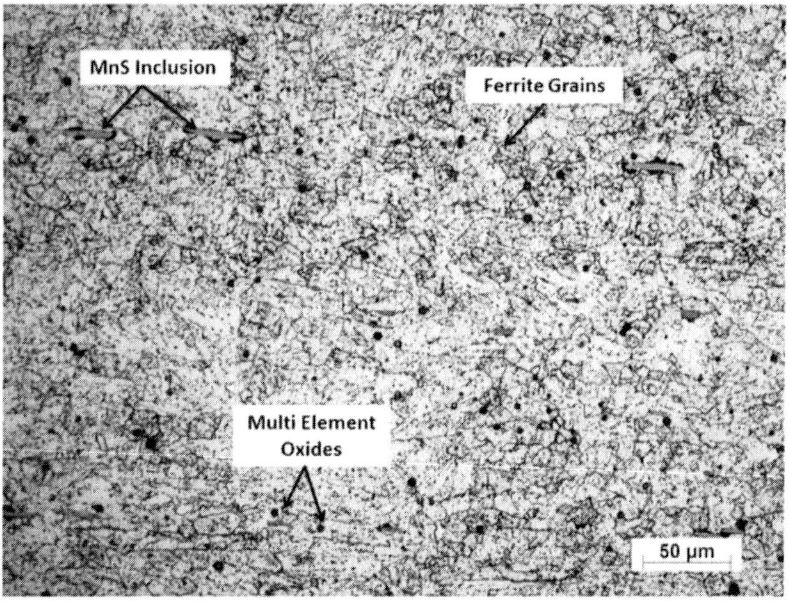

(a)

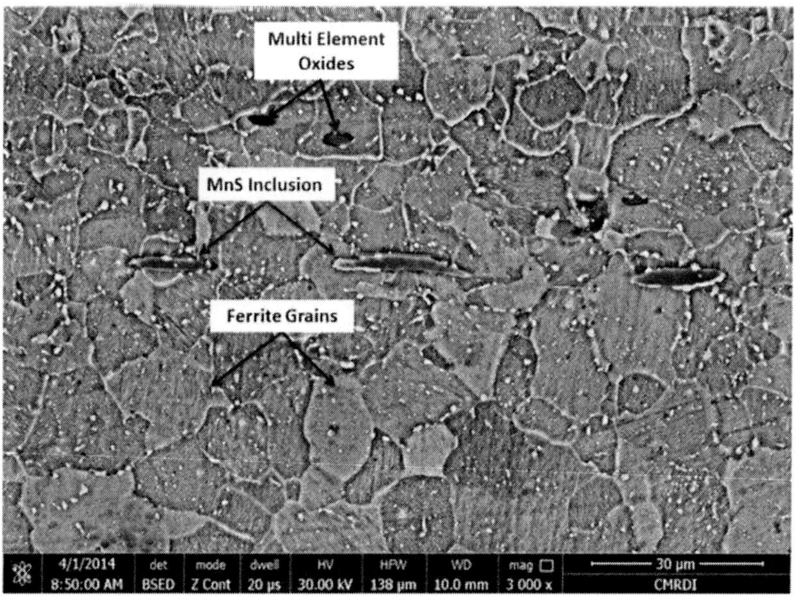

(b)

Figure 1. (**a**) Optical image of microstructure (magnification 200×) and (**b**) scanning electron microscopy (SEM) image of microstructure (magnification 3000×) of the as-received (AR) sample.

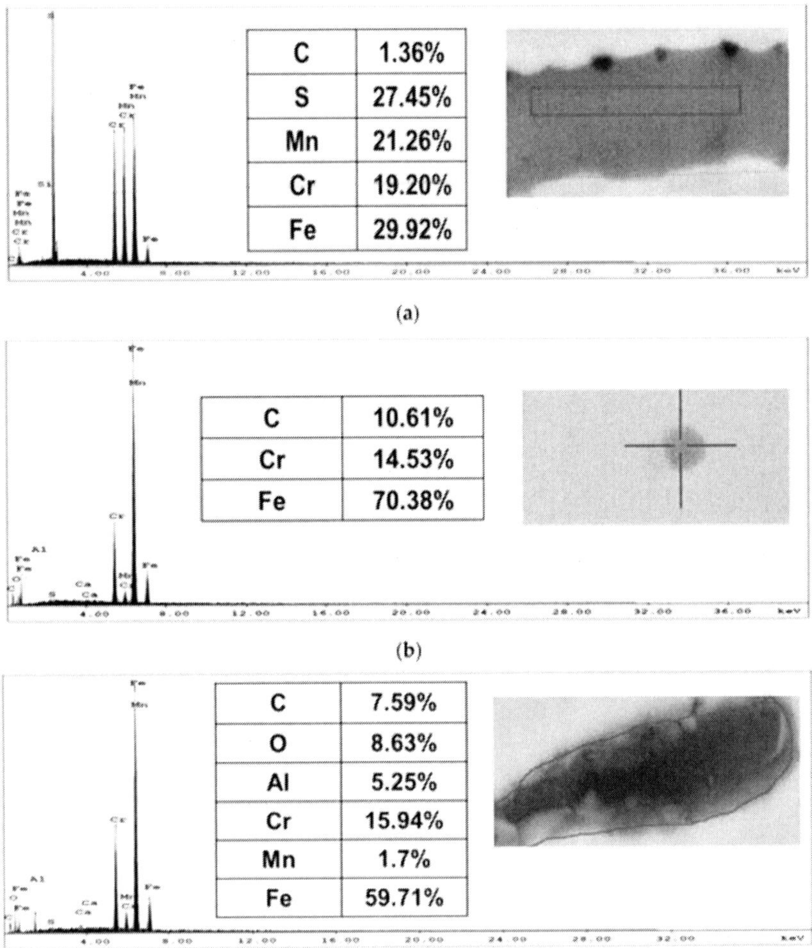

C	1.36%
S	27.45%
Mn	21.26%
Cr	19.20%
Fe	29.92%

(a)

C	10.61%
Cr	14.53%
Fe	70.38%

(b)

C	7.59%
O	8.63%
Al	5.25%
Cr	15.94%
Mn	1.7%
Fe	59.71%

(c)

Figure 2. Energy dispersive X-ray (EDX) analyses of AR AISI416: (**a**) MnS inclusions; (**b**) carbide particles; and (**c**) multi-element oxides.

3.1.2. Microstructure of the CHT Specimens

Martensitic structures appeared in the CHT specimens (oil quenched from austenite), as shown in Figure 3. Carbide particles, multi-element oxides and MnS inclusions were observed as well. In addition, these phases were present in both AR and CHT specimens as detected by the XRD spectra shown in Figure 4.

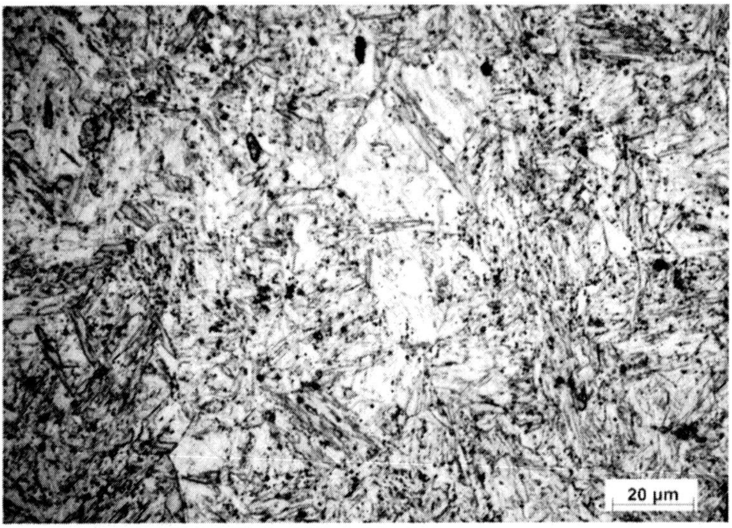

Figure 3. Optical image of microstructure (magnification 200×) of conventionally hardened AISI 416 stainless steel.

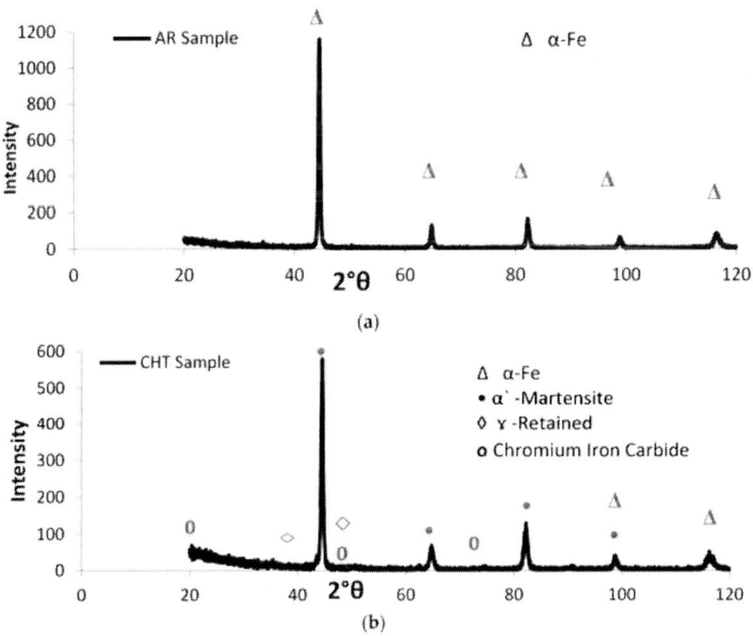

Figure 4. X-ray diffraction pattern of (**a**) untreated AISI 416 and (**b**) conventionally heat treated (CHT) samples.

3.1.3. Microstructure of the LHT Specimens

Figure 5 shows typical optical micrographs of LHT samples treated at laser power of 1000 W. The surface layer of all specimens was free of cracks and pores. The EDX data of such samples were analyzed in Figure 6 for the condition indicated. The XRD spectra for the specimens under different processing conditions are shown in Figure 7. All LHT specimens were composed mainly of martensite, with retained austenite and small traces of the ferrite phase. Extra investigation via transmission electron microscopy (TEM) is planned for future work.

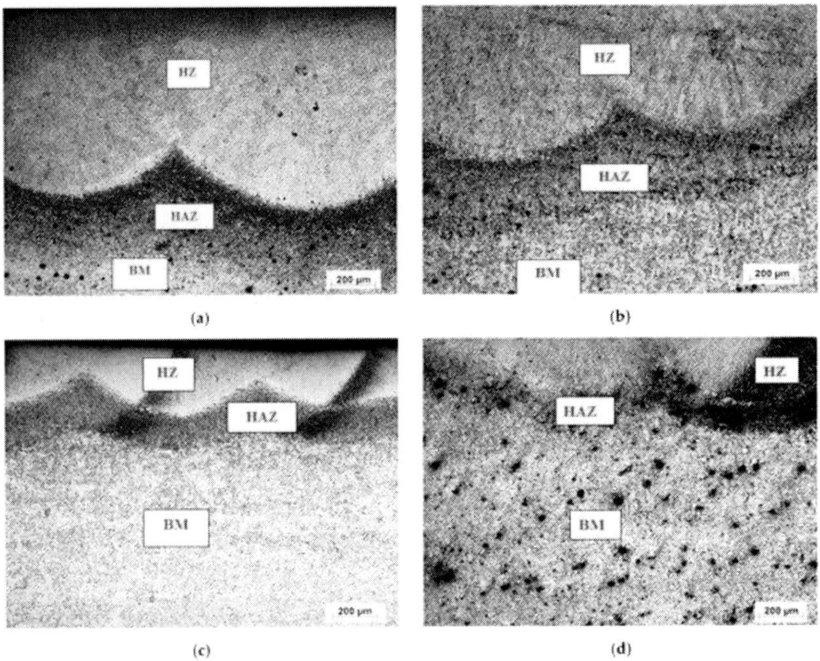

Figure 5. Typical microstructure features (magnification 50×) of laser heat treated (LHT) AISI 416 stainless steel at a constant power of 1000 W and different laser scanning speeds: (**a**) 0.5; (**b**) 1; (**c**) 2; and (**d**) 3 $m \cdot min^{-1}$.

Additional information of both hardened zones (HZ) and heat affected zones (HAZ) was gained from the light micrographs of the laser surface treated specimens at a higher magnification using SEM, as presented in Figure 8.

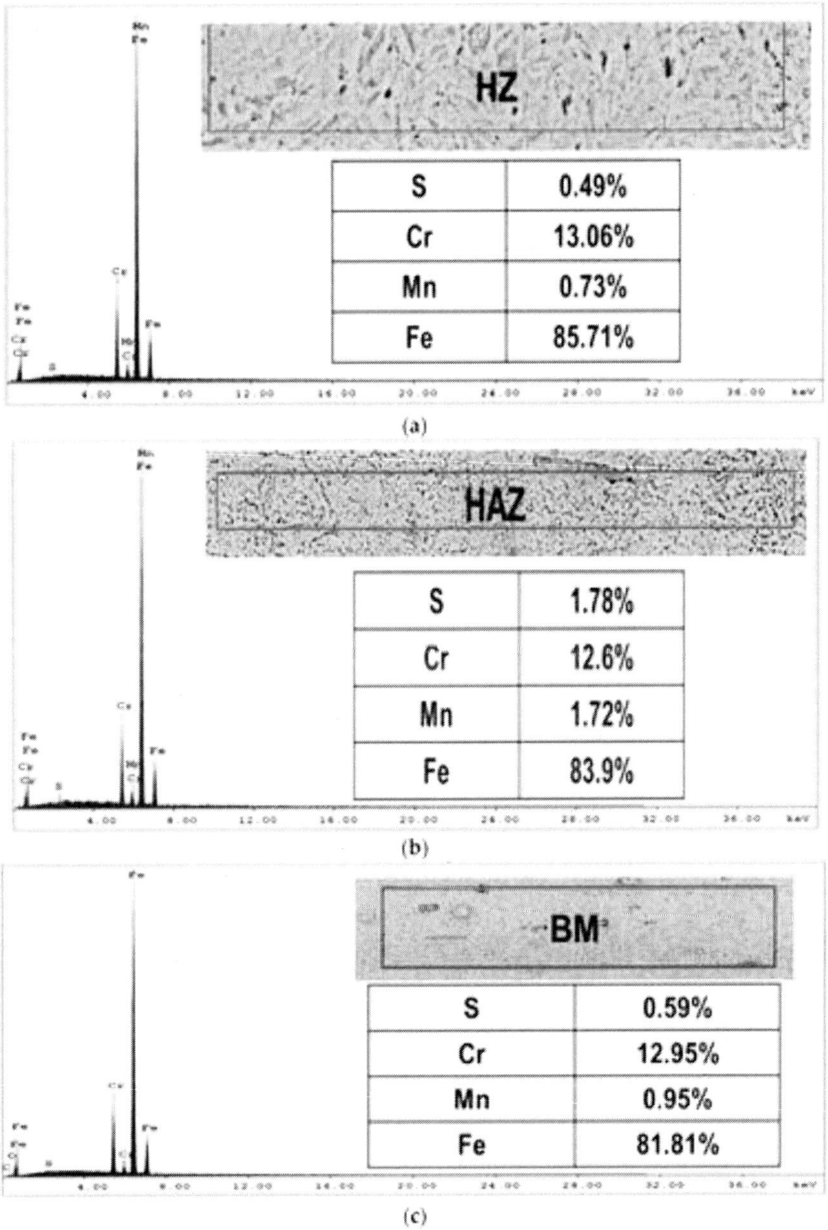

Figure 6. EDX analysis of the LHT P-1-0.5 sample: (**a**) hardened zone (HZ); (**b**) heat affected zone (HAZ); and (**c**) base metal (BM).

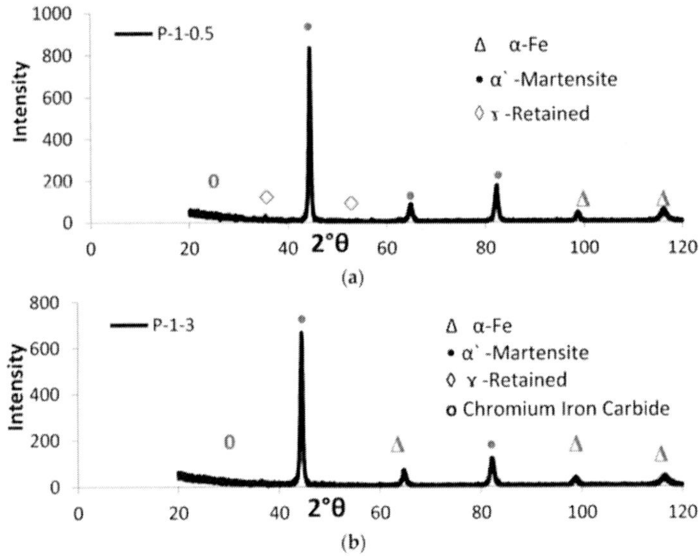

Figure 7. XRD spectra for various LHT AISI 416 specimens processed at laser power 1000 W and scanning speeds of (**a**) 0.5 and (**b**) 3 m·min^{-1}.

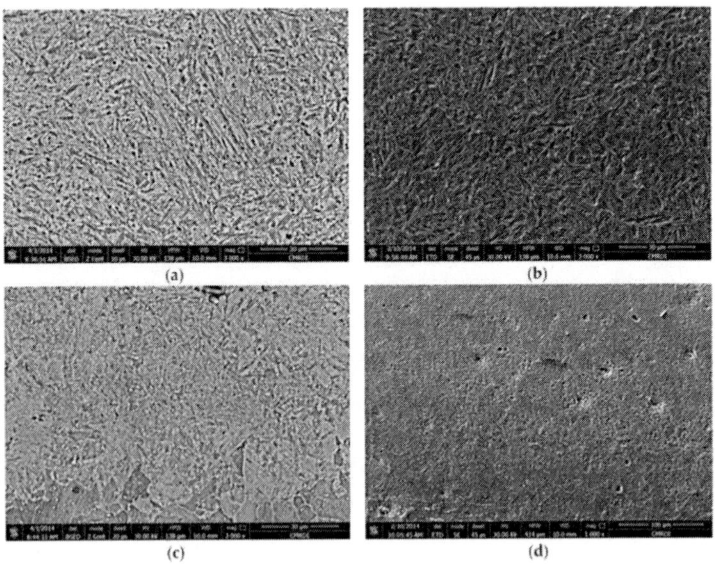

Figure 8. SEM microstructure (magnification 1000×): (**a**) HZ; (**b**) HAZ of sample P-07-0.5 and (**c**) HZ; (**d**) HAZ of sample P-1-0.5.

As mentioned before, the optimal smoothness of track profiles was acquired by adopting an overlap of 25%. No defects were recorded, as seen in Figure 9. The obtained results coincided with previous reports, for example, [25]. The microstructure consisted of mixture of martensite and retained austenite.

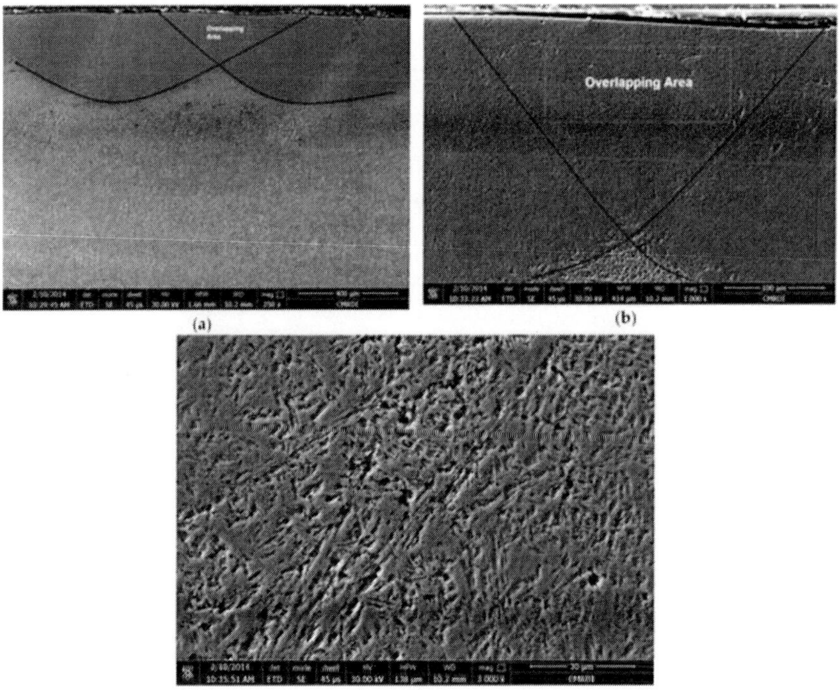

Figure 9. SEM micrographs of the highlighted overlapping area of sample P-1-2, at magnifications of (**a**) 250×; (**b**) 1000×; (**c**) 3000×.

3.2. Hardened Depth

The hardened depths were evaluated under the optical microscope at magnification 50× from the highest point to the deepest point in the hardened zone, as shown in Figure 10. The hardened depths were measured at exactly the same location (in the middle of the hardened layer) in all samples. The hardened depth due to laser hardening was directly affected by the laser power and beam spot diameter (power density W·mm^{-2}) [26]. As the power density increased and the transverse speed decreased the hardened depth and width increased.

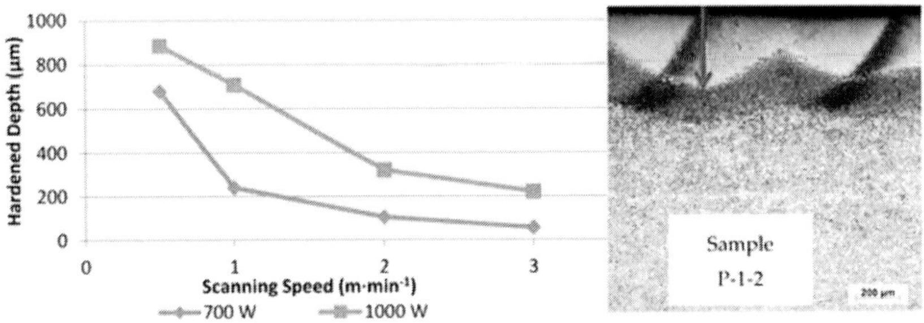

Figure 10. Variations in the depth of hardened regions with respect to laser power and laser scanning speed.

3.3. Macrohardness

The hardness value of the base material was in the range of 155–160 HV, and the hardness value of the CHT samples was in the range of 500–514 HV. Both cases showed uniformity of hardness distribution. On the other hand, laser treated samples exhibited variations in hardness distribution, being in the range of 440–510 HV. Heat input values were calculated to combine the effect of laser power with the scanning speed and the laser beam diameter into one value. Heat input was calculated from the following equation H=p/d.v where: H: heat input (J·mm^{-2}), p: laser power (J·s^{-1}), d: beam diameter (mm), and v: laser scanning speed (mm·s^{-1}) [4].

It was indicated that at constant laser scanning speed, samples treated at higher laser powers (1000 W) had higher hardness values than samples treated at lower laser powers (700 W). The present experimental findings (as shown in Figure 11) showed that the obtained surface hardness strongly depended on the values of heat input.

3.4. Microhardness Distribution

Figure 12 and Figure 13 display the Vickers microhardness depth profiles for laser hardened samples. All profiles could be divided into three zones corresponding to the microstructure, the hardened zone (highlighted in figures), the heat-affected zone, and the base metal zone, respectively.

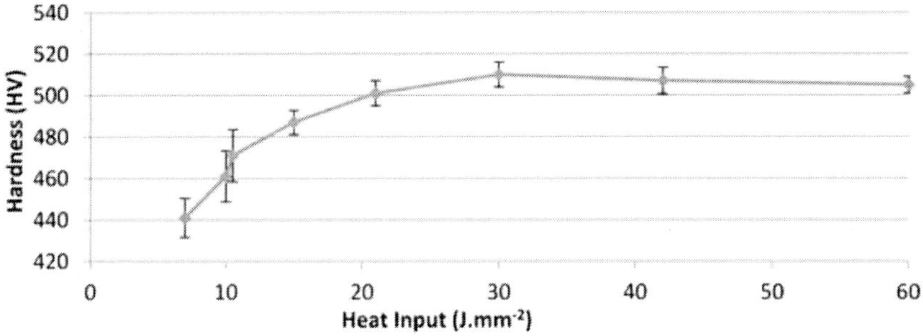

Figure 11. Heat input values versus average hardness values of laser surface treated AISI 416 samples. Error bars represent standard deviation.

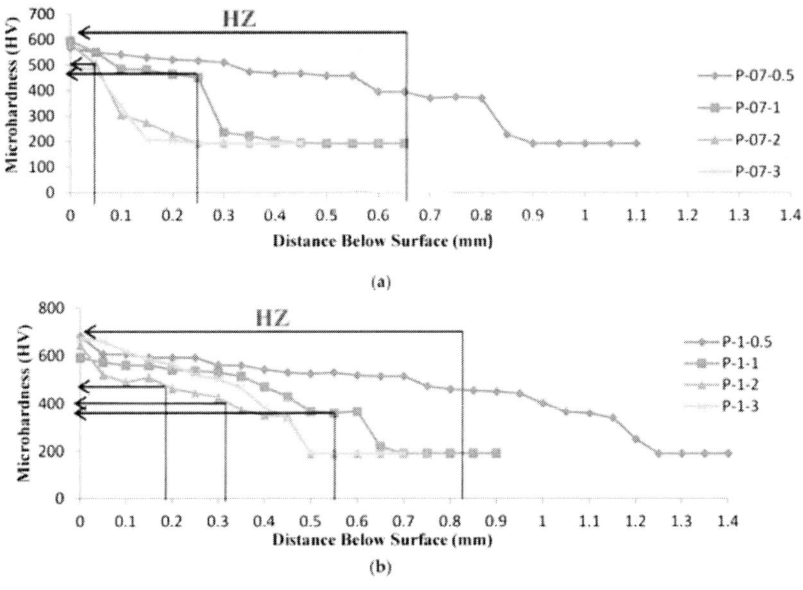

Figure 12. The microhardness profiles of the laser-hardened specimens treated at laser powers of: (**a**) 700 W; and (**b**) 1000 W, corresponding to different laser scanning speeds (0.5, 1, 2 and 3 m·min^{-1}).

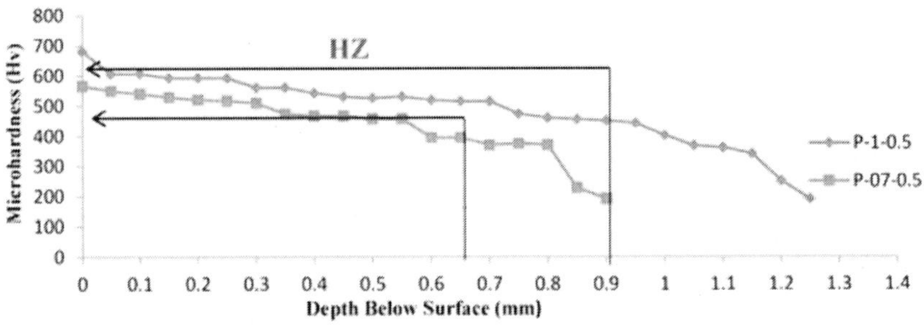

Figure 13. The microhardness profiles of laser treated samples processed at two different laser powers (700 and 1000 W) at a constant laser scanning speed of 0.5 m·min^{-1}.

3.5. Wear Behavior

The loss of weight as a result of wear is shown in Figure 14 with error bars representing standard deviation. In this figure, the wear resistance of the LHT samples was compared with those of the CHT samples. Up to a laser scanning speed of 1 m·min^{-1}, the wear resistance of laser treated samples, at both powers, was comparable to conventionally hardened samples. On the other hand, LHT samples exhibited a decrease in wear resistance as laser scanning speed increased. It was demonstrated from the curve of the weight loss versus the heat input (Figure 14) that the weight loss decreased with an increase in heat input. This was reflected through the surface hardness values after the laser surface treatment (Figure 11).

3.6. Impact Results

Figure 15 presents impact energy values for CHT and LHT samples at different laser hardening conditions. In this work, the tempering of the CHT samples at two different temperatures of 500 and 600 °C was performed in order to compare the impact energy of the LHT samples with the hardened plus tempered martensite stainless steel, as normally treated.

3.7. Corrosion Results

Figure 16a,b shows typical potentiodynamic polarization curves for AR, CHT, and LHT at laser powers of 700 W and 1000 W, respectively. The corrosion potentials obtained from these curves were presented in Table 2.

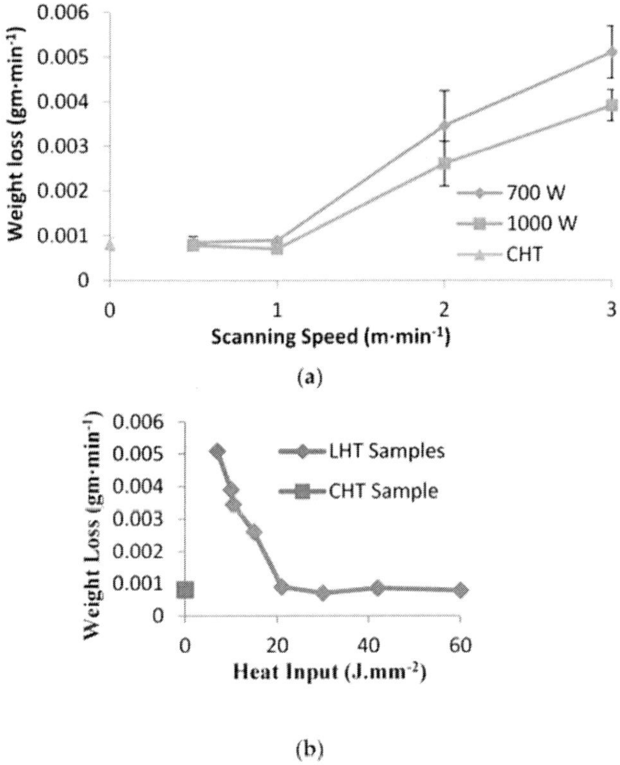

Figure 14. Wear behaviors of CHT and LHT samples, (**a**) weight loss vs. scanning speed; (**b**) weight loss vs. heat input.

4. DISCUSSION

4.1. Microstructural Analysis

Laths of martensite were formed in the conventionally treated samples, due to the slower cooling rate from austenite compared to laser surface treatment, as discussed below. As for the laser treated samples, the microstructures observed in this work were in accord with those mentioned previously [21]. The morphology of the cross section of the laser affected region could be divided into three regions based on their different microstructures: (i) the hardened zone, (ii) the heat affected zone and (iii) the base metal, respectively.

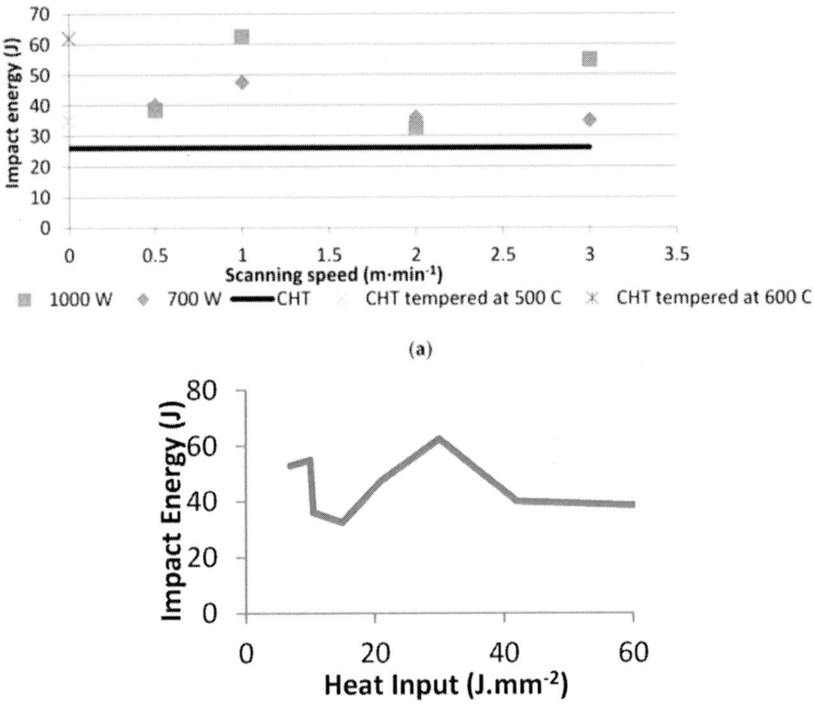

(b)

Figure 15. The impact energy profiles of CHT and LHT samples, (**a**) impact energy vs. laser powers and laser scanning speeds; (**b**) impact energy vs. heat input.

SEM showed that the HZ consisted of martensite containing a small amount of retained austenite (identified with the help of microhardness measurement). The HAZ consisted of martensite, carbide particles, MnS inclusions, and traces of initial ferrite grain structures. The results were in good agreement with those reported from XRD analysis.

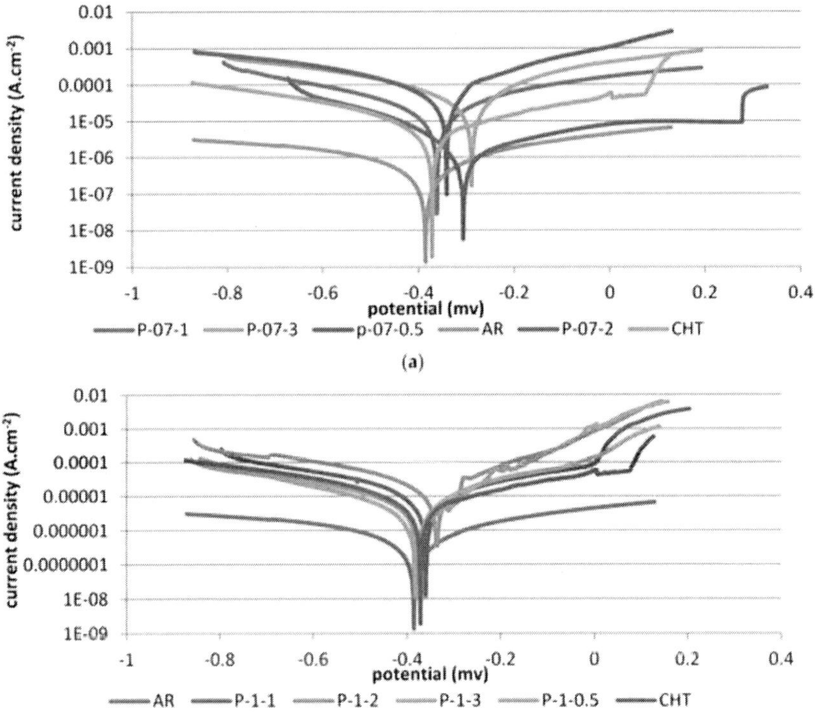

(b)

Figure 16. Potentiodynamic polarization curves of AR 416, CHT, and LHT samples treated at four different laser scanning speeds at laser powers of: (**a**) 700 W; and (**b**) 1000 W.

Table 2. Corrosion potentials determined from potentiodynamic polarization curves.

Condition	AR	CHT	P-07-0.5	P-07-1	P-07-2	P-07-3	P-1-0.5	P-1-1	P-1-2	P-1-3
Corrosion potential (mV vs. SRE)	–385	–372	–340	–359	–307	–328	–380	–361	–335	–378

The principal observation from the EDX data of the HZ, HAZ and BM was that there were higher sulfur and manganese contents in the HAZ as compared to the HZ and BM; this could be attributed to the high laser power and slow scanning speed used, because in this condition, partial melting may have occurred and resulted in diffusion of sulfur and manganese from the hardened zone toward the heat affected zone.

The present data recorded a shallower hardened zone accompanying the high laser scanning speeds, and the opposite occurred at slower scan speeds. This was understood as a high laser scanning speed leading to less energy being applied to the substrate.

4.2. Hardness and Wear Behavior

The variations in heat input values may have been the reason for the differences in the hardness values [27]. The hardness values increased with increasing heat input, or decreased with increasing laser scanning speed, as shown in Figure 11. Changing the surface absorption coefficient of the laser may cause differences in absorbed laser beam energy values, especially in the case of multi-pass heating with overlapping tracks [3]. Selective hardening advantageously traps the diffused free carbon in the steel via both rapid heating and quenching; this makes the steel harder than in normal phase transformations [21].

At high heat input above 21 $J \cdot mm^{-2}$, there was sufficient heat to obtain a complete austenitization process before quenching, which meant that the carbon dissolved completely in the austenite. Therefore, the martensite thus formed exhibited higher hardness values. On the other hand, at heat inputs lower than 21 $J \cdot mm^{-2}$, carbon dissolution in austenite was incomplete before cooling occurred, and this austenite transformed into martensite with lower hardness values. Also it could be seen that, for heat input <21 $J \cdot mm^{-2}$, the energy absorbed by the surface was not large enough to heat the material surface to the required quenching temperature (above 960 °C) [28], as had been noted by Tianmin et al. [16]. Therefore, hardness values decreased below 500 HV.

The microstructure of the hardened zone consisted mainly of martensite with microhardness values of approximately HV 700. The heat affected zone, which contained primary ferrite structures, exhibits microhardness values of approximately HV 450, which was higher than that of HV 200 in the AISI 416 martensitic stainless steel alloy substrate, as a result of martensitic transformation which occurs during laser treatment. Microhardness decreased significantly from the hardened surface to the substrate. A difference in the resulting hardened depth at constant laser scanning speed and different powers was observed, from the data presented in Figure 12 and Figure 13. By taking a cutoff at HV 500, it was evident that depth values of 0.7 mm and 0.3 mm were obtained at a laser scanning speed of 50 $cm \cdot min^{-1}$ and powers of 1000 and 700 W, respectively.

The variation of surface hardness with heat input reported in Figure 11 supported the wear behavior of the specimens under different laser parameters. When the hardness level increases, the wear resistance also increases [29,30]. Furthermore, as seen in Figure 12 and Figure 13, the depth of hardening was reduced at lower laser powers and at higher laser scanning speeds (lower heat input conditions). The more weight loss observed in Figure 14, which was associated with lower laser power and higher laser scanning speeds, may have been also partially related to shallower hardened depth, which was likely to be entirely worn–out during the wear test. The weight loss measured in this case could include both the hardened zone as well as the tougher heat affected zone (HAZ) which contained some untransformed ferrite. This might explain the increasing weight loss with the decreasing heat input of the processing laser.

It is noteworthy that at scanning speed 1 $m \cdot min^{-1}$ for both laser powers, the corresponding samples showed comparable weight losses, in spite of the considerable difference of hardened depth in both cases (700 vs. 240 μm for 1000 and 700 W, respectively). This anomaly may suggest that the worn surface layer was less than 240 μm in both cases, and therefore the effect of laser power appeared insignificant.

Based on the mentioned experimental results (Figure 11 and Figure 14), the wear resistance not only depended on the hardened depth, but also on other factors, such as the amount of heat input and the hardness level. Additionally, compressive stresses generated from laser hardening also add to factors controlling abrasive wear resistance [3]. The high stresses induced during rapid cooling of the surface hardened layer tended to create fine and highly dislocated martensite, with some level of stable austenite.

4.3. Impact Toughness

The measured impact energy for laser treated samples actually represented the overall effects contributed by both the hardened zones and the tough interior base metal (structure somewhat similar to a composite material [15]), unlike conventionally heat treated samples. The same level of wear resistance was achieved in both conditions. Clearly the impact energy tended to increase between the heat input range of 15–30 $J \cdot mm^{-2}$ (Figure 15). This observation was in accord with previously reported results although that the laser type was a Q-switched pulsed laser, in contrast with the present continuous laser [31]. The reason for this increase was due to the residual compressive stress induced by a laser surface treatment which was able to restrain the growth of the crack. However, the decrease of the impact energy for a heat input larger than 30 $J \cdot mm^{-2}$ needs more study to be sure that it is a recurring phenomenon.

The improved toughness of the laser-hardened specimens was also attributed to the refined microstructures with lower MnS and carbide particles concentrations along the prior austenite grain boundaries in the laser-hardened zone (as identified by EDX, Figure 6) [15].

The present data also showed that CHT samples tempered at 500 °C and 600 °C had impact energies of 35 and 62 J, respectively. The reason for these different values was that during tempering at 500 °C, a secondary hardening may have occurred in the stainless steel, due to carbide precipitations [32]. CHT samples tempered at 600 °C had comparable impact energy with LHT samples, but had much lower hardness levels (280 Hv) than the hardness levels achieved in LHT samples (510 Hv). A remarkable achievement in the impact energy (more than twice) could be noticed at a laser scanning speed of 1 $m \cdot min^{-1}$ and a laser power of 1000 W, as compared to CHT conditions (after tempering at 500 °C). Lower wear resistance is expected to accompany the lower hardness of the tempered CHT specimens.

It should be pointed out that the data presented here concerning impact was rather limited, and mainly directed towards comparing the impact toughness levels of the conventional and the laser treated samples. However, we conclude drawn that the results have some practical significance for impact toughness, as our objective was to find the heat input values that would lead to optimum mechanical and physical properties (hardness, wear resistance, impact toughness, and corrosion resistance).

4.4. Corrosion Resistance

The corrosion performance of AISI 416 martensitic stainless steel is essentially determined by two factors: (a) the amount and size of carbide particles and manganese sulfide inclusions on the grain boundaries of the hardened zone; and (b) the refinement of microstructure [20]. Chromium carbides are rich in this type of steel, because chromium is a dominant carbide-forming element. The higher the carbide level, the more the chromium is bounded in carbides, which results in less chromium in the solid solution for the formation of the passive film. Furthermore, the corrosion resistance increases with an increasing degree of microstructure refinement.

It could be observed from Figure 16 that corrosion potentials for all samples were very similar, and showed that LHT treatment did not change the fundamental nature of the passive film. Therefore from this standpoint, LHT of AISI 416 is capable of enhancing corrosion resistance by reducing the amount and size of manganese sulfide inclusions and carbide particles in the hardened zone, as compared to the CHT samples.

Data in Table 2 shows that LHT improved the corrosion potentials of grade AISI 416 stainless steel by 80 and 70 mV over that verified for the AR and CHT samples, respectively. These results agreed with the findings reached by Yong [21]. The corrosion potential of some LHT samples shifted towards an optimum direction. According to the polarization curves in Figure 16, it could be supposed that the optimum condition which provided the best corrosion resistance was P-07-200 at a heat input of 10.5 $J \cdot mm^{-2}$, and this could be attributed to a matrix having a microstructure consisting of the appropriate proportions of martensite, finer carbides, retained austenite, and lower sulfur and different oxides. Such a microstructure displayed a promising combination of toughness and strength, and contained less weak sites for corrosion attack.

5. CONCLUSIONS

1) Laser surface treatment of AISI 416 martensitic stainless steel considerably increased the surface hardness from 155 to 510 HV. The obtained hardness values were found to strongly depend on the heat input values. The hardness values increased with increasing heat input, or decreased with the increasing laser scanning speed.
2) The microhardness distribution in the laser affected region showed a stepwise profile. The average microhardness in the hardened zone reached HV 700, and in the heat affected zone it reached HV 450. The size of the shallowest hardened layer was 57 μm, and was associated with the lowest heat input (7 $J \cdot mm^{-2}$), while the deepest hardened layer was 885 μm and was achieved with the highest heat input value (60 $J \cdot mm^{-2}$), both with the lowest laser scanning speed of 0.5 $m \cdot min^{-1}$.
3) The laser affected region could be divided into two zones based on different microstructures: (i) the hardened zone and (ii) the heat affected zone.
4) As for heat input above 21 $J \cdot mm^{-2}$, heat was sufficient to obtain complete austenitization, and the carbon completely dissolved in austenite and caused an increase in hardness values and vice versa.
5) Wear results showed that up to scanning speed of 1 $m \cdot min^{-1}$, the wear resistance of samples processed at both laser powers was comparable to the conventionally hardened samples. On the other hand, at higher speeds, LHT samples showed a decrease in wear resistance as the scanning speed increased; this may have partially been related to a shallower hardened depth, which might have entirely worn out during the wear test. The weight loss decreased

with the increase in heat input. This was reflected through the surface hardness values after laser surface treatment.

6) The wear resistance depended on the hardened depth, the amount of heat input, hardness levels, and compressive stresses.

7) A significant result concerning the impact toughness and hardness values of laser heat treated samples, as compared to conventional heat treated samples, was reported. At the same level of hardness, laser treated samples experienced more than double the improvement in impact toughness. The reason for this increase was due to residual compressive stress induced by a laser surface treatment, which was able to restrain the growth of cracks.

8) The CHT samples tempered at 600 °C had comparable impact energy with LHT samples, but at much lower hardness levels (280 Hv vs. 510 Hv), with correspondingly higher wear resistance in LHT samples.

9) The corrosion potentials of the LHT samples improved by 80 and 70 mV over those recorded for the AR and CHT samples respectively.

10) The optimum condition for wear resistance, impact toughness and corrosion resistance was recorded at a heat input value of 21 $J \cdot mm^{-2}$. This may not have been the maximum value for each property, but it was the general optimum, consideringall of the properties as a whole.

This paper threw light on the significance of laser treated samples as compared to conventional treated samples, particularly in regards to impact toughness, wear resistance, and corrosion resistance, under the conditions specified.

ACKNOWLEDGMENTS

This research was supported by the Central Metallurgical Research and Development Institute (CMRDI) and Faculty of Engineering, Cairo University. Laser surface treatment, metallographic analysis, and wear and corrosion tests were carried out in the Central Metallurgical Research and Development Institute (CMRDI). Microhardness distribution and impact test were performed in the Faculty of Engineering, Cairo University.

AUTHOR CONTRIBUTIONS

S.R. Al-Sayed and A.A. Hussein conceived and designed the experiments; S.R. Al-Sayed and partialy H. Elgazzar performed the experiments; S.R. Al-Sayed, A.A. Hussein, A.A. Nofal and S.I. Hassab Elnaby analyzed all the data; A.A. Nofal contributed reagents/materials/analysis tools; S.R. Al-Sayed and A.A. Hussein wrote the paper.

CONFLICTS OF INTEREST

The authors declare no conflict of interest.

REFERENCES

1. Budinski, K.G. Surface Engineering for Wear Resistance; Prentice Hall: Englewood Cliffs, NJ, USA, 1988.
2. Purushothaman, D.B.; Gengusamynaidu, B.; Karupuudaiyar, R.B. Experimental studies on the microstructure and hardness of laser-treated steel specimens. Trans. Can. Soc. Mech. Eng. 2012, 36, 241–258.
3. Kusinski, J.; Kac, S.; Kopia, A.; Radziszewska, A.; Rozmus-Górnikowska, M.; Major, B.; Major, L.; Marczak, J.; Lisiecki, A. Laser modification of the materials surface layer—A review paper. Bull. Pol. Acad. Sci. Tech. Sci. 2012, 60, 711–728.
4. Steen, W.M. Laser Surface Treatment. In Laser Material Processing; Springer: London, UK, 1991; pp. 172–232.
5. Majumdara, J.D.; Nath, A.K.; Manna, I. Studies on laser surface melting of tool steel-Part II: Mechanical properties of the surface. Surf. Coat. Technol. 2010, 204, 1326–1329.
6. Lovejoy, P.T. Structure and constitution of wrought martensitic stainless steels. In Handbook of Stainless Steels, 1st ed.; Packner, D., Bernstein, I.M., Eds.; McGraw-Hill: New York, NY, USA, 1977; p. 800.
7. Berns, H. Manufacture and application of high nitrogen steels. ISI Int. 1996, 36, 909–914.
8. Krishna, B.V.; Bandyopadhyay, A. Surface modification of AISI 410 stainless steel using laser engineered net shaping (LENSTM). Mater. Des. 2009, 30, 1490–1496.
9. Bommi, V.C.; Mohan, M.K.; Prakash, S. Surface Modification of Martensitic Stainless steel Using Metal Working CO2Laser. In Proceedings of the International Symposium of Research Students on Materials Science and Engineering, Chennai, India, 20–22 December; 2004.; pp. 1–30.
10. Ready, J.F.; Farson, D.F.; Feeley, T. Surface Treatment: Heat Treatment. In LIA Hand book of Laser Material Processing; Laser Institute of America; Springer: Berlin/Heidelberg, Germany; New York, NY, USA, 2001.
11. Loa, K.H.; Chenga, F.T.; Kwoka, C.T.; Man, H.C. Effects of laser treatments on cavitation erosion and corrosion of AISI 440C martensitic stainless steel. Mater. Lett. 2003, 58, 88–93.
12. Kwok, C.T.; Man, H.C.; Cheng, F.T. Cavitation erosion and pitting corrosion behaviour of laser surface-melted martensitic stainless steel UNS S42000. Surf. Coat. Technol. 2000, 126, 238–255.
13. Klar, E.; Samal, P.K. Powder Metallurgy Stainless Steels: Processing, Microstructures, and Properties; ASM International: Materials Park, OH, USA, 2007; pp. 5–18.

14. Ion, J.C.; Moisio, T. Laser surface modification of a 13.5% Cr, 0.6% C steel. J. Mater. Sci. 1991, 26, 43–48.
15. Tsay, L.W.; Chang, Y.M.; Torng, S.; Wu, H.C. Improved Impact Toughness of 13Cr Martensitic Stainless Steel Hardened by Laser. J. Mater. Eng. Perform. 2002, 11, 422–427.
16. Tianmin, S.; Meng, H.; Yuen, T.H. Impact wear behavior of laser hardened hypoeutectoid 2Cr13 martensite stainless steel. Wear 2003, 255, 444–455.
17. Escudero, M.L.; Bello, J.M. Laser surface treatment and corrosion behaviour of martensitic stainless AISI 420 steel. Mater. Sci. Eng. A 1992, 158, 227–233.
18. Ingelgem, Y.V.; Vandendae, I.; van den Broek, D.; Hubin, A.; Vereecken, J. Influence of laser surface hardening on the corrosion resistance of martensitic stainless steel. Electrochim. Acta 2007, 52, 7796–7801.
19. Conde, A.; Colac, R.; Vilar, R.; de Damborenea, J. Corrosion behaviour of steels after laser surface melting. Mater. Des.2000, 21, 441–445.
20. Kwok, C.T.; Lo, K.H.; Cheng, F.T.; Man, H.C. Effect of processing conditions on the corrosion performance of laser surface-melted AISI 440C martensitic stainless steel. Surf. Coat. Technol. 2003, 166, 221–230.
21. Yong, T.T.; Sudesh, L.W.; Gary, K.L.N.; Daniel, J.B. Investigation into the influence of laser melting on the sulphide inclusions in AISI 416 stainless steel. Corros. Sci. 2011, 53, 3950–3955.
22. Charschan, S.S. Lasers in Industry; Van Nostrand Reinhold Company: New York, NY, USA, 1972.
23. Rana, J.; Goswami, G.L.; Jha, S.K.; Mishra, P.K.; Prasad, B.V.S.S.S. Experimental studies on the microstructure and hardness of laser-treated steel specimens. Opt. Laser Technol. 2007, 39, 385–393.
24. Jeffrey, A. ASM Handbook Volume 8: Mechanical Testing and Evaluation; Kuhn, H., Medlin, D., Eds.; ASM International: Materials Park, OH, USA, 2004; Volume 8.
25. Navasa, C.; Condea, A.; Fernándeza, B.J.; Zubirib, F.; Damborenea, J. Laser coatings to improve wear resistance of mould steel. Surf. Coat. Technol. 2005, 194, 136–142.
26. Asibu, E.K., Jr. Laser surface modification. In Principles of Laser Materials Processing; John Wiley & Sons: Hoboken, NJ, USA, 2009; pp. 409–427.
27. Katsamas, A.I.; Haidemenopoulos, G.N. Surface hardening of low-alloy 15CrNi6 steel by CO2 laser beam. Surf. Coat. Technol. 1999, 115, 249–255.
28. National Technical Committee for Standardization of Heat Treatment. Applied Standard Handbook of Metal Heat Treatment; Machinery Industry: Beijing, China, 1994.
29. Selvan, J.S.; Subramanian, K.; Nath, A.K. Effect of laser surface hardening on En18 (AISI 5135) steel. J. Mater. Process. Technol. 1999, 91, 249–255.
30. Hussein, H.T.; Kadhim, A.; Al-Amiery, A.A.; Kaghum, A.A.H.; Mohamad, A.B. Enhancement of the Wear Resistance and Microhardness of Aluminum Alloy by Nd:YaG Laser Treatment. Sci. World J. 2014, 2014, 1–5.
31. Weiguang, X.; Lei, L.; Yanpeng, W.; Aimin, Z.; Yacong, G.; Chenguang, H.; Hongxiang, Y.; Lingchen, Z. Impact toughness of a gradient hardened layer of Cr5Mo1V steel treated by laser shock peening. Acta Mech. Sin. 2016, 32, 301–308.
32. Ulrike, T.; Veronika, C.; Ute, S.; Michael, J.H. ASM Handbook Volume 9: Metallography and Microstructures; ASM International: Materials Park, OH, USA, 2004.

CHAPTER 5

Synthesis, Structural Study and Spectroscopic Characterization of a Quinolin-8-Yloxy Derivative with Potential Biological Properties

Elida Romano[1], María V. Castillo[1], Jorgelina L. Pergomet[2], Juan Zinczuk[2], Silvia A. Brandán[1]*

[1]*Faculty of Biochemistry, Chemistry and Pharmacy of the National University of Tucumán, Tucumán, R. Argentina*

[2]*Chemistry Rosario Institute (CONICET-UNR), Faculty of Biochemistry and Pharmacists Sciences, Rosario, R. Argentina*

ABSTRACT

We have prepared the (5-chloro-quinolin-8-yloxy) acetic acid and characterized it by using infrared, Raman and multidimensional nuclear magnetic resonance spectroscopies. The density functional theory (DFT) together with the 6-31G* and 6-311++G** basis sets were used to study its structure and vibrational properties. Three stable conformations of the compound were theoretically determined in gas phase and probably these conformations are present in the solid phase. The harmonic vibrational wavenumbers for the optimized geometries were calculated at the same theory levels. For a complete assignment of the observed bands in the vibrational spectra, the DFT calculations were combined with Pulay's scaled quantum mechanical force field (SQMFF) methodology in order to fit the theoretical wavenumber values to the experimental ones. Besides, the force constants of the three conformers of (5-chloro-quinolin-8-yloxy) acetic acid were calculated and compared with those obtained by us for the 2-(quinolin-8-yloxy) acetic acid. In addition, the characteristics of the electronic delocalization of those structures were performed by using natural bond orbital (NBO), while the corresponding topological properties of electronic charge density are analysed by employing Bader's atoms in molecules theory (AIM).

Keywords: (5-Chloro-Quinolin-8-Yloxy) Acetic Acid; Vibrational Spectra; Molecular Structure; Force Field; DFT Calculations

1. INTRODUCTION

Heterocyclic compounds that contain the (quinolin-8- yloxy) moiety exhibit a wide range of biological properties [1-5], such as the 2-(quinolin-8-yloxy) acetohydrazones that have antiamoebic activities [5] for which, the structural and vibrational study of these types of compounds are of great chemical and pharmaceutical importance. Recently, a complete vibrational analysis of all observed bands in the vibrational spectra of 2-(quinolin- 8-yloxy)-acetic acid [6] was performed by means of the DFT calculations combined with Pulay's scaled quantum mechanical force field (SQMFF) methodology [7-9]. In that study, the normal mode calculations were accomplished by using a generalized valence force field (GVFF) [8,9] considering three different structures for the anhydrous and monohydrated compounds. In the present work, we report an experimental and theoretical vibrational study of (5-chloro-quinolin-8-yloxy) acetic acid (CQA) by means of the B3LYP calculations using 6-31G* and 6-311++G** basis sets. For a complete assignment of the compound, the DFT calculations were combined with the SQMFF methodology [7-9] in order to fit the theoretical wavenumber values to the experimental ones. So far, the crystal and molecular structure of (5-chloro-quinolin-8- yloxy) acetic acid has not been determined. First, we have synthesized and characterized the compound and then, the optimized geometries and frequencies for the normal modes of vibration of CQA considering three different stable structures were calculated in order to carry out a complete assignment of all observed bands in the infrared and Raman spectra. Here, the normal mode calculations and the force fields for those structures were obtained by using the transferable scaling factors of Rauhut and Pulay [9] while the harmonic force constants were subsequently scaled to reproduce as well as possible the experimental frequencies. Thus, a complete assignment of the compound in terms of the potential energy distribution was performed. Furthermore, when the forces constants values of this compound were compared with those corresponding to the 2-(quinolin-8-yloxy)- acetic acid [6] numerous changes in the structural, topological and vibrational properties attributed to the chloro atom were observed. Additionaly, the nature of the different rings and bonds of the three studied structures of (5-chloro-quinolin-8-yloxy) acetic acid were analyzed by means of the NBO studies [10-12] while the topological properties of electronic charge density were determined employing the Bader's atoms in molecules theory (AIM) [13].

2. EXPERIMENTAL METHODS

2.1. Synthesis

(5-chloroquinolin-8-yloxy) acetic acid was obtained according to Cho et al. [1] using the following procedure.

2.1.1. (5-Chloroquinolin-8-Yloxy) Acetic Acid Methyl Ester

A mixture of 5-chloro 8-hydroxyquinoline (3.6 g, 20 mmol), methyl bromoacetate (3.7 g, 24 mmol) and K_2CO_3 (5.52 g, 40 mmol) in acetone (50 mL) was heated for 3 h under reflux, filtered and concentrated in vacuo. The residue was partitioned between ethyl acetate and brine, and the organic layer was dried with $MgSO_4$, filtered, and concentrated in vacuo. The residue was purified by crystallization from isopropyl ether to give (5-chloroquinolin-8-yloxy) acetic acid methyl ester 3584 g (71%) m.p. 102.1°C - 102.3°C.

2.1.2. (5-Chloroquinolin-8-Yloxy) Acetic Acid (CQA)

A mixture of ester ((1.26 g, 5 mmol) and $LiOH{\cdot}H_2O$ (352 mg, 8.40 mmol) in 75 mL of $THF/CH_3OH/H_2O$ (1:1:1, 6) was stirred and heated at reflux for 1 h and concentrated. The aqueous layer was washed with ether and adjusted to pH 3 with 1 N HCl. The precipitate was filtered and dried to give (5-chloroquinolin-8-yloxy) acetic acid (1085 g, 91%) m.p. 219.5°C - 221°C.

2.2. NMR Spectra

1H NMR (300 MHz, $CDCl_3$) δ: 4.92 (2H, s, O-CH_2); 7.08 (1H, d, J = 8.48, H_7), 7.64 (1H, d, J = 8.48, H_6); 8,47 (1H, d, J = 8.58, H_4); 7.70 (1H, dd, J = 4.18, J = 8.58, H_3); 8.94 (1H, d, J = 4.18, H_2).

^{13}C NMR (75 MHz, $CDCl_3$) δ: 65.94 (CH_2); 110.54 (C7); 121.69 (C5); 123.53 (C3); 126.66 (C4a); 127.07 (C6); 132.64 (C4); 140.55 (C8a); 150.28(C2); 153.54 (C8); 170.34 (CO_2H) (Atom numbering according to naphthalene).

2.3. Infrared and Raman Spectra

The infrared spectrum of the solid substance was recorded in KBr pellets in the wavenumbers range from 4000 to 400 cm^{-1} with a FT-IR Perkin Elmer spectrometer, provided with a Globar source and a DGTS detector. FT-Raman spectrum of the crystalline solid was recorded on a Bruker RFA 106/S FT-Raman instrument using the 1064 nm excitation line from an Nd: YAG laser in the region of 4000 - 0 cm^{-1}. Two hundred scans were accu mulated at 4 cm^{-1} resolution using a laser power of 150 mW.

3. COMPUTATIONAL DETAILS

The potential energy curves associated with the internal rotation described by the C21-C18-O17-C10 dihedral angle for CQA were studied at the B3LYP/6-31G* and 6-311++G** theory levels. With both calculations, two conformations stable, named, C_I and C_{II}, according to the position of the OH group in relation to the N atom of the ring were obtained. Another plane structure was also considered (C_{III}) in agreement with the experimental structure of the 8-(carboxymethoxy) quinolinium nitrate monohydrate compound [14]. The structures and labelling of the atoms for all conformers of CQA can be seen in **Figure 1**. The electronic charge density topological analysis for those structures were performed by using the AIM200 program package [15] while the NBO calculations were obtained by means of the NBO 3.1 [16] program, as implemented in the GAUSSIAN 03 package [17]. The natural internal coordinates for the compound have been defined as those reported in the literature [7,8,18] and are listed in the Supporting Material as Table S1. The resulting force fields were transformed to "natural" internal coordinates by using the MOLVIB program [19,20]. Then, following the SQMFF procedure [7-9,21], the harmonic force fields for those structures were evaluated at the B3LYP/6-31G* level. The potential energy distribution components (PED) higher than or equal to 10% were subsequently calculated with the resulting SQMFF. The nature of all vibration modes was carried out by means of the Gauss View program [22].

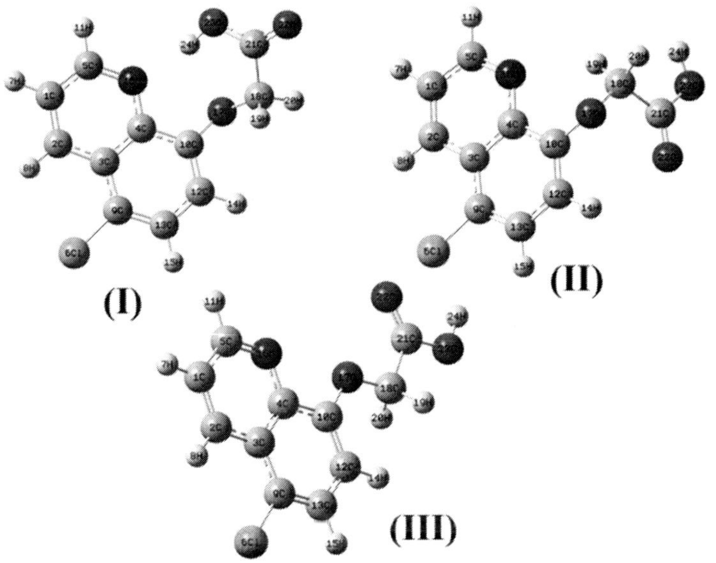

Figure 1. Theoretical structures and labelling of the atoms for the stable C_I, C_{II} and C_{III} conformers of (5-chloro-quinolin-8-yloxy) acetic acid.

4. RESULTS AND DISCUSSION

4.1. Geometry

Table S2 (Supporting Material) shows the comparison of the total energies for the compound, and the corresponding dipole moment values for the C_I, C_{II} and C_{III} conformers of C1 symmetries. With both methods, the energy of the C_I is lower than the other ones, as was observed in the compound without chlorine [6], and the potential energy difference between the C_I and C_{II} forms, by using the B3LYP/6-31G* and B3LYP/6-311++G** methods, are −43.80 and −46.43 kJ/mol, respectively. When the C_I conformer is compared with the planar ones the potential energy difference values slightly by using both basis sets decrease up to −27.80 and −32.36 kJ/mol, respectively.

Also, in this case, the high values of the dipolar moments for the C_I structures could probably explain its stabilities, as was observed in similar molecules [6,23-25].

Table 1 shows a comparison of the calculated geometrical parameters for all structures of CQA, with the ones corresponding to that observed from X-ray diffraction for the 8-(carboxymethoxy) quinolinium nitrate monohydrate compound [14] by means of the root mean of square deviations (rmsd) values. Note that the 6-311++G** basis set reproduce reasonably well the theoretical bond lengths for the C_I structure (0.019 Å) and the bond angles for the C_{III} structure (2.00°), as is expected because this last conformer is planar as that compared compound [14]. Besides, the calculation predicts that the C_I and C_{III} conformers are the most stable and probably both are present in the crystalline state.

The stabilities of the C_{II} and C_{III} structures of CQA, in relation to the C_I structure, were investigated by using the natural atomic charges [26-29] and the results are given in Table S3. Note that the natural charges values (NPA) corresponding to the C21 atoms have the most positive values in all structures while the most negative values correspond to the O23 atoms. In this compound, as in 2-(quinolin-8-yloxy)-acetic acid [6], the higher NPA charges observed on the O17 and O22 atoms in the C_{II} structure justified the repulsion between those atoms. This way, the C_{II} structure is more unstable than the corresponding to the C_I and C_{III} conformers. The bond orders, expressed by Wiberg's index for all conformers of CQA are given in Table S4. The bond order values for the O23 atoms in all conformers have the lowest values (belong to the COO groups) than the other ones (the O17 and O22 atoms belongs to the heterocyclic rings), while the bond orders for the N16 atoms have higher values.

4.2. NBO Study

The stability of the three structures of CQA was also investigated by means of NBO calculations [10-12]. The second order perturbation energies $E^{(2)}$ (donor $\rightarrow$ acceptor) that involve the most important delocalization all conformers of CQA are given in Tables S5, S6. The contributions of the stabilization energies for the $DET_{s\rightarrow s*}$ charge transfers are similar to those obtained for the 2-(quinolin-8-yloxy)-acetic acid compound [6].

Here, the delocalizations $DET_{LP\rightarrow s*}$ for the C_{II} structure has lower values than the $DET_{s\rightarrow s*}$delocalizations, and the calculated total energy values favours to the C_I and C_{III} conformers, which structures are the most stable in the gas phase. Thus, the total energy values clearly show the higher stability of the C_I conformer however, by using the 6-311++G** basis set, the C_{II} and C_{III}conformers have the same ones approximately total energy values.

4.3. AIM Analysis

Furthermore, the three structures of CQA were analysed by means of Bader's charge electron density topological analysis [13]. The calculated electron density, (r) and the Laplacian values, Ñ^{2}r(r) in the bond critical points (BCPs) and ring critical points (RCPs) for those structures are shown in Table S7. The BCP has the typical properties of the closed–shell interaction and for this, the value of ρ(r) is relatively low, the relationship $|l_1|/l_3$ is <1 and Ñ^{2}r(r) is positive indicating that the interaction is dominated by the charge contraction away from the interatomic surface toward each nucleus. Note that the results are clearly dependent of the size basis set. Thus, the analysis shows two BCPs and RCPs in the C_I structure when the 6-31G* basis set is employed but, only one BCP and RCP when the 6-311++G** basis set is used. On the contrary, for the C_{II} structure with both basis sets only one BCP and RCP are observed while for the C_{III} structure there is one BCP with the 6-31G* basis set. These results, togheter to the NBO analysis, justify the stabilities of the C_I and C_{II} structures due to the presence of short intramolecular O17-H24 and N16-H24 bonds, in the first case and only a N16-H19 bonds in the second one, as observed in Table S7.

4.4. Vibrational Analysis

The Figures 2, 3 show the registered infrared and Raman spectra for the compound in solid phase. In this study, in accordance with the NBO and AIM results, the three structures for the compound in gas phase were considered. The CQA's structures have C_1 symmetries and 66 normal vibration modes, all active in the infrared and Raman spectra. Probably, the three species are present in the solid phase because the comparison of each vibrational spectrum with the corresponding experimental one is very different among them, as observed in **Figure 4**, however, a comparison between the average calculated infrared spectra (from B3LYP/6-31G* level for the C_I, C_{II} and C_{III} conformers) by using average wavenumbers and intensities with the corresponding experimental one demonstrate a good correlation, as observed in Figure S1.

Thus, the resulting IR spectrum reproduces rather well some bands of the experimental spectrum, especially in the 2000 - 400 cm^{-1} region. The assignment of the experimental bands to the expected normal vibration modes were made on the basis of the PED in terms of symmetry coordinates and taking into account the corresponding assignment of related molecules [6,30-34]. Tables 2, S8-S10 show the experimental and calculated frequentcies, potential energy distribution based on the 6-31G* basis set, and assignment for the C_I, C_{II} and C_{III} structures of CQA. In a similar way as observed in the molecular packing of the 2-(2'-furyl)-1Himidazole [23], 2-(2'- furyl)-4,5-1H-dihydroimidazole [27] and tolazoline hydrochloride molecules [35,36], the broad band with several shoulders between 2800 and 1900 cm^{-1} can be attributed to the N-H and O-H hydrogen bondings formed by the spatial arrangement of molecules in the lattice crystal.

Table 1. Calculated geometrical parameters for the conformers of (5-chloro-quinolin-8-yloxy) acetic acid.

Parameter	6-31G*			6-311++G**			Exp.
	C_I	C_{II}	C_{III}	C_I	C_{II}	C_{III}	
				Bond length (Å)			
C3-C4	1.432	1.437	1.434	1.430	1.434	1.432	1.402 (3)
C4-N16	1.362	1.363	1.357	1.360	1.365	1.354	1.368 (2)
C5-N16	1.319	1.318	1.318	1.317	1.318	1.315	1.317 (2)
C4-C10	1.429	1.431	1.439	1.427	1.431	1.437	1.411 (3)
C9-Cl6	1.758	1.762	1.763	1.758	1.758	1.763	
C10-O17	1.367	1.374	1.357	1.367	1.377	1.356	1.362 (2)
C18-O17	1.433	1.413	1.407	1.432	1.412	1.406	1.424 (2)
C18-C21	1.537	1.531	1.520	1.535	1.530	1.518	1.498 (3)
C21-O22	1.210	1.202	1.203	1.205	1.202	1.196	1.203 (2)
C21-O23	1.331	1.362	1.358	1.329	1.363	1.359	1.303 (2)
RMSD	0.020	0.027	0.025	0.019	0.027	0.025	
				Bond angle (degrees)			
C4-N16-C5	118.7	118.4	118.2	118.9	118.2	118.5	122.98 (19)
C4-C10-C12	119.9	120.1	119.8	119.9	120.3	119.7	118.90 (19)
C4-C10-O17	117.6	121.8	115.3	117.8	121.8	115.5	114.39 (17)
C12-C10-O17	122.4	117.9	124.9	122.2	117.7	124.8	126.71 (18)
C10-O17-C18	119.2	118.5	117.8	119.7	118.4	118.0	117.04 (15)
O17-C18-C21	112.5	112.3	108.6	113.1	112.4	109.0	108.78 (16)
C18-C21-O23	117.4	114.0	108.8	117.7	113.9	108.7	110.57 (19)
C18-C21-O22	119.3	125.2	127.2	119.2	125.3	127.4	125.03 (19)
O22-C21-O23	123.3	120.8	124.0	123.0	120.8	123.9	124.40 (2)
RMSD	4.05	4.65	2.01	4.24	4.71	2.00	
				Dihedral angles (degrees)			
C21-C18-O17-C10	−112.8	−76.7	−180.0	−112.4	−76.7	−180.0	−178.52 (17)
C4-C10-O17-C18	110.4	−63.4	180.0	107.8	−62.0	180.0	176.72 (17)
C12-C10-O17-C18	−73.2	121.9	0.0	−76.4	123.3	0.0	−3.3 (3)
RMSD	67.3	167.0	2.8	69.4	166.7	2.8	

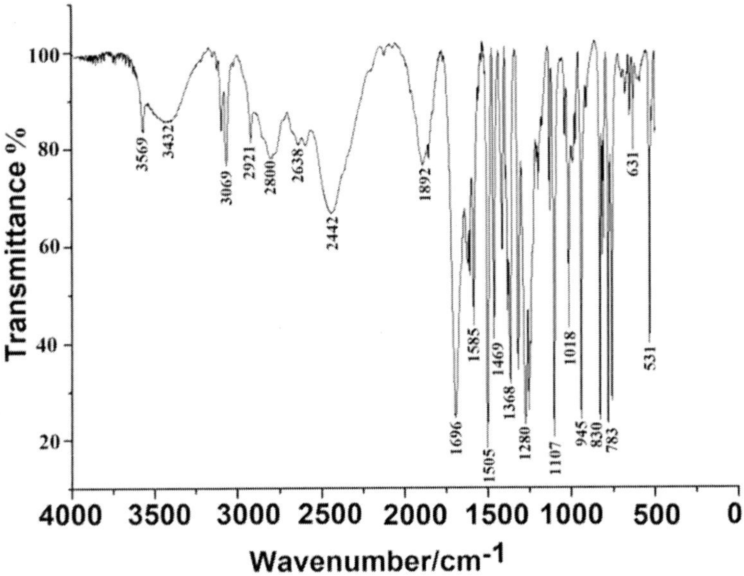

Figure 2. Experimental infrared spectrum of the solid (5- chloro-quinolin-8-yloxy) in KBr pellets.

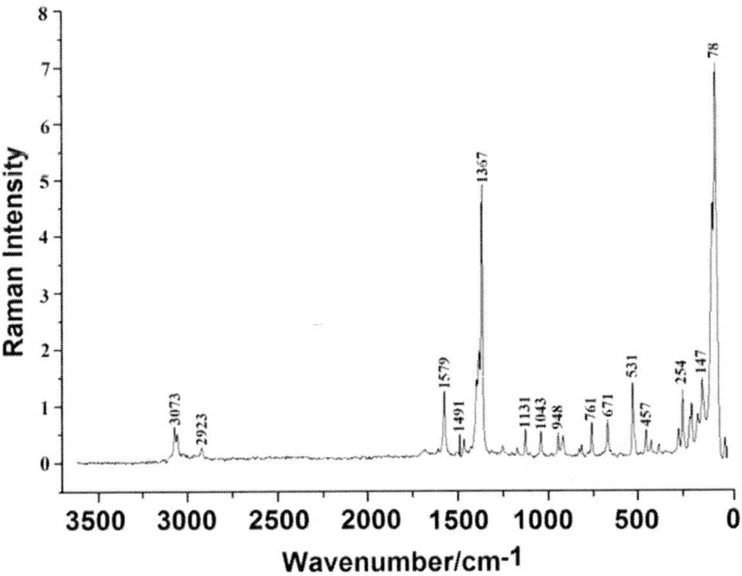

Figure 3. Experimental raman spectrum of the solid (5- chloro-quinolin-8-yloxy)of in a glass capillary.

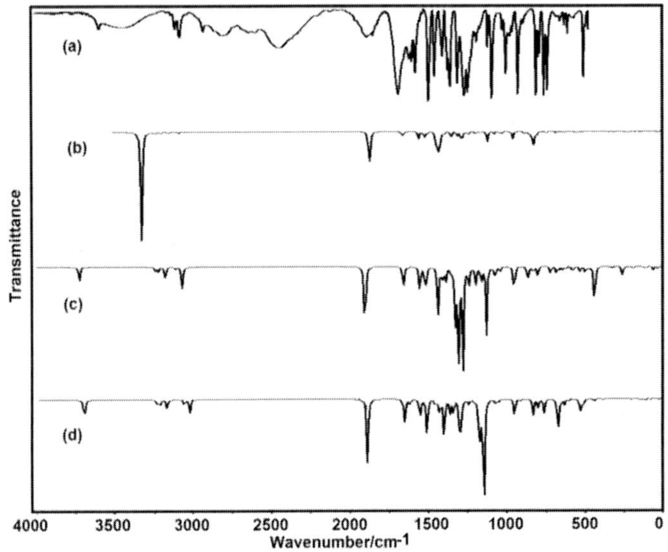

Figure 4. Comparison between: a) the experimental infrared spectrum of the solid (5-chloro-quinolin-8-yloxy) with, the theoretical spectra corresponding to the b) conformer C_I; c) conformer C_{II}, and; d) conformer C_{III} by using B3LYP/6-31G* level.

Here, the B3LYP/6-31G* calculations were considered because the used scale factors are defined for this basis set. The SQM force fields for this compound can be obtained upon request. Below we discuss the assignment of the most important groups.

4.4.1. Assignments

4.4.1.1. OH Modes

The weak bands in the IR spectra at 3569, 3432 and 3149 cm^{-1}, in accordance to the values reported for similar molecules [30,33,34], can be assigned to the O-H stretching vibrations of the three conformers, as observed in **Table 2**. The width and shape of these bands shows clearly the typical spectroscopic signature of an H-bond interaction (inter or intramolecular). In accordance to that reported value for the benzoic acid [33,34] and to the theoretical calculations, the corresponding deformation modes of this group are assigned, for those conformers, to the strong IR bands at 1368, 1280 and 1254 cm^{-1}, for the C_I, C_{III} and C_{II}conformers, respectively. The out-ofplane deformation mode (tOH or gOH) in the 4-hydroxybenzoic acid dimer [28] is calculated at 877 cm^{-1} while in CQA that mode is assigned for the C_I conformers to the strong band in the IR spectrum at 760 cm^{-1} and to the very weak band in the same spectrum at 504 cm^{-1} for the C_{III} conformer while for the C_{II} conformers that mode is assigned to the shoulder in the Raman spectrum at 409 cm^{-1}. This last band in the C_p conformer of 2-(quinolin-8-yloxy)-acetic acid [6] is assigned to the weak IR band at 497 cm^{-1}.

Table 2. Observed and calculated wavenumbers (cm^{-1}) and assignment for the (5-chloro-quinolin-8-yloxy) acetic acid.

Experimental[a]		C_I		C_{II}		C_{III}	
IR solid	Raman solid	SQM[b]	Assignment	SQM[b]	Assignment	SQM[b]	Assignment
3569 w				3558	ν(O23-H24)		
3432 w,br						3541	ν(O23-H24)
3149 vvw	3145 vvw	3175	ν(O23-H24)				
3125 vvw	3130 vvw	3100	ν(C2-H8)	3109	ν(C12-H14)	3102	ν(C12-H14)
3100 w		3096	ν(C13-H15)	3097	ν(C2-H8)	3096	ν(C2-H8)
	3098 vvw	3085	ν(C1-H7)	3092	ν(C13-H15)	3086	ν(C13-H15)
3069 w	3073 w	3081	ν(C12-H14)	3079	ν(C1-H7)	3076	ν(C1-H7)
	3057 w	3056	ν(C5-H11)				
3025 vvw	3028 vw			3036	ν(C5-H11)	3037	ν(C5-H11)
	3006 vvw	3006	ν_a CH_2				
	2981 vw			2974	ν_a CH_2		
	2966 vw	2946	ν_s CH_2				
	2943 vvw			2931	ν_s CH_2	2931	ν_a CH_2
2921 vw	2923 vw					2895	ν_s CH_2
1892 m,br				1823	ν(C21-O22)		
1854 m						1814	ν(C21-O22)
1696 s	1687 vw	1792	ν(C21-O22)				
1626 m	1646 vw	1613	ν(C2-C3)	1613	ν(C2-C3)	1614	ν(C2-C3)
1608 m	1610 vw	1599	ν(C1-C2)	1600	ν(C1-C2)	1601	ν(C13-C9)
1585 m	1579 m	1575	ν(C13-C9)	1570	ν(C13-C9)	1572	ν(C13-C12)
1505 vs	1491 vw	1509	βC1-H7	1508	βC1-H7	1513	βC1-H7
1469 s	1470 vw	1482	βC5-H11	1480	βC5-H11	1485	βC5-H11
	1431 vw			1446	δ CH_2	1453	δ CH_2
1423 sh		1433	δ CH_2			1431	*wag* CH_2
1417 m		1408	βC2-H8	1408	βC2-H8		
1395 sh	1398 m	1402	ν(C5-N16)	1403	ν(C5-N16)	1407	ν(C10-C4)
1386 m	1386 m	1374	*wag* CH_2	1390	*wag* CH_2	1405	βC2-H8
1368 s	1367 s	1353	δ OH			1350	ν(C3-C4), ν(C12-C10)
1324 s		1340	ν(C3-C4)	1336	ν(C3-C4)	1320	ν(C10-O17)
	1317 vvw			1305	ν(C12-C10)		
1280 s	1278 vvw	1305	ν(C12-C10)	1279	ρCH_2	1289	δ OH
1254 s	1255 vw	1269	ρCH_2	1254	δ OH	1259	ν(C4-N16)
1251 sh	1253 w	1251	ν(C4-N16)	1250	ν(C4-N16)	1252 1251	ν(C5-N16) ρCH_2
1238 sh		1238	ν(C10-C4)	1234	ν(C10-C4)		
	1219 vw	1221	ν(C21-O23)			1212	βC13-H15

1202 w	1202 vvw	1207	βC13-H15, ν(C3-C9)	1207	βC13-H15, ν(C3-C9)		
1176 vw	1174 vw	1166	βC12-H14	1167	βC12-H14	1180	βC12-H14
1133 m	1131 w	1131	R_1(A2)	1136	R_1(A2)	1143	ν(C1-C2)
1107 s	1105 vvw	1083	ν(C13-C12)	1112 1086	ν(C21-O23) ν(C13-C12)	1128 1104	ν(C13-C12), δ OH, ν(C21-O23)
1042 vw	1043 w	1037	ν(C1-C5)	1036	ν(C1-C5)	1035	ν(C1-C5)
1018 m	1012 vvw			1029	ν(C18-O17)	1029	ν(C18-O17)
996 w		998	ν(C18-O17)				
981 w		993	γC1-H7	991	γC2-H8	991	γC2-H8
954 sh	967 sh	963	γC5-H11	967	w CH_2	962	w CH_2
948 sh	948 vw	961	w CH_2	960	γC5-H11	959	γC5-H11
945 s		937	γC13-H15	939	γC12-H14	926	R_1(A2)
913 vw	919 vw	925	ν(C9-C16)	921	R_1(A1)	915	γC13-H15
836 sh		851	ν(C18-C21)	847	γC13-H15	884	ν(C18-C21)
830 s	833 vw	839	γC12-H14	838	ν(C18-C21)	818	βR_3(A2)
820 m	817 vw	817	R_2(A1)	816	R_3(A2)	818	γC12-H14
806 m		812	γC2-H8	809	γC1-H7	809	γC1-H7
783 s	785 vw	778	τR_1(A2)	774	τR_1(A2)	777	τR_1(A1)
760 s	761 w	737	τOH			742	R_1(A1)
701 vw		720	ν(C10-O17)				
680 vw				699	ν(C10-O17)		
	671 w	665	R_2(A1)	666	γC10-O17	655	R_2(A1)
654 vw	655 vvw	649	γC10-O17	650	R_2(A1)		
631 w	633 vvw	624	δ COO	624	τR_1(A1)	632 620	τR_1(A2) δ COO
613 vw	613 vvw					619	γCOO
606 vw	600 vvw	601	τR_1(A1)	595	γC9-C16	597	R_2(A2)
590 vw		572	R_1(A2)	569	ρC10-O17, R_2(A2)	585	γC10-O17
531 m	531 m	534	γCOO	544	γCOO		
520 vw	522 sh	528	R_2(A2)	522	R_2(A2)	526	ν(C3-C9)
504 vw		503	R_1(A1)	490	δ COO	508 495	R_3(A1) τOH
	488 vvw	469	τR_3(A1)	480	R_3(A1)		
	457 m	457	τR_1(A1)	460	τR_2(A1)	463	τR_2(A2)
	431 vw			415	τR_3(A2)	439	ρCOO
	409 sh	416	τR_1(A2)	399	τOH	418	τR_1(A2)
	386 vw	388	γC9-C16	385	ν(C9-C16)	378	ν(C9-C16)
	356 vvw	335	ρCOO	330	ρCOO	346	γC9-C16
	278 w	312	ρC10-O17	314	δ CCO		

254 s	240	δ CCO	254	δ COC	252	β C9-C16
216 w	225	β C9-C16	219	β C9-C16	223	ρ C10-O17
208 m	207	δ COC	196	ρ C10-O17	199	δ CCO
173 w	168	Butt	174	Butt	188	Butt
147 m	135	τR_2 (A2)	123	τR_2 (A2)	135	τR_3 (A1)
92 s	88	CCOC, w COO	102	τR_3 (A1)	116 109	w Ring τR_2 (A1)
78 vs	87	w COO	54	w COO	80	δ COC
	48	w Ring	36	CCOC	33	CCOC
20 vw	22	CCOC	24	w Ring	25	w COO

4.4.1.2. CH Modes

The C-H stretching modes can be clearly assigned, for their positions, to the group of bands in the 3125 - 3025 cm^{-1} region of the IR and Raman spectra. The in-plane deformation modes of the phenyl ring, in similar compounds are observed between 1267 and 1042 cm^{-1} [31] while in this compound those modes are assigned to the IR bands between 1505 and 1176 cm^{-1}, as observed in **Table 2**. On the other hand, the six expected out-of-plane deformations of C-H group for the three conformers of CQA are assigned to the IR and Raman bands located between 981 and 806 cm^{-1}.

4.4.1.3. CH_2 Modes

The group of bands in the 3006 - 2923 cm^{-1} region of the Raman spectra are assigned to the antisymmetric and symmetric stretching modes of this group in agreement to similar compounds [26,32]. The scissoring modes for the three conformers of CQA can be assigned to the shoulder in the IR spectrum and to the weak Raman band respectively at 1423 and 1431 cm^{-1}. The shoulder in the IR spectrum and the weak Raman band at 1423 cm^{-1} and the band of medium intensity at 1386 cm^{-1} are assigned to the wagging modes of this group, as predicted by calculation, while the strong bands and the shoulder at 1280, 1254 and 1251 cm^{-1} are assigned to the rocking modes. The twisting mode for all CQA conformers are associated with the shoulders in the IR spectrum at 954 and 948 cm^{-1} as was observed in the 2-(quinolin-8-yloxy)-acetic acid [6].

4.4.1.4. COO Modes

The C=O stretching mode, for the C_I conformer of CQA is calculated at 1792 cm^{-1}, it is at lower wavenumber than the other conformers (1823 and 1814 cm^{-1}, C_{II} and C_{III}conformers, respectively). On the other hand, for the benzoic acid the C=O stretching mode is predicted between 1786 and 1608 cm^{-1} while the C-O stretching mode are assigned between 1359 and 1334 cm^{-1} [33,34]. Hence, the IR bands of the media intensities and the strong IR band at 1892, 1854 and 1696 cm^{-1} are clearly assigned to the C=O stretching modes, as predicted by the calculations while the the Raman and IR bands respectively at 1219 and 1107 cm^{-1} are assigned to the C-O stretching mode corresponding to the three conformers of CQA. In the 4-hidroxybenzoic acid dimer [30] the two g(COO) modes are predict at 772 cm^{-1}, whereas the two d(COO) modes at 771 and 620 cm^{-1}. Here, those modes are predict at different wavenumbers, hence the IR bands at 613 and 531 cm^{-1} are associated with the g(COO) modes while the IR bands at 631 and 504 cm^{-1} are assigned to the d(COO) modes, as observed in **Table 2**. The rocking modes of COO groups are predicted between 439 and 335 cm^{-1}, as in similar compounds [30,33,34], hence, they are assigned to the IR bands at 431 and 356 cm^{-1}. The SQM clearly predict the twisting modes between 87 and 25 cm^{-1}, accordingly, these modes are associated to the very strong and very weak Raman bands at 78 and 20 cm^{-1}, respectively.

4.4.1.5. Skeletal Modes

Here, the skeletal stretching modes in the three CQA's conformers are predicted strongly mixed among them (Tables S8-S10). In agreement to the values reported for similar molecules [30-34] and our theoretical results, the IR bands of the media intensities at 1626, 1608 and 1585 cm^{-1} are mainly associated to a C=C stretchings. Also, the IR bands at 1368, 1324, 1202, 1133, 1107, 1042, 913 and 830 cm^{-1}, and the shoulders in the same spectrum at 1395, 1238 and 836 cm^{-1} are associated to the C-C stretching modes, as observed in **Table 2**. The strong band and the shoulders in the IR spectrum respectively at 1254, 1395 and 1251 $cm^{-1}$are associated to the C-N stretching modes. A very important observation in this compound is related to the C-Cl stretchings mode because for the C_I conformer is predicting at higher wavenumber than the other ones. Thus, the IR band at 813 cm^{-1} is assigned to that mode for the C_I conformer while the very weak Raman band at 386 cm^{-1} are assigned to those modes for the remains conformers. The six phenyl ring deformations and torsions modes for the three CQA's conformers, as the calculation predicts and taking into account the assignments for similar molecules [30-34], are assigned in the expected regions, as observed in **Table 2**. Finally, the butterfly modes are predicted by the calculations between 188 and 168 cm^{-1}, for this, they are assigned to the weak Raman band at 173 cm^{-1}, as observed in **Table 2**.

4.5. Force Field

The calculated forces constants for the three CQA's conformers are given in **Table 3** and they were calculated by means of the scaling procedure of Pulay et al. [7,8] with the MOLVIB program [19,20]. The C=O stretching force constants values for the C_{II} and C_{III} conformers (14.16 and 14.10 mdyn·Å^{-1}, respectively) are higher than the corresponding to the C_I structure (13.62 mdyn·Å^{-1}) because the calculated C=O distances in those conformers are lower (1.202 and 1.203 Å, respectively) than the corresponding to C_I conformer (1.210 Å). Also, these same reasons justify the higher f(C-O) force constant value in the C_I structure (6.33 mdyn·Å^{-1}) in relation to the others (see **Table 1**). The formation of a H bond (OH-O) in the C_I structure through the OH group and the lower C-O-H angle value in this structure (112.4°) justifies the lower f(nOH) force constant value in this structure (6.11 mdyn·Å^{-1}) in reference to the others. The differences between the f(O=C=O), f(H-C-H) and f(C-O-H) force constant values for the C_I structure in relation to the others are also attributed to the geometrical parameters. On the other hand, the analysis of the force constants for all conformers of this compound with the values for 2-(quinolin-8-yloxy) acetic acid [6] show that the values are higher in CQA, with exeption of the f(C-O) force constant. An explanation can be probably due to that the chloro atom increase the topological properties of the RCPs in the C_I conformer and as conesquence decrease the O-H distance increasing the C-O distance in this conformer in relation to the 2-(quinolin-8-yloxy) acetic acid [6].

4.6. HOMO-LUMO Energy Gap

The frontier molecular HOMO and LUMO orbitals were calculated for 2-(quinolin-8-yloxy) acetic acid and compared with the corresponding values for the (5-chloroquinolin-8-yloxy) acetic acid [37], as observed in Table S11. The results show clearly that both orbitals are mainly localized on the rings, indicating that the HOMOLUMO are mostly the π-antibonding type orbitals and that the values of the energy separation between those orbitals are higher in the 2-(quinolin-8-yloxy) acetic acid than in CQA. This large HOMO-LUMO gap for the 2- (quinolin-8-yloxy) acetic acid automatically means high excitation energies for many excited states, a good stability and a high chemical hardness. For these reasons, the presence of a Cl atom in CQA increases the reactivity compared to the 2-(quinolin-8-yloxy) acetic acid.

4.7. NMR Analysis

Tables 4, 5 show a comparison between the experimental and calculated chemical shifts for the ^{1}H and ^{13}C nuclei, respectively. The calculated chemical shifts for the H nuclei show a reasonable agreement in relation to experimental values with observed RMSD values between 0.578 and 0.174 ppm, while the chemical shifts for the carbon nuclei show higher RMSD values (7.325 and 0.254 ppm). The calculated ^{1}H chemical and ^{13}C shifts show a good concordance for the three conformers when the 6-311++G** basis set is used. Thus, **Table 4** shows that the calculated ^{13}C chemical shifts with the GIAO method using the 6-311++G** basis set are in accordance with the experimental values. In general, the calculated shifts for the ^{13}C nuclei are higher than the corresponding experimental values.

Table 3. Comparison of scaled internal force constants for the three conformers of (5-chloro-quinolin-8-yloxy) acetic acid.

Force constant	B3LYP/6-31G*				
	(5-chloro-quinolin-8-yloxy) acetic acid			2-(8-quinolinyloxy) acetic acid[b]	
	C_I	C_{II}	C_{III}	C_I	C_{II}
f(C=O)	13.62	14.16	14.10	12.53	13.07
f(C-O)	6.33	5.93	6.15	6.73	5.50
f(O-H)	6.11	7.71	7.65	5.55	7.10
f($C-H_2$)	5.31	5.23	5.10	4.87	4.80
f(C-H)	5.68	5.67	5.66	5.17	5.16
f(C-Cl)	3.51	3.49	3.42	-	-
f(C-N)	7.90	7.77	7.84	7.26	7.12
f(C-C)	6.65	6.59	6.61	6.22	6.17
f(O=C=O)	1.33	1.24	1.26	1.33	1.23
f(H-C-H)	0.91	0.87	0.88	0.84	0.80
f(C-O-H)	1.01	0.79	0.77	0.91	0.70
f(C-O-C)	0.99	1.16	1.46	0.98	1.18

Units are mdyn·$Å^{-1}$ for stretching and stretching/stretching interaction and mdyn·Å·rad^{-2} for angle deformations; [a]This work; [b]Anhydrous from Ref [6].

5. CONCLUSION

The (5-chloro-quinolin-8-yloxy) acetic acid was synthesized and characterized by infrared, Raman and NMR spectroscopic techniques. The presence of C_I, C_{II} and C_{III} conformers was detected in both spectra, and a complete assignment of the vibrational modes was accomplished. The

B3LYP/6-31G* and B3LYP/6-311++G** calculations suggest the existence of three conformers for CQA in the gas phase and, probably, the three are present in the solid state. An SQM/B3LYP/6-31G* force field was obtained for the three structures of CQA after adjusting the force constants obtained theoretically to minimize the difference between the observed and calculated wavenumbers. Also, the principal force constants for the stretching and deformation modes of CQA were determined. The NBO and AIM analysis confirm the O-H and N-H bonds in the three conformers of CQA while the HOMO-LUMO study shows that the Cl atom increases the reactivity of CQA, as compared with 2-(quinolin-8-yloxy) acetic acid.

ACKNOWLEDGEMENTS

Elida Romano, María V. Castillo, Jorgelina L. Pergomet, Juan Zinczuk, Silvia A. Brandán (Consejo de Investigaciones, Universidad Nacional de Tucumán) and CONICET (Consejo Nacional de Investigaciones Científicas y Técnicas, R. Argentina). The authors thank Prof. Tom Sundius for his permission to use MOLVIB and the Dr. Jorge Güida for the Raman spectrum.

REFERENCES

1. S. Y. Cho, J. H. Ahn, J. D. Ha, S. K. Kang, J. Y. Baek, S. S. Han, E. Y. Shin, S. S. Kim, K. R. Kim, H. G. Cheon and J. K. Choi, "Protein Tyrosine Phosphatase 1B Inhibitors: Heterocyclic Carboxylic Acids," Bulletin of the Korean Chemical Society, Vol. 24, No. 10, 2003, pp. 1455-1464.
2. B. P. Kennedy and C. Ramachandran, "Protein Tyrosine Phosphatase-1B in Diabetes," Biochemical Pharmacology, Vol. 60, No. 7, 2000, pp. 877-883.
3. N. Moller, L. Iversen, H. Andersen, J. McCormack, "Protein Tyrosine Phosphatases (PTPs) as Drug Targets: Inhibitors of PTP-1 for the Treatment of Diabetes," Current Opinion in Drug Discovery & Development, Vol. 3, No. 5, 2000, pp. 527-540.
4. Hubele, "Use of Quinoline Derivatives for the Protection of Cultivated Plants from Herbicides,"US Patent No. 4902340, 1990.
5. F. Hayat, A. Salahuddin, J. Zargan and A. Azam, "Synthesis, Characterization, Antiamoebic Activity and Cytotoxicity of Novel 2-(Quinolin-8-Yloxy) Acetohydrazones and Their Cyclized Products (1,2,3-Thiadiazole and 1,2,3-Selenadiazole Derivatives)," European Journal of Medicinal Chemistry, Vol. 45, No. 12, 2010, pp. 6127-6134.
6. G. R. Argañaraz, E. Romano, J. Zinczuk and S. A. Brandán, "Structural and Vibrational Study of 2-(Quinolin- 8-yloxy)-Acetic Acid Based on FT-IR-Raman Spectroscopy and DFT Calculations," Journal of Chemistry and Chemical Engineering, Vol. 5, No. 8, 2011, pp. 747-758. [Citation Time(s):10]
7. P. Pulay, G. Fogarasi, F. Pang and E. J. Boggs, "Systematic ab Initio Gradient Calculation of Molecular Geometries, Force Constants, and Dipole Moment Derivatives," Journal of the American Chemical Society, Vol. 101, No. 10, 1979, pp. 2550-2560.
8. P. Pulay, G. Fogarasi, G. Pongor, J. E. Boggs and A. Vargha, "Combination of Theoretical ab Initio and Experimental Information to Obtain Reliable Harmonic Force Constants. Scaled Quantum Mechanical (QM) Force Fields for Glyoxal, Acrolein, Butadiene, Formaldehyde,

and Ethylene," Journal of the American Chemical Society, Vol. 105, No. 24, 1983, pp. 7037-7047.
9. G. Rauhut and P. Pulay, "Transferable Scaling Factors for Density Functional Derived Vibrational Force Fields," The Journal of Physical Chemistry, Vol. 99, No. 39, 1995, pp. 3093-3100.
10. E. Reed, L. A. Curtis and F. Weinhold, "Intermolecular Interactions from a Natural Bond Orbital, Donor-Acceptor Viewpoint," Chemical Review, Vol. 88, No. 6, 1988, pp. 899-926.
11. J. P. Foster and F. J. Weinhold, "Natural Hybrid Orbitals," Journal of the American Chemical Society, Vol. 102, No. 24, 1980, pp. 7211-7218.
12. E. Reed and F. Weinhold, "Natural Localized Molecular Orbitals," Chemical Physics, Vol. 83, No. 4, 1985, pp. 1736-1740.
13. R. F. W. Bader, "Atoms in Molecules: A Quantum Theory," Oxford University Press, Oxford, 1990.
14. F. Sun, L. Chen, H. C. Fang, X. M. Lin and Y. P. Cai, "8-(Carboxymethoxy)Quinolinium Nitrate Monohydrate," Acta Crystallographica, Vol. 64, 2008, p. 1641.
15. F. Biegler-König, J. Schönbohm and D. Bayles, "AIM 2000—A Program to Analyze and Visualize Atoms in Molecules," Journal of Computational Chemistry, Vol. 22, No. 5, 2001, pp. 545-559.
16. E. D. Glendening, J. K. Badenhoop, A. D. Reed, J. E. Carpenter and F. Weinhold, "Natural Bond Orbital: NBO 3.1," University of Wisconsin, Madison, 1996.
17. M. J. Frisch, G. W. Trucks, H. B. Schlegel, G. E. Scuseria, M. A. Robb, J. R. Cheeseman, J. A. Montgomery, T. Vreven, K. N. Kudin, J. C. Burant, J. M. Millam, S. S. Iyengar, J. Tomasi, V. Barone, B. Mennucci, M. Cossi, G. Scalmani, N. Rega, G. A. Petersson, H. Nakatsuji, M. Hada, M. Ehara, K. Toyota, R. Fukuda, J. Hasegawa, M. Ishida, T. Nakajima, Y. Honda, O. Kitao, H. Nakai, M. Klene, X. Li, J. E. Knox, H. P. Hratchian, J. B. Cross, C. Adamo, J. Jaramillo, R. Gomperts, R. E. Stratmann, O. Yazyev, A. J. Austin, R. Cammi, C. Pomelli, J. W. Ochterski, P. Y. Ayala, K. Morokuma, G. A. Voth, P. Salvador, J. J. Dannenberg, V. G. Zakrzewski, S. Dapprich, A. D. Daniels, M. C. Strain, O. Farkas, D. K. Malick, A. D. Rabuck, K. Raghavachari, J. B. Foresman, J. V. Ortiz, Q. Cui, A. G. Baboul, S. Clifford, J. Cioslowski, B. B. Stefanov, G. Liu, A. Liashenko, P. Piskorz, I. Komaromi, R. L. Martin, D. J. Fox, T. Keith, M. A. Al-Laham, C. Y. Peng, A. Nanayakkara, M. Challacombe, P. M. W. Gill, B. Johnson, W. Chen, M. W. Wong, C. Gonzalez and J. A. Pople, "Gaussian 03, Revision B.01," Gaussian, Inc., Pittsburgh, 2003.
18. G. Fogarasi, X. Zhou, P. Taylor and P. Pulay, "The Calculation of Ab Initio Molecular Geometries: Efficient Optimization by Natural Internal Coordinates and Empirical Correction by Offset Forces," Journal of the American Chemical Society, Vol. 114, No. 21, 1992, pp. 8191-8201.
19. T. Sundius, "Molvib—A Flexible Program for Force Field Calculations," Journal of Molecular Structure, Vol. 218, No.1-2, 1990, pp. 321-326.
20. T. Sundius, "Scaling of Ab Initio Force Fields by MOLVIB," Vibrational Spectroscopy, Vol. 29, No. 1-2, 2002, pp. 89-95.
21. F. Kalincsák and G. Pongor, "Extension of the Density Functional Derived Scaled Quantum Mechanical Force Field Procedure," Spectrochimica Acta Part A: Molecular and Biomolecular Spectroscopy, Vol. 58, No. 5, 2002, pp. 999-1011.
22. B. Nielsen and A. J. Holder, "Gauss View, User's Reference," GAUSSIAN, Inc., Pittsburgh, 1997-1998.
23. E. Ledesma, S. A. Brandán, J. Zinczuk, O. Piro, J. J. L. González, A. B. Altabef, "Structural and Vibrational Study of 2-(2'-Furyl)-1H-Imidazole," Journal of Physical Organic Chemistry, Vol. 21, No. 12, 2008, pp. 1086-1097.
24. E. Romano, A. B. Raschi, A. Benavente and S. A. Brandán, "Structural Analysis, Vibrational Spectra and Coordinated Normal of 2R-(−)-6-Hydroxytremetone," Spectrochimica Acta Part A: Molecular and Biomolecular Spectroscopy, Vol. 84, No. 1, 2011, pp. 111-116.
25. S. A. Brandán, G. Benzal, et al., "Theoretical and Experimental Atudy of the Vibrational Apectra of 1,5-Simethylcytosine," Vibrational Spectroscopy, Vol. 46, No. 2, 2008, pp. 89-99.

26. E. Ledesma, J. Zinczuk, A. B. Altabef, J. J. López-González and S. A. Brandán, "Synthesis and Vibrational Analysis of N-(2'-Furyl)-Imidazole," Journal of Raman Spectroscopy, Vol. 40, No. 8, 2009, pp. 1004- 1010.
27. J. Zinczuk, A. E. Ledesma, S. A. Brandán, O. E. Piro, J. J. López-González, A. B. Altabef, "Structural and Vibrational Study of 2-(2'-Furyl)-4,5-1H-Dihydroimidazole," Journal of Physical Organic Chemistry, Vol. 22, No. 12, pp. 1166-1167.
28. S. A. Brandán, E. Erog˘lu, A. E. Ledesma, O. Oltulu and O. B. Yalçınkaya, "A New Vibrational Study of Acetazolamide Compound Based on Normal Coordinate Analysis and DFT Calculations," Journal of Molecular Structure, Vol. 993, No. 1-3, 2011, pp. 225-231.
29. L. C. Bichara, H. E. Lanús, C. G. Nieto and S. A. Brandán, "Density Functional Theory Calculations of the Molecular Force Field of l-Ascorbic Acid, Vitamin C," The Journal of Physical Chemistry A, Vol. 114, No. 14, 2010, pp. 4997-5004.
30. S. A. Brandán, F. M. López and M. Montejo, J. J. L. González, A. B. Altabef, "Theoretical and Experimental Vibrational Spectrum Study of 4-Hydroxybenzoic Acid as Monomer and Dimer," Spectrochimica Acta Part A: Molecular and Biomolecular Spectroscopy, Vol. 75, No. 5, 2010, pp. 1422-1434.
31. E. Ledesma, C. Contreras, J. Svoboda, A. Vektariane and S. A. Brandán, "Theoretical Structures and Experimental Vibrational Spectra of Isomeric Benzofused thieno [3,2-b] Furan Compounds," Journal of Molecular Structure, Vol. 967, No. 1-3, 2010, pp. 159-165.
32. D. Contreras and M. Montejo, J. J. L. González, J. Zinczuk and S. A. Brandán, "Structural and Vibrational Analyses of 2-(2-Benzofuranyl)-2-Imidazoline," Journal of Raman Spectroscopy, Vol. 42, No. 1, 2011, pp. 108-116.
33. J. Antony, G. V. Helden, G. Meijer and B. Schmidt, "Anharmonic Midinfrared Vibrational Spectra of Benzoic Acid Monomer and Dimer," Journal of Chemical Physics, Vol. 123, No. 1, 2005, pp. 14305-14311.
34. D. Contreras, A. E. Ledesma, H. E. Lanús, J. Zinczuck and S. A. Brandán, "Hydration of L-Tyrosine in Aqueous Medium. An Experimental and Theoretical Study by Mixed Quantum Mechanical/Molecular Mechanics Methods," Vibrational Spectroscopyc, Vol. 57, No. 1, 2011, pp. 108- 115.
35. S. Ghose and J. K. Dattagupta, "Crystal Structure of Tolazoline Hydrochloride (Priscoline), an α-Adrenergic Antagonist," Journal of the Chemical Society, Perkin Transactions 2, Vol. 200, No. 1-2, 1989, pp. 599-601.
36. D. Contreras, A. E. Ledesma, J. Zinczuk and S. A. Brandán, "Vibrational Study of Tolazoline Hydrochloride by Using FTIR-Raman and DFT Calculations," Spectrochimica Acta Part A: Molecular and Biomolecular Spectroscopy, Vol. 79, No. 5, 2011, pp. 1710-1714.
37. Romano, M. V. Castillo, J. L. Pergomet, J. Zinczuk and S. A. Brandán, "Synthesis and Structural and Vibrational Analysis of (5,7-Dichloro-Quinolin-8-Yloxy) Acetic Acid," Journal of Molecular Structure, Vol. 1018, No. 27, 2012, pp. 149-155.

CHAPTER 6

Synthesis and Quasi-Static Compressive Properties of Mg-AZ91D-Al_2O_3Syntactic Foams

David B. Newsome, Benjamin F. Schultz *, J.B. Ferguson and Pradeep K. Rohatgi

Materials Science and Engineering Department, University of Wisconsin-Milwaukee, 3200 N. Cramer St., Milwaukee, WI 53211, USA

ABSTRACT

Magnesium alloys have considerably lower density than the aluminum alloy matrices that are typically used in syntactic foams, allowing for greater specific energy absorption. Despite the potential advantages, few studies have reported the properties of magnesium alloy matrix syntactic foams. In this work, Al_2O_3 hollow particles of three different size ranges, 0.106–0.212 mm, 0.212–0.425 mm, and 0.425–0.500 mm were encapsulated in Mg-AZ91D by a sub-atmospheric pressure infiltration technique. It is shown that the peak strength, plateau strength and toughness of the foam increases with increasing hollow sphere wall thickness to diameter (t/D) ratio. Since t/D was found to increase with decreasing hollow sphere diameter, the foams produced with smaller spheres showed improved performance—specifically, higher energy absorption per unit weight. These foams show better performance than other metallic foams on a specific property basis.

Keywords: syntactic foam; metal matrix composite; energy absorption; compressive properties

1. INTRODUCTION

Metal matrix syntactic foams (MMSFs) are metal matrix composites containing a high volume fraction of hollow reinforcements (typically ~50%) giving the composite characteristics similar to a metal foam. The compressive stress–strain curves of metal foams exhibit an initial elastic region up to a plateau stress which is maintained until the material becomes fully dense (typically at ~50% strain) [1]. Consequently, metal foams can absorb a large amount of energy during compression before the stress rapidly increases. The shape of MMSF stress–strain curves differ slightly from the typical behavior of metal foams, in that in many cases, the initial elastic region ends in a peak stress, after which the stress drops and then gradually increases until densification (defined herein as the past-peak strain at which the stress returns to the initial peak stress) [2,3]. The energy absorbed before densification may be determined by integrating the stress–strain curve up to the densification strain. An Ashby style log–log chart showing the specific peak strength *vs.* the specific energy absorption for a variety of open-celled and syntactic foams in the open literature is shown in Figure 1a. To eliminate errors based on differing definition of material properties, the peak strength and energy absorption properties were determined directly from stress–strain plots provided in each publication. This figure shows the potential superiority of MMSFs over metal foams in terms of both specific peak strength and specific energy absorption capacity.

Although magnesium alloys and their composites have gained popularity due to their light weight and attractive properties, comparatively few studies on magnesium based metal matrix syntactic foams have been published. One of the main difficulties in the development of magnesium based composites by liquid metallurgy processes is the reactivity of magnesium with the typical hollow reinforcements available (including SiO_2, fly ash, Al_2O_3 *etc.*). These reactions can have a deleterious effect leading to the breakage and infiltration of the spheres. Weise *et al.*, used a cold chamber high pressure die casting technique to infiltrate a Mg alloy into preforms of 3M S60HS hollow glass bubbles (soda-lime-borosilicate glass) [20]. The combination of high infiltration pressure and the fracture of the spheres due to their reaction with the Mg alloy resulted in complete infiltration of the hollow spheres.

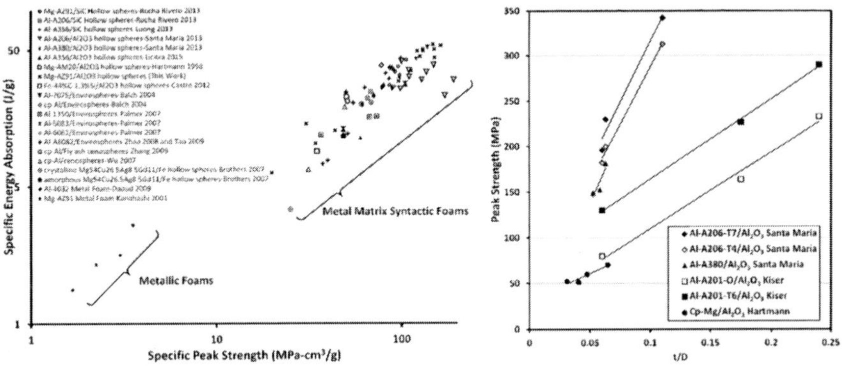

(a) (b)

Figure 1. (**a**) Log–log plot of specific peak strength and specific energy absorption for different types of foams [2,4,5,6,7,8,9,10, 11,12,13,14, 15,16,17,18]; (**b**) Effect of reinforcement wall thickness to diameter ratio (*t/D*) on the peak strength of various aluminum alloy and commercially pure magnesium-hollow Al_2O_3 sphere metal matrix syntactic foams [2,5,10,19].

Under certain processing conditions, the reaction between the magnesium and the reinforcement can be limited, or can even have a beneficial effect such as to improve bonding between the spheres and the matrix. Daoud *et al.*, synthesized Mg-ZC63- 25 vol% Fly Ash (61 wt % SiO_2, 25 wt % Al_2O_3, 5 wt % Fe_2O_3, 1.6 wt % MgO and 1 wt % TiO_2) composites via a stir casting method [21]. The magnesium alloy was shown to have reacted with the spheres to form a surface layer of MgO, which aided in wetting and incorporation of the spheres in the Mg-ZC63 matrix. Despite the reaction, the spheres remained largely intact and empty during the stir mixing and casting process.

In a study by Hartmann *et al.*, magnesium based alloy-Al_2O_3 hollow sphere syntactic foams were successfully made by a sub-atmospheric pressure infiltration technique designed to limit the amount of contact between the magnesium alloy and the Al_2O_3 spheres, thus limiting reaction [10]. In their work, a pressure of 0.5–2.9 bar (typically 0.5 bar) was used to infiltrate commercially pure (cp)-Mg, Mg-AM20, Mg-AM50 and Mg-AZ91 into a preform of hollow sintered alumina spheres with *t/D* ratios varied between 0.03–0.06. Their work suggests that a threshold pressure exists beyond which, cracking and infiltration of the spheres occurs. Defouw *et al.*, used a pressure infiltration technique to synthesize pure Mg and Mg-AZ91-carbon microsphere syntactic foams [22]. In that study, the density of the foams was varied by adjusting the infiltration pressure from 200 to 800 kPa relative to vacuum. Foam densities from

0.7 to 1.03 g/cc were achieved with microstructures containing varied amounts of filled spheres and porosity due to lack of infiltration. Higher density foams were stronger, reaching peak strengths of 53 and 25 MPa for the AZ91 and pure Mg foams respectively. The overall energy absorption was 3–9 J/g independent of the density of the foam.

Brothers *et al.*, synthesized a low density, low melting point amorphous $Mg_{60}Cu_{21}Ag_7Gd_{12}$ alloy syntactic foam incorporating a network of crystalline iron hollow spheres by pressure infiltration [6]. Differential scanning calorimetry studies of the amorphous alloy and foam showed little to no deterioration of the glass-forming ability of the alloy as a result of the infiltration process. Selected specimens were annealed to crystallize the matrix and both amorphous and crystalline syntactic foams were tested in compression. Amorphous foams exhibited a significantly larger peak strength and stiffness compared to the crystalline foams. The incorporation of the ductile iron spheres significantly improved the compressive failure strain and energy absorption compared to the monolithic amorphous Mg alloy in compression.

Recently, the quasistatic and high strain rate compressive mechanical properties of SiC hollow sphere reinforced Mg-AZ91 syntactic foams have been studied [4,23]. Rocha Rivero found that the incorporation of SiC hollow spheres (1 mm nominal diameter, 70 μm wall thickness) resulted in refinement of the microstructure surrounding the spheres in comparison to the unreinforced alloy. EDS of the matrix surrounding the spheres showed the presence of small amounts of Silicon, however this did not seem to have a significant effect on the intermetallics formed (chiefly $Mg_{17}Al_{12}$, β-phase) [4]. In both studies, Mg-AZ91-SiC syntactic foams exhibited high specific peak strengths and specific energy absorption under quasistatic conditions. Rocha Rivero found that under high strain rate conditions up to 726/s, the Mg-AZ91-SiC syntactic foams do not exhibit strain rate dependence suggesting that the strain rate sensitivity of these materials are governed by the properties of the matrix, which shows strain dependence only at higher strain rates (>1000/s) [4,24]. Further testing of these materials by Anantharaman *et al.*, at higher strain rates (1330–2300/s) did in fact show strain rate dependence, where peak strength increased with increasing strain rate [23].

One of the key attributes of the reinforcement that influences the properties of MMSFs is their wall thickness (t) to diameter (D) ratio. A cross-section of the published peak strength data for a variety of Al alloy and cp-Mg based MMSFs is plotted *versus* the t/D ratio of the reinforcement in Figure 1b, and shows a trend of increasing peak strength as the t/D ratio increases. The mechanical properties of the matrix are also known to influence the peak strength of MMSFs, as demonstrated in Figure 1b for aluminum alloys A201, A206 and A380. While Hartmann *et al.*, examined the effects of the hollow sphere *t/D* ratio (ranging from 0.03 to 0.06) on the peak strength of cp-Mg and several Mg alloys, they unfortunately only report the full range of data for the cp-Mg syntactic foam. However, their work suggests that the use of high strength magnesium alloys will have a similar complementary effect to the peak strength of syntactic foams as is the case for aluminum alloys.

In the present study, high strength magnesium alloy AZ91D/Al_2O_3 syntactic foams with *t/D* ratios ranging from 0.06–0.1 are synthesized by a sub-atmospheric pressure infiltration technique, and the mechanical properties and sphere/matrix interface are investigated.

2. RESULTS AND DISCUSSION

Representative micrographs of the as-cast composites are shown in Figure 2. The hollow particles appear to be uniformly distributed, and fully encapsulated by the metal matrix with little to no visible porosity in the matrix between the neighboring hollow particles.

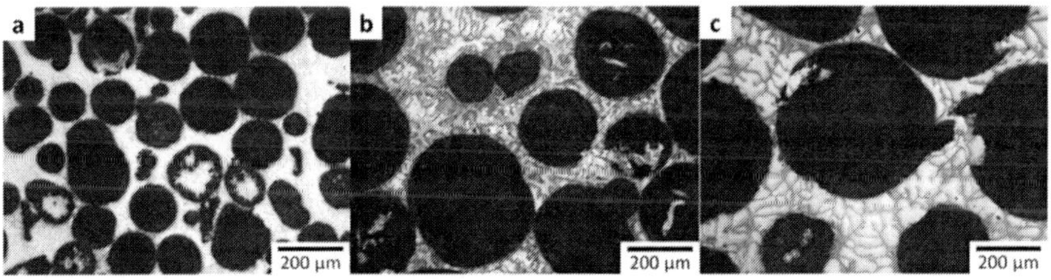

Figure 2. Representative microstructures of Mg AZ91D-Al_2O_3 syntactic foams containing hollow particles with diameters (**a**) 0.106–0.212 mm; (**b**) 0.212–0.425 mm and (**c**) 0.425–0.500 mm.

As can be seen in Figure 2a, a portion of the spheres were broken and their interiors were infiltrated at this pressure. The bulk densities of the reinforcements in each size range are presented in Table 1. It is probable that a portion of the infiltrated spheres shown in Figure 2a had been fractured prior to processing, as the flotation technique employed for eliminating fractured spheres for the larger sphere sizes is not capable of separating spheres with a bulk density greater than 1.483 g/cm^3 (the density of chloroform).

Table 1. Measured hollow sphere density for differing size ranges.

Sieve Size Range (mm)	Hollow Sphere Bulk Density (g/cc)
0.106–0.212	2.03
0.212–0.425	1.33
0.425–0.85	1.24

The Al_2O_3 hollow particles used in this investigation were on average spherical, however, especially in the case of the finest reinforcement size, the shape of the exterior and interior walls exhibit some asymmetry. The thickness of the wall therefore can vary for each hollow particle. For these reasons (along with the difficulty in sectioning and polishing the spheres along the equatorial plane), quantifying the t/D ratio by traditional imaging techniques can be difficult. One method to determine the t/D ratio has been derived by considering the sphere geometry and the average density of the sphere, ρ_{sphere}, as shown in Equation (1) where ρAl2O3 is the density of Al_2O_3 and d is the internal diameter of the hollow sphere [5].

$$\rho_{sphere} = \rho_{Al_2O_3} \frac{\left(\frac{4}{3}\right)\pi\frac{D^3}{8} - \left(\frac{4}{3}\right)\pi\frac{d^3}{8}}{\left(\frac{4}{3}\right)\pi\frac{D^3}{8}} = \rho_{Al_2O_3}\left(1 - \frac{d^3}{D^3}\right) \quad (1)$$

The t/D ratio of the hollow particles may then be determined from the densities of Al_2O_3 and hollow particles by Equation (2).

$$\frac{t}{D} = \frac{1}{2}\left[1 - \left(1 - \frac{\rho_{sphere}}{\rho_{Al_2O_3}}\right)^{\frac{1}{3}}\right] \quad (2)$$

Figure 3 shows that though the average hollow sphere wall thickness increases with increasing sphere diameter, the t/D ratio decreases asymptotically.

Representative quasi-static engineering stress–strain curves are shown in Figure 4 for each size range. Each curve is typical of the quasistatic compression of MMSFs, in that an initial peak is reached (peak stress), followed by a sharp drop, and subsequent densification/shearing events that correspond to the shear of the MMSF at the critical shear stress, and crushing and densification of the hollow particles [5]. The smallest size reinforcements (0.106–0.212 mm) exhibited a significantly higher strength than the larger size reinforcements, though the densification strain (the past-peak strain at which the stress returns to the initial peak stress value) was roughly the same for each size range.

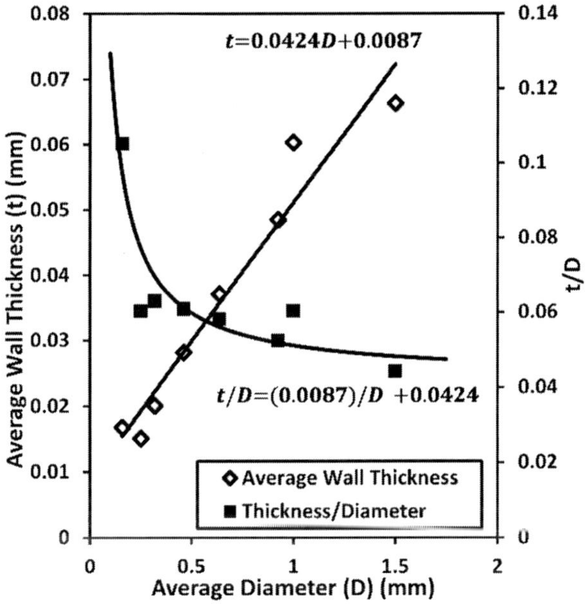

Figure 3. Dependence of average wall thickness and *t*/*D* ratio on reinforcement diameter.

The measured peak strength, plateau strength, densification strain and energy absorption are reported in Table 2. The energy absorption (toughness) of the MMSFs was determined by calculating the area under the stress–strain curve up to the densification strain (the strain at which the stress reached the magnitude of the initial peak stress and subsequently continues to rise). The plateau strength reported in this work is the average of all of the measured stress data points between the strain corresponding to the initial peak and the densification strain. The specific energy absorption and specific peak strength of the MMSFs produced in this study are plotted on Figure 1a (shown with an "*x*" marker) for the purpose of comparison with metal foams of different compositions produced in other studies. The MMSFs containing the smallest Al_2O_3 hollow spheres, (0.106–0.212 mm) exhibited the best combination of specific peak strength and specific energy absorption, outperforming a variety of MMSFs and metal foams produced in other studies [2,4,5,6,7,8,9,10,11,12,13,14,15,16,17,18].

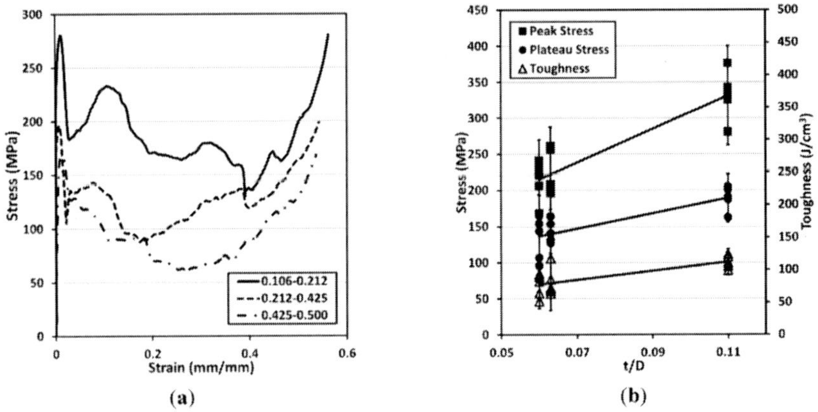

Figure 4. (**a**) Typical compressive stress–strain curves for Mg AZ91D-Al_2O_3 syntactic foams; (**b**) Peak strength (filled diamonds), plateau strength (filled squares) and toughness (open triangles) *vs.* t/D ratio. All properties shown increase with increasing t/D ratio. Error bars are shown (two standard deviations) based on the average value of each property.

Figure 4b shows a plot of each measured property in relation to the t/D ratio. Each measured property was observed to increase with increasing t/D ratio. Also of note, the peak strength of the MMSFs exceeds the yield strength of the matrix alloy Mg-AZ91D (160 MPa) [25]. This indicates that there is load transfer from the matrix to the reinforcement as a result of good bonding between the Mg-AZ91D alloy and the Al_2O_3 spheres. The peak stress in this case corresponds to the concurrent permanent deformation of the matrix and spheres at the critical shear stress. After this point, the compressive stress–strain curves for the MMSFs become non-linear, corresponding to plastic deformation of the matrix and further fracture of the spheres [5,26]. Closer examination of the interface between the Al_2O_3 spheres and the matrix via scanning electron microscope (SEM) (Topcon, Hasunuma-cho, Itabashi-Ku, Tokyo, Japan) in Figure 5 reveals that in many cases, the magnesium alloy has infiltrated pores in the walls of the hollow particles. The results of energy dispersive X-ray spectroscopy (EDS) (Advanced Analysis Technologies, Dane, WI, USA) dot mapping of aluminum (Al) and magnesium (Mg) in the magnified regions of Figure 5b,c are shown in the insets to the right of each image respectively. A good interface is observed between the surfaces of the spheres and the surrounding matrix suggesting that bonding has occurred. This combination of mechanical and chemical bonding between the matrix and the hollow particles is therefore expected to be responsible for the improvement in strength observed in these MMSFs over that of the base alloy.

Table 2. Summary of quasi-static compression data for Mg-AZ91D/Al_2O_3 syntactic foams.

Hollow Sphere Size Range (mm)	Peak Stress (MPa)	Plateau Stress (MPa)	Toughness (J/cm³)	Densification Strain (%)	Density (g/cm³)
	342	162	99	60%	2.27
	325	200	120	59%	2.20
0.106–0.212	280	187	106	56%	2.15
	376	205	124	59%	2.31
	332	190	111	57%	2.21
	261	132	85	61%	1.59
	208	141	68	48%	1.90
0.212–0.425	199	153	118	58%	1.98
	256	164	72	44%	1.91
	196	127	64	54%	2.10
	241	165	93	56%	1.85
	221	154	89	58%	1.82
0.425–0.500	230	144	82	56%	1.83
	168	96	52	54%	1.75
	206	107	64	58%	1.83

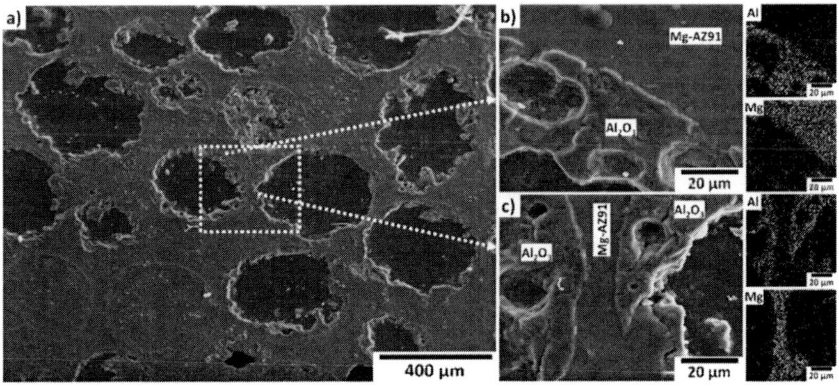

Figure 5. Scanning electron microscope (SEM) Micrographs and energy dispersive X-ray spectroscopy (EDS) dot maps of Mg-AZ91D/Al_2O_3 (0.425–0.5 mm) syntactic foam; (**a**) polished cross-section showing filled and empty hollow particles; (**b**) upper portion of left side sphere (within dashed box in Figure 5a) showing physical and mechanical bonding at the Mg/Al_2O_3 interface; (**c**) region between spheres (within dashed box in Figure 5a) showing physical and mechanical bonding at the Mg/Al_2O_3interface.

3. EXPERIMENTAL SECTION

Syntactic foams composed of Mg alloy AZ91D (Magnesium Elektron, West Des Moines, IA, USA) incorporating approximately 50 vol% hollow Al_2O_3 hollow particles (ALODUR white bubble alumina) (C-E Minerals Processing Inc., Newell, WV, USA) were synthesized. The nominal compositions of the alloy and the hollow particles are presented in Table 3 (information provided by suppliers). The hollow particles were supplied by C-E Minerals (C-E Minerals Processing Inc., Newell, WV, USA) in standard size (diameter) ranges of 0–0.5, 1–3 and 1–5 mm. The spheres were further sorted by size using an Allen Bradley Sonic Sifter (Allen Bradley Co., Milwaukee, WI, USA) to produce several size ranges although only 0.106–0.212, 0.212–0.425 and 0.425–0.500 mm were used to synthesize syntactic foams. After sorting, the Al_2O_3 particles were floated in chloroform ($CHCl_3$) (Thermo Fisher Scientific, Waltham, MA, USA) which has a density of 1.483 g/cm^3, causing the defective spheres to sink and the intact spheres to float, thus allowing them to be collected. The spheres with size range 0.106–0.212 were not floated in chloroform due to their higher bulk density. The bulk density of the hollow particles for each size range was determined by measuring the mass of a fixed amount of spheres and the corresponding volume of water displaced by these same spheres. The size ranges evaluated in this study along with their measured densities are shown in Table 1.

Table 3. Nominal composition of matrix and reinforcements.

Material	Component	Nominal Content (wt %)
AZ91D	Mg	88–91
	Al	8.3–9.7
	Mn	0.13 min
	Zn	0.35–1.0
	Si	0.50 max
	Cu	0.1 max
	Ni	0.03 max
Al_2O_3 Hollow Sphere	Al_2O_3	98.8
	SiO_2	0.8
	Na_2O	0.1
	MgO	0.05
	Fe_2O_3	0.03
	CaO	0.03

A sub-atmospheric pressure infiltration apparatus was used to produce the MMSFs in this work. A 13.5 mm × 13.5 mm steel tube with one end welded shut was tap-packed with hollow particles to a height of 70–90 mm. An ingot of AZ91D alloy was placed above the hollow spheres, separated by a 2 mm thick layer of zirconia felt (Zircar Zirconia Inc., Florida, NY, USA). The felt was used as a reaction barrier between the magnesium alloy melt and the spheres prior to infiltration and served as a filter to remove the oxide layer from the liquid melt. The crucible containing the preform and ingot was heated in a quartz chamber under vacuum to 650 °C and held for 1 h at which time the alloy had fully melted and uniformly sealed the inner perimeter of the crucible. Argon gas (Airgas, Radnor Township, PA, USA) was then rapidly introduced into the heated quartz chamber, reaching a sub-atmospheric pressure of 0.4 bar within 10 s, thereby forcing the molten alloy into the evacuated spaces between the hollow particles. The quartz chamber containing the sample was then removed from the tube furnace and the steel crucible quenched immediately in room temperature water. The density of the composites was measured using a Mettler Toledo AT261 Delta Range Microbalance (Mettler Toledo, Columbus, OH, USA) equipped with a density measurement apparatus (Archimedes method) (Mettler Toledo, Columbus, OH, USA). The specimens were first lightly coated with vacuum grease to prevent infiltration of surface pores during the density measurement. Microstructural analysis was performed with a Nikon Eclipse TS100 microscope (Nikon, Brighton, MI, USA) equipped with an automated stage and Clemex Professional Image analysis software (Clemex Technologies, Inc., Longueuil, PQ, Canada), and a Topcon SM300 scanning electron microscope (SEM) (Topcon, Hasunuma-cho, Itabashi-Ku, Tokyo, Japan) equipped with energy dispersive X-ray spectroscopy (EDS) (Advanced Analysis Technologies, Dane, WI, USA) was used to observe the interfaces between the AZ91D matrix and Al_2O_3 spheres. To facilitate imaging, the specimens were sputter coated with gold/palladium using a Denton Vacuum Desk II sputter coater (Denton Vacuum LLC, Moorestown, NJ, USA).

Quasi-static compression testing was performed in accordance with ASTM C365-94 (ASTM International, West Conshohocken, PA, USA) on specimens having dimensions of approximately 13 mm × 13 mm × 12 mm (thickness). Testing was carried out using a SATEC Model 50Ud Universal Testing Machine (Instron, Norwood, MA, USA) at constant crosshead speed with an initial strain rate of 10^{-3} s^{-1} and a self-leveling platen. Strains were calculated from the crosshead displacement, and were corrected for deflection of the load frame. The quasi-static compression curves typically exhibited an initial peak followed by a lower plateau stress and later densification. Compression was stopped when the stress slightly exceeded the magnitude of the initial peak stress.

4. CONCLUSIONS

The microstructure and quasi-static mechanical properties of Mg-AZ91D-Al_2O_3 hollow sphere syntactic foams have been characterized for foams of three different microsphere sizes, specifically 0.106–0.212, 0.212–0.425 and 0.425–0.5. The peak strength, plateau strength and toughness of the AZ91D-Al_2O_3 syntactic foams increase with increasing *t/D* ratio. Because *t/D* was found to increase with decreasing sphere diameter, the foams produced with finer hollow particles result in improved performance. Good bonding between the matrix and the reinforcement as observed by SEM suggests that the increased strength of the MMSFs relative to the base alloy yield strength is as a result of load transfer during compression.

ACKNOWLEDGMENTS

This research was supported by the U.S. Army-TARDEC through TACOM R&D Contract# W56HZV-08-C-0716. The authors would also like to acknowledge and thank C-E Minerals for providing the ALODUR alumina hollow sphere material. Disclaimer: Reference herein to any specific commercial company, product, process, or service by trade name, trademark, manufacturer, or otherwise, does not necessarily constitute or imply its endorsement, recommendation, or favoring by the United States Government or the Department of the Army (DoA). The opinions of the authors expressed herein do not necessarily state or reflect those of the United States Government or the DoA, and shall not be used for advertising or product endorsement purposes.

AUTHOR CONTRIBUTIONS

David B. Newsome prepared the syntactic foam specimens and performed the density measurements, compression testing and metallography. Benjamin F. Schultz performed SEM characterization and analyzed the compression data. J.B. Ferguson performed further analysis of the compression and density data. Pradeep K. Rohatgi provided general guidance. All authors contributed in writing the article.

CONFLICTS OF INTEREST

The authors declare no conflict of interest.

REFERENCES

1. Evans, A.G.; Hutchinson, J.W.; Ashby, M.F. Multifunctionality of cellular metal systems. Prog. Mater. Sci. 1998, 43, 171–221.
2. Santa Maria, J.A.; Schultz, B.F.; Ferguson, J.B.; Rohatgi, P.K. Al-Al2O3 syntactic foams—Part I: Effect of matrix strength and hollow sphere size on the quasi-static properties of Al-A206/Al2O3 syntactic foams. Mater. Sci. Eng. A2013, 582, 415–422.
3. Majlinger, K.; Orbulov, I.N. Characteristic compressive properties of hybrid metal matrix syntactic foams. Mater. Sci. Eng. A 2014, 606, 248–256.
4. Rocha Rivero, G.A.; Schultz, B.F.; Ferguson, J.B.; Gupta, N.; Rohatgi, P.K. Compressive properties of Al-A206/SiC and Mg-AZ91/SiC syntactic foams. J. Mater. Res. 2013, 28, 2426–2435.
5. Santa Maria, J.A.; Schultz, B.F.; Ferguson, J.B.; Gupta, N.; Rohatgi, P.K. Effect of hollow sphere size and size distribution on the quasi-static and high strain rate compressive properties of Al-A380-Al2O3 syntactic foams. J. Mater. Sci. 2014, 49, 1267–1278.
6. Brothers, A.H.; Dunand, D.C.; Zheng, Q.; Xu, J. Amorphous Mg-based metal foams with ductile hollow spheres. J. Appl. Phys. 2007, 102.
7. Luong, D.D.; Strbik, O.M., III; Hammond, V.H.; Gupta, N.; Cho, K. Development of high performance lightweight aluminum alloy/SiC hollow sphere syntactic foams and compressive characterization at quasi-static and high strain rates. J. Alloys Compd. 2013, 550, 412–422.
8. Licitra, L.; Luong, D.D.; Strbik, O.M.; Gupta, N. Dynamic properties of alumina hollow particle filled aluminum alloy A356 matrix syntactic foams. Mater. Des. 2015, 66, 504–515.
9. Castro, G.; Nutt, S.R. Synthesis of syntactic steel foam using gravity-fed infiltration. Mater. Sci. Eng. A 2012, 553, 89–95.
10. Hartmann, M.; Reindel, K.; Singer, R.F. Fabrication and properties of syntactic magnesium foams. In MRS Proceedings; Cambridge University Press: Cambridge, UK, 1998; Volume 521, pp. 211–216.
11. Balch, D.K.; O'Dwyer, J.G.; Davis, G.R.; Cady, C.M.; Gray, G.T., III; Dunand, D.C. Plasticity and damage in aluminum syntactic foams deformed under dynamic and quasi-static conditions. Mater. Sci. Eng. A 2005, 391, 408–417.
12. Palmer, R.A.; Gao, K.; Doan, T.M.; Green, L.; Cavallaro, G. Pressure infiltrated syntactic foams—Process development and mechanical properties. Mater. Sci. Eng. A 2007, 464, 85–92.
13. Tao, X.F.; Zhang, L.P.; Zhao, Y.Y. Al matrix syntactic foam fabricated with bimodal ceramic microspheres. Mater. Des.2009, 30, 2732–2736.
14. Tao, X.F.; Zhao, Y.Y. Compressive behavior of Al matrix syntactic foams toughened with Al particles. Scr. Mater. 2009, 61, 461–464.
15. Zhang, Q.; Lee, P.D.; Singh, R.; Wu, G.; Lindley, T.C. Micro-CT characterization of structural features and deformation behavior of fly ash/aluminum syntactic foam. Acta Mater. 2009, 57, 3003–3011.
16. Wu, G.H.; Dou, Z.Y.; Sun, D.L.; Jiang, L.T.; Ding, B.S.; He, B.F. Compression behaviors of cenosphere-pure aluminum syntactic foams. Scr. Mater. 2007, 56, 221–224.
17. Daoud, A. Effect of fly ash addition on the structure and compressive properties of 4032-fly ash particle composite foams. J. Alloys Compd. 2009, 487, 618–625.

18. Kanahashi, H.; Mukai, T.; Yamada, Y.; Shimojima, K.; Mabuchi, M.; Aizawa, T.; Higashi, K. Experimental study for the improvement of crashworthiness in AZ91 magnesium foam controlling its microstructure. Mater. Sci. Eng. A 2001, 308, 283–287.
19. Kiser, M.; He, M.Y.; Zok, F.W. Mechanical response of ceramic microballoon reinforced aluminum matrix composites under compressive loading. Acta Mater. 1999, 47, 2685–2694.
20. Weise, J.; Zanetti-Bueckmann, V.; Yezerska, O.; Schneider, M.; Haesche, M. Processing, properties and coating of micro-porous syntactic foams. Adv. Eng. Mater. 2007, 9, 52–56.
21. Daoud, A.; El-khair, M.T.A.; Abdel-Aziz, M.; Rohatgi, P. Fabrication, microstructure and compressive behavior of ZC63 Mg-microballoon foam composites. Compos. Sci. Technol. 2007, 67, 1842–1853.
22. DeFouw, J.D.; Rohatgi, P.K. Low density magnesium matrix syntactic foams. In Supplemental Proceedings: Materials Fabrication, Properties, Characterization, and Modeling, Volume 2; Minerals, Metals and Materials Society: San Diego, CA, USA, 2011; pp. 797–804.
23. Anantharaman, H.; Shunmugasamy, V.C.; Strbik, O.M.; Gupta, N.; Cho, K. Dynamic properties of silicon carbide hollow particle filled magnesium alloy (AZ91D) matrix syntactic foams. Int. J. Impact Eng. 2015, 82, 14–24.
24. Mukai, T.; Kanahashi, H.; Yamada, Y.; Shimojima, K.; Mabuchi, M.; Nieh, T.G.; Higashi, K. Dynamic compressive behavior of an ultra-lightweight magnesium foam. Scr. Mater. 1999, 41, 365–371.
25. Kaufman, J.G. Magnesium Alloy Database; Knovel: New York, NY, USA, 2011.
26. Orbulov, I.N.; Ginsztler, J. Compressive characteristics of metal matrix syntactic foams. Compos. Part A: Appl. Sci. Manuf. 2012, 43, 553–561.

CHAPTER 7

The Microwave-Assisted Green Synthesis of TiC Powders

Hui Wang [1,2,*], Wencheng Zhu [1], Yanchun Liu [3], Lingke Zeng [1] and Luyi Sun [2,*]

[1]*College of Materials Science & Engineering, South China University of Technology, Guangzhou 510640, China*

[2]*Department of Chemical & Biomolecular Engineering and Polymer Program, Institute of Materials Science, University of Connecticut, Storrs, CT 06269, USA*

[3]*Guangzhou Redsun Gas Appliances Co., Ltd., Guangzhou 510460, China*

ABSTRACT

Titanium carbide (TiC) is an important engineering material and has found widespread applications. Currently, TiC is typically synthesized through carbothermal reduction, requiring a high temperature (ca. 1700–2300 °C) and long reaction time (ca. 10–20 h), which is not eco-friendly. During a conventional reaction path, anatase TiO_2 (A-TiO_2) was first converted to rutile TiO_2 (R-TiO_2), which was subsequently reduced to TiC. Herein, we explored the synthesis of TiC powders with the assistance of microwave heating. In particular, we achieved the conversion of A-TiO_2, which was more reactive than R-TiO_2 for the carbothermal reduction, to TiC, which was directly due to quick microwave heating. As such, the carbothermal reduction started at a much lower temperature of ca. 1200 °C and finished within 30 min when reacting at 1400 °C, leading to significant energy saving. This study shows that microwave-assisted synthesis can be an effective and green process for preparing TiC powders, which is promising for future large-scale production. The influence of the reaction temperature, the reaction duration, and the carbon content on the synthesis of TiC powders was investigated.

Keywords: TiC powders; microwave-assisted synsthesis

1. INTRODUCTION

Titanium carbide (TiC) is an important engineering material because of its superior hardness, high melting temperature, good electrical conductivity, and outstanding abrasion resistance [1,2,3]. Therefore, TiC powders have been used extensively in cutting tools, grinding wheels, polishing pastes, etc. [3,4,5,6].

Currently, TiC powders are mainly synthesized by the carbothermal reduction of TiO_2, and the heat during the process of reaction is usually provided by an external heating system. The synthesis is typically in the temperature range of 1700–2300 °C for 10–20 h [7]. The size of the synthesized TiC powders not only depends on the size of raw materials, but also on the synthesis conditions. Therefore, it is difficult to prepare uniform TiC powders by conventional heating processes [8,9].

As an alternative heating technology, microwave heating has advantages such as high thermal efficiency, selective heating, quick heating, and short processing time. It has been implemented in many industrial applications in recent years [10,11,12]. The major difference between microwave heating and conventional heating is the inverse temperature profile inside microwave-heated samples [13,14]. The center of the sample becomes hotter than the surface, which is exposed to the colder furnace atmosphere [15]. Many materials including ceramics, polymers, and metallic powders can be directly exposed to microwaves [16]. In microwave carbothermal reduction process, two factors might be favorable for lowering the onset reaction temperature. One is the existence of thermal and non-thermal effects from microwave heating [17]. The other is the uniform heating achieved by the energy directly delivered to the starting materials via molecular-level interactions under an electromagnetic field [18,19]. Microwave-assisted synthesis is attractive for the preparation of TiC because it is faster and more effective than conventional heating. Carbon, one of the reactants for the synthesis of TiC, is a very effective absorber of microwave radiation, offering extra facilitation of the reaction. Cross and coworkers [20,21] made some early explorations of the synthesis of TiC via microwave heating and proved that it was viable and indeed more effective to synthesize TiC via microwave heating.

In this report, microwave heating was adopted to synthesize TiC powders by carbothermal reduction of TiO_2. The influence of reaction temperature and carbon content on the phase composition of the produced TiC powders was systematically investigated. In particular, we aimed to achieve the direct conversion of anatase TiO_2 (A-TiO_2), which was more reactive than rutile TiO_2 (R-TiO_2), for the carbothermal reduction to TiC by taking advantage of quick microwave heating.

2. EXPERIMENTAL

2.1. Materials

Both A-TiO_2 powders (99%, ~40 nm in diameter) and carbon black (99%, ~10 nm in diameter) were purchased from Tuoyi Co. (Guangzhou, China). Ethanol was acquired from Fuyu Fine Chemical Co. (Fuzhou, China). Polyacrylic acid (PAA) was obtained from KeMiOU Chemical Co. (Tianjin, China). All the chemicals were used as received without further purification.

2.2. Synthesis

First, a 20.0 wt % A-TiO_2 ethanol dispersion was ultrasonicated for 30 min to form a uniform system, which was mixed with polyacrylic acid (PAA/TiO_2 = 0.5 wt %) to stabilize the dispersion of A-TiO_2 in ethanol. Subsequently, a pre-determined amount of carbon black was mixed with the above dispersion. The reaction precursor was obtained after evaporating ethanol.

The synthesis was carried out in a KL-2D-16 microwave furnace (Kailin Microwave Equipment Co., Guangzhou, China). During reaction, the microwave frequency was 2450 ± 50 MHz, and microwave power was 6000 W. Thermocouples were used to measure the temperatures during reaction. The entire reaction was carried out under argon protection.

The carbothermal reduction of TiO_2 is a complex process. Under a conventional heating process, before the reduction reaction, A-TiO_2 is first converted to R-TiO_2. With the rising of temperature, the reduction consists of a series of intermediate reactions, producing various intermediate products and phases (Equations (1)–(4)). The general procedures of the carbothermal reduction reaction between carbon black and R-TiO_2 through conventional heating to generate TiC are as follows [22,23]:

$$2R - TiO_2 + C = Ti_2O_3 + CO; \quad (1)$$

$$Ti_2O_3 + C = 2TiO + CO; \quad (2)$$

$$TiO + 2C = TiC + CO. \quad (3)$$

Total reaction:

$$TiO_2 + 3C = TiC + 2CO. \quad (4)$$

In this work, we aimed to achieve a direct reduction of A-TiO_2 to TiC under the assistance of microwave heating, as illustrated in Figure 1. Such a direction reduction route is expected to occur at a lower temperature, thus saving energy and time. In order to investigate the reduction process of TiO_2 by microwave heating, we monitored the phase transition of the reaction products at various temperatures and reaction durations by X-ray diffraction, and studied the influence of carbon content on the synthesis of the TiC powders.

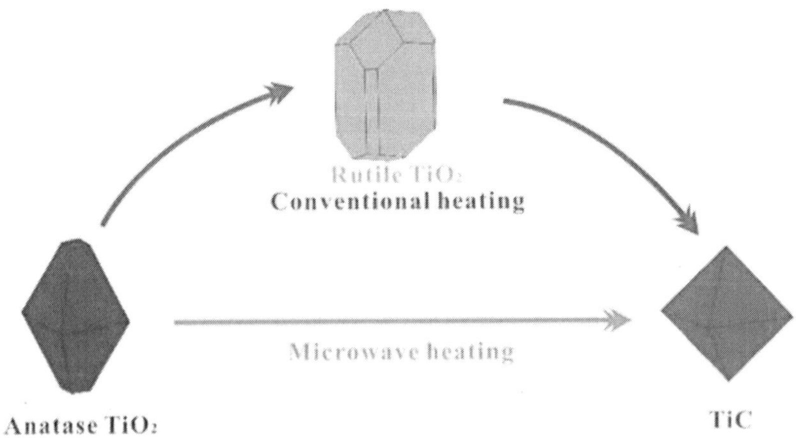

Figure 1. Schematic of different reaction routes between conventional and microwave heating.

2.3. Characterization

Phase composition of the reaction products was determined by X-ray diffraction on a PANalytical X-ray diffractometer (XRD, monochromated Cu Kα radiation) at 25 °C. The morphology of the powders was imaged by scanning electron microscopy (SEM) on a Nova S-430 microscope operated at 20 kV, which was equipped with an energy-dispersive X-ray spectrometer (EDS).

3. RESULTS AND DISCUSSION

3.1. Effect of Temperature and Reaction Time

The XRD patterns of the samples synthesized at various temperatures (1100, 1200, 1300, and 1400 °C) and durations of reaction (10 and 30 min) at a TiO_2 and C molar ratio of 1:3.6 are presented in Figure 2.

Figure 2A displays the XRD patterns of the samples reacted at 1100 °C for 10 and 30 min. It shows that TiC phase was not formed at 1100 °C for 10 min of reaction, but a small amount of TiC was generated after 30 min of reaction. The conversion from A-TiO_2 to R-TiO_2 was observed at this temperature. A number of studies reported that A-TiO_2 began to transform to R-TiO_2 at a temperature up to 610 °C [24,25]. With an increase in reaction time, more R-TiO_2 was generated, as evidenced by stronger R-TiO_2 phase diffraction peaks in the XRD pattern. The XRD patterns of the samples synthesized at 1200 °C (Figure 2B) showed that most A-TiO_2 was converted to be R-TiO_2; meanwhile, Ti_2O_3 started to form after 10 min of reaction at 1200 °C. After 30 min of reaction at 1200 °C, the TiC phase started to appear as evidenced by the corresponding XRD peaks. The TiC phase could be clearly observed on the XRD pattern when the reaction temperature was raised to 1300 °C. With an increase in reaction time, a higher concentration of TiC was generated, as supported by the more intensive diffraction peaks of TiC (Figure 2C). According to the XRD pattern, the virtually pure TiC phase was synthesized after 30 min of reaction at 1400 °C (Figure 2D).

Based on the above observations, the reaction mechanism is proposed as follows. At 1100 °C, the main reaction was the transformation from A-TiO_2 to R-TiO_2, but this is a relatively slow process:

$$A - TiO_2 \rightarrow R - TiO_2 \,. \tag{5}$$

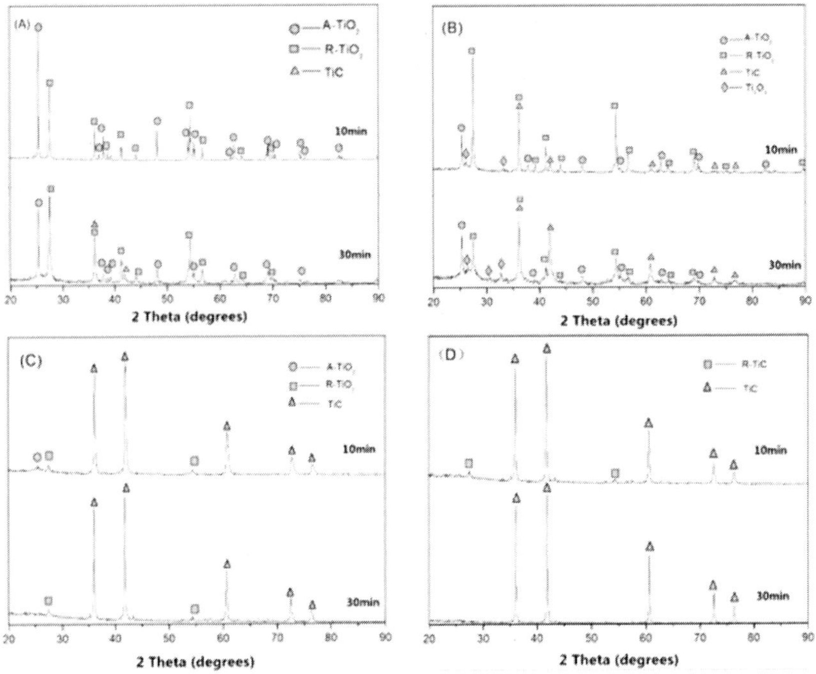

Figure 2. X-ray diffractometer (XRD) patterns of the synthesized samples at various temperatures: (**A**) 1100 °C; (**B**) 1200 °C; (**C**) 1300 °C; (**D**) 1400 °C.

It was reported that the activity of A-TiO_2 with C was higher than that of R-TiO_2 [26,27,28], so the rate of Reaction (6) is faster than that of Reaction (1):

$$2A - TiO_2 + C \rightarrow Ti_2O_3\,(s) + CO\,(g)\,. \tag{6}$$

During conventional heating, A-TiO_2 tends to transform to R-TiO_2 before the reduction reaction. This is one of the key reasons that a much higher temperature; thus, much more energy is required to convert the less active R-TiO_2 to TiC. If one can quick heat to directly convert A-TiO_2 to TiC, it is much more favorable in terms of energy consumption. Microwave heating can help achieve this process. Because the rate of microwave heating is very fast, it takes a very short time to raise the temperature from 1100 to 1200 °C. Therefore, when the temperature was quickly increased to 1200 °C, some A-TiO_2 remained. Such unconverted A-TiO_2 can quickly react with C to form TiC at relatively lower temperatures compared with the reduction reaction temperature of R-TiO_2 to TiC. Figure 2C shows that, after 30 min of reaction at 1300 °C, there was still some R-TiO_2 left but no A-TiO_2. As such, a high heating rate is very desirable to directly reduce A-TiO_2 to TiC, which allows for the synthesis of TiC at lower temperatures while saving energy. After 30 min of reaction at 1300 °C, most TiO_2 was reacted to form TiC, which began to be the dominating phase.

With an increase in reaction temperature and time, it was observed that the *(200)* peak of the synthesized TiC shifted from 41.90° to 41.70° (Figure 3A,B). The lattice constant of the TiC (1300 °C, 10 min) and TiC (1400 °C, 30 min) were calculated to be 4.321 Å and 4.324 Å, lower than the standard value of 4.327 Å. The results show that a higher temperature and longer reaction time are beneficial for the growth of TiC, which was also reported by Preiss et al. [29,30].

SEM images of the samples synthesized at various temperatures (1100, 1200, 1300, and 1400 °C) for 30 min are shown in Figure 4. With an increase in reaction temperature, the size of the powders became larger, changing from ca. 0.2 μm at 1100 °C to ca. 0.4 μm at 1200 °C. When the reaction temperature was increased to 1300 °C, the TiC powders exhibited a pseudocubic morphology (Figure 4C) with a particle size of ca. 0.7 μm. It should be noted that, although no other diffraction peaks were observed in Figure 2D, amorphous carbon should still exist because excessive carbon was added to ensure a complete carbothermal reduction, because carbon black acts as both the carbon source and the media to transform microwave dielectric to heat.

3.2. Effect of Carbon Content

TiO_2 is a poor microwave absorber, while carbon black absorbs microwave very effectively. During the reaction, carbon black not only participates in the reaction but also serves as a medium to absorb microwave radiation to help heat the mixture of the reactants. Its content was reduced along the reaction; therefore, excessive carbon black was added during the reactions. In order to study the role of carbon content in the process of the reduction of TiO_2, mixtures of TiO_2 and C at various molar ratios (1:3.0; 1:3.2; 1:3.4, and 1:3.6) were reacted at 1400 °C for 30 min via microwave heating.

Table 1 presents the results calculated from the data of the XRD and EDS measurements of the products from the samples starting at various TiO_2 and C mole ratios. The results showed that, with an increasing amount of carbon in the mixture, both the concentration and lattice constant of TiC increased, which suggests that a higher concentration of C is favorable for the growth of TiC.

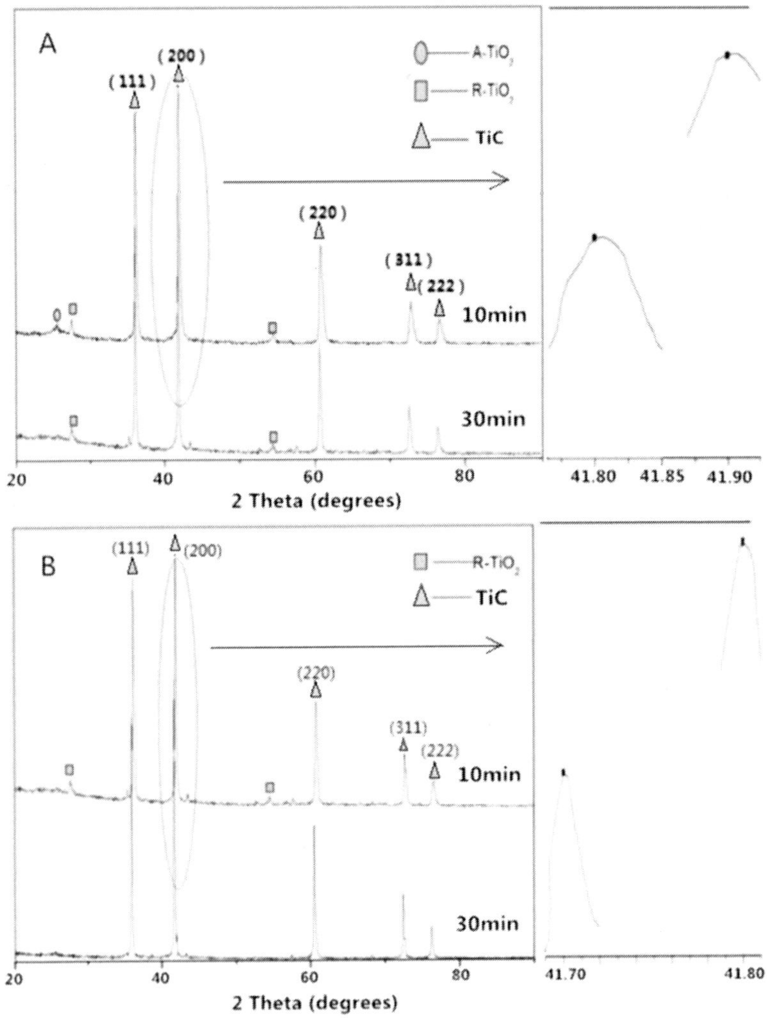

Figure 3. XRD patterns of the samples synthesized at (**A**) 1300 °C and (**B**) 1400 °C.

4. CONCLUSIONS

In conclusion, our experimental results showed that pure and pseudocubic TiC phase could be synthesized by directly reducing A-TiO_2 instead of R-TiO_2 with the assistance of quick microwave heating. The lattice constant of TiC increases with an increasing reaction temperature and time. Carbon black acts as both the carbon source and the media to transform microwave dielectric to heat. A high ratio of C to TiO_2 in the starting materials is favorable for improving the conversion rate and quality of TiC powders. Compared with the carbothermal reduction using conventional heating to synthesize TiC powders, which typically requires a temperature of 1700–2400 °C for 10 to 24 h, TiC powders could be synthesized at a temperature of as low as 1200 °C during microwave heating, and the carbothermal reduction can be finished within 30 min at 1400 °C. Therefore, the preparation of TiC powders via microwave heating is much more energy-effective and thus promising for large-scale production.

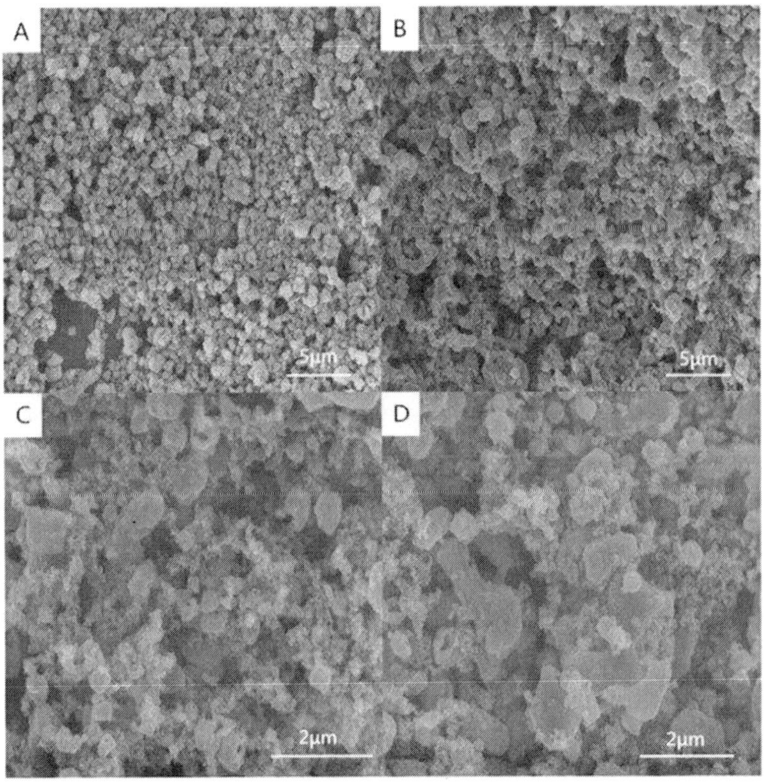

Figure 4. SEM images of the products synthesized at various temperatures for 30 min: (**A**) 1100 °C; (**B**) 1200 °C; (**C**) 1300 °C; (**D**) 1400 °C.

Table 1. Product composition after reaction at 1400 °C for 30 min under various TiO_2 and C ratios.

TiO_2:C (mol:mol/wt:wt)	Composition of Product (wt %)			TiC (200) Peak Position (Degree)	Lattice Constant of TiC (Å)
	TiO_2	C	TiC		
1:3.0/1:0.45	15.5	4.2	80.3	41.92	4.319
1:3.2/1:0.48	6.7	5.7	87.6	41.86	4.320
1:3.4/1:0.51	2.4	8.3	89.3	41.78	4.322
1:3.6/1:0.54	0	9.4	90.6	41.70	4.324

ACKNOWLEDGMENTS

This work was financially supported by the National High-tech R&D Program (863 Program) of China and the Science and Technology Planning Project of Guangdong Province, China.

AUTHOR CONTRIBUTIONS

Hui Wang and Lingke Zeng conceived and designed the experiments; Wencheng Zhu performed the experiments; Lingke Zeng and Luyi Sun analyzed the data; Yanchun Liu contributed in experiments and characterizations; Wang Hui and Luyi Sun wrote the paper.

CONFLICTS OF INTEREST

The authors declare no conflict of interest.

REFERENCES

1. Shin, Y.; Li, X.S.; Wang, C.; Coleman, J.R.; Exarhos, G.J. Synthesis of hierarchical titanium carbide from titania-coated cellulose paper. Adv. Mater. 2004, 16, 1212–1215.
2. Pierson, H.O. Handbook of Refractory Carbides & Nitrides: Properties, Characteristics, Processing and Apps; William Andrew: Norwich, NY, USA, 1996.
3. Toth, L. Transition Metal Carbides and Nitrides; Elsevier: Amsterdam, The Netherlands, 2014.
4. Koc, R. Kinetics and phase evolution during carbothermal synthesis of titanium carbide from ultrafine titania/carbon mixture. J. Mater. Sci. 1998, 33, 1049–1055.

5. El-Eskandarany, M.S. Structure and properties of nanocrystalline TiC full-density bulk alloy consolidated from mechanically reacted powders. J. Alloys Compd. 2000, 305, 225–238.
6. Ushakov, A.; Karpov, I.; Lepeshev, A.; Krushenko, G. Physicochemical properties of nanomodifiers based on electric arc titanium nitride powder for polymer nanocomposite materials. Tekhnol. Met. 2011, 3, 16–21.
7. Kappe, C.; Dallinger, D.; Murphree, S. Practical Microwave Synthesis for Organic Chemists: Strategies, Instruments, and Protocols; Wiley VCH Verlag GmbH & Co., KGaA: Weinheim, Germany, 2009.
8. Choi, Y.; Rhee, S.-W. Effect of aluminium addition on the combustion reaction of titanium and carbon to form TiC. J. Mater. Sci. 1993, 28, 6669–6675.
9. Koc, R.; Folmer, J. Carbothermal synthesis of titanium carbide using ultrafine titania powders. J. Mater. Sci. 1997, 32, 3101–3111.
10. Thakur, S.K.; Kong, T.S.; Gupta, M. Microwave synthesis and characterization of metastable (Al/Ti) and hybrid (Al/Ti + SiC) composites. Mater. Sci. Eng. A 2007, 452, 61–69.
11. Rajkumar, K.; Aravindan, S. Microwave sintering of copper-graphite composites. J. Mater. Process. Technol. 2009, 209, 5601–5605.
12. Leparoux, S.; Vaucher, S.; Beffort, O. Assessment of microwave heating for sintering of Al/SiC and for in-situ synthesis of TiC. Adv. Eng. Mater. 2003, 5, 449–453.
13. Campanone, L.; Zaritzky, N. Mathematical analysis of microwave heating process. J. Food Eng. 2005, 69, 359–368.
14. Ciacci, T.; Galgano, A.; Di Blasi, C. Numerical simulation of the electromagnetic field and the heat and mass transfer processes during microwave-induced pyrolysis of a wood block. Chem. Eng. Sci. 2010, 65, 4117–4133.
15. Rosa, R.; Veronesi, P.; Leonelli, C. A review on combustion synthesis intensification by means of microwave energy. Chem. Eng. Process. Process Intensif. 2013, 71, 2–18.
16. Mishra, R.R.; Sharma, A.K. Microwave–material interaction phenomena: Heating mechanisms, challenges and opportunities in material processing. Compos. A Appl. Sci. Manuf. 2016, 81, 78–97.
17. Porcelli, M.; Cacciapuoti, G.; Fusco, S.; Massa, R.; d'Ambrosio, G.; Bertoldo, C.; De Rosa, M.; Zappia, V. Non-thermal effects of microwaves on proteins: Thermophilic enzymes as model system. FEBS Lett. 1997, 402, 102–106.
18. Moshtaghioun, B.; Poyato, R.; Cumbrera, F.; de Bernardi-Martin, S.; Monshi, A.; Abbasi, M.; Karimzadeh, F.; Dominguez-Rodriguez, A. Rapid carbothermic synthesis of silicon carbide nano powders by using microwave heating. J. Eur. Ceram. Soc.2012, 32, 1787–1794.
19. Zhang, H.; Li, F.; Jia, Q.; Ye, G. Preparation of titanium carbide powders by sol-gel and microwave carbothermal reduction methods at low temperature. J. Sol-Gel Sci. Technol. 2008, 46, 217–222.
20. Binner, J.G.P.; Hassine, N.A.; Cross, T.E. The possible role of the pre-exponential factor in explaining the increased reaction rates observed during the microwave synthesis of titanium carbide. J. Mater. Sci. 1995, 30, 5389–5393.
21. Hassine, N.A.; Binner, J.G.P.; Cross, T.E. Synthesis of refractory metal carbide powders via microwave carbothermal reduction. Int. J. Refract. Met. Hard Mater. 1995, 13, 353–358.
22. Holt, J.; Munir, Z. Combustion synthesis of titanium carbide: Theory and experiment. J. Mater. Sci. 1986, 21, 251–259.
23. Zeng, L.K.; Liu, Y.C.; Zhu, W.C.; Liu, P.A.; Wang, H.; Cheng, X.S.; Liang, Q.Y. Investigation on the continuous microwave synthesis of nano titanium carbide powder. Adv. Mater. Res. 2014, 1064, 66–70.
24. Chen, Y.-F.; Lee, C.-Y.; Yeng, M.-Y.; Chiu, H.-T. The effect of calcination temperature on the crystallinity of TiO2nanopowders. J. Cryst. Growth 2003, 247, 363–370.
25. Colmenares, J.; Aramendia, M.; Marinas, A.; Marinas, J.; Urbano, F. Synthesis, characterization and photocatalytic activity of different metal-doped titania systems. Appl. Catal A Gen. 2006, 306, 120–127.

26. Lucarelli, L.; Nadtochenko, V.; Kiwi, J. Environmental photochemistry: Quantitative adsorption and FTIR studies during the TiO2-photocatalyzed degradation of orange II. Langmuir 2000, 16, 1102–1108.
27. Diamandescu, L.; Vasiliu, F.; Tarabasanu-Mihaila, D.; Feder, M.; Vlaicu, A.; Teodorescu, C.; Macovei, D.; Enculescu, I.; Parvulescu, V.; Vasile, E. Structural and photocatalytic properties of iron-and europium-doped TiO2 nanoparticles obtained under hydrothermal conditions. Mater. Chem. Phys. 2008, 112, 146–153.
28. Jung, Y.-S.; Kim, D.-W.; Kim, Y.-S.; Park, E.-K.; Baeck, S.-H. Synthesis of alumina-titania solid solution by sol-gel method. J. Phys. Chem. Solids 2008, 69, 1464–1467.
29. Preiss, H.; Berger, L.-M.; Schultze, D. Studies on the carbothermal preparation of titanium carbide from different gel precursors. J. Eur. Ceram. Soc. 1999, 19, 195–206.
30. Cao, Y.; Zhang, H.; Li, F.; Lu, L.; Zhang, S. Preparation and characterization of ultrafine ZrB2–SiC composite powders by a combined sol-gel and microwave boro/carbothermal reduction method. Ceram. Int. 2015, 41, 7823–7829.

CHAPTER 8

Binary Alkali-Metal Silicon Clathrates by Spark Plasma Sintering: Preparation and Characterization

Igor Veremchuk [1], Matt Beekman [2], Iryna Antonyshyn [1], Walter Schnelle [1], Michael Baitinger [1], George S. Nolas [3,*] and Yuri Grin [1,*]

[1]*Max-Planck-Institut für Chemische Physik fester Stoffe, 01187 Dresden, Germany*

[2]*Department of Physics, California Polytechnic State University, San Luis Obispo, CA 93407, USA*

[3]*Department of Physics, University of South Florida, Tampa, FL 33620, USA*

ABSTRACT

The binary intermetallic clathrates $K_{8-x}Si_{46}$ (x = 0.4; 1.2), $Rb_{6.2}Si_{46}$, $Rb_{11.5}Si_{136}$ and $Cs_{7.8}Si_{136}$ were prepared from M_4Si_4 (M = K, Rb, Cs) precursors by spark-plasma route (SPS) and structurally characterized by Rietveld refinement of PXRD data. The clathrate-II phase $Rb_{11.5}Si_{136}$ was synthesized for the first time. Partial crystallographic site occupancy of the alkali metals, particularly for the smaller Si_{20} dodecahedra, was found in all compounds. SPS preparation of $Na_{24}Si_{136}$ with different SPS current polarities and tooling were performed in order to investigate the role of the electric field on clathrate formation. The electrical and thermal transport properties of $K_{7.6}Si_{46}$ and $K_{6.8}Si_{46}$in the temperature range 4–700 K were investigated. Our findings demonstrate that SPS is a novel tool for the synthesis of intermetallic clathrate phases that are not easily accessible by conventional synthesis techniques.

Keywords: spark plasma synthesis; Zintl phase; intermetallic clathrate

1. INTRODUCTION

Binary phases of alkali metals with silicon have been prepared by thermal decomposition of the monosilicides M_4Si_4 (M = Na, K, Rb, Cs) for over six decades [1,2,3,4]. Structural investigations of the polycrystalline products revealed the existence of clathrate-I $M_{8-x}Si_{46}$ and clathrate-II $M_{24-x}Si_{136}$ silicides [5,6]. In both clathrate types, the alkali-metal atoms are enclosed in polyhedral cages formed in the tetrahedrally bonded silicon framework. In the clathrate-I crystal structure, the unit cell contains two Si_{20} and six Si_{24} cages, while in the clathrate-II structure there are sixteen Si_{20} and eight Si_{28} cages present in the unit cell. In the clathrate phases prepared by thermal decomposition the filling fraction of M in the silicon cages depends explicitly on both the reaction conditions and the size of the M atom relative to the size of the cage [6] (Figure 1).

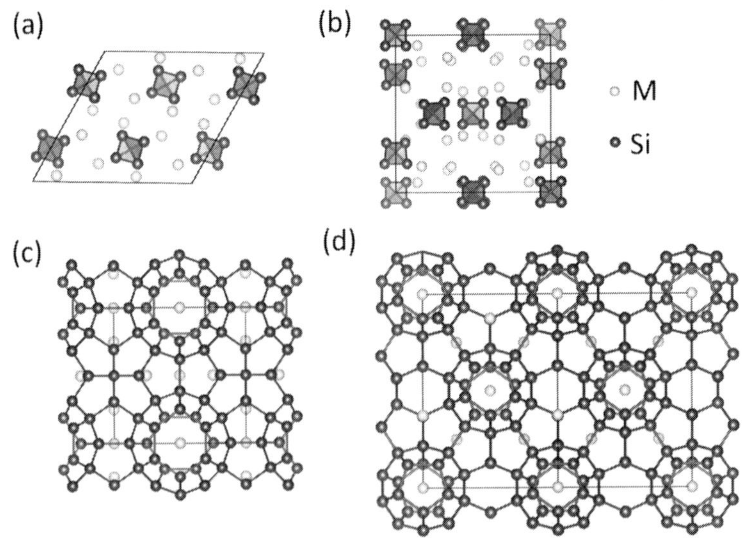

Figure 1. Crystal structures of (**a**) Na_4Si_4; (**b**) M_4Si_4 (M = K, Rb, or Cs); (**c**) $M_{8-x}Si_{46}$; and (**d**) $M_{24-x}Si_{136}$.

Alternatively, redox techniques have been developed to obtain metastable clathrate phases [7,8,9,10] including Ge and Si allotropes with empty cages [11,12]. Defect-free single crystals of alkali metal silicon clathrates large enough for structure investigations have only been obtained in exceptional cases [4]. For example, crystals of K_8Si_{46} were grown on α-Si exposed to potassium vapor [13], while crystals of ternary clathrate-II phases $Rb_8Na_{16}Si_{136}$ and $Cs_8Na_{16}Si_{136}$ form in closed ampoules directly from the melt [14,15]. Single-crystals suitable for measurement of intrinsic transport properties were first prepared by the electrochemically-driven transformation of Na_4Si_4 to $Na_{24}Si_{136}$ during the spark plasma treatment (SPS) [16,17]. Another approach—the kinetically controlled thermal decomposition (KCTD) method—can also provide relatively large single crystals [18,19]. Recently, a thermal decomposition at lower temperatures was employed for the preparation of $Na_{24-x}(Si_yGe_{1-y})_{136}$ and proved to be useful for the synthesis of silicon clathrates [20]. Extensions of the SPS and KCTD approaches have also recently reported, including use of multi-phase precursors or simultaneous ion-exchange and electrochemical reactions to prepare ternary or quaternary clathrates [21,22,23].

The unique advantage of SPS for the synthesis of clathrates by producing scalable, compact bulk materials also continues to be of great interest [4]. In this work we discuss both the preparation and consolidation of alkali-metal silicon clathrates by SPS. We report on the preparation and crystal structure of different clathrate phases (Table 1) as well as SPS densification of $K_{7.6}Si_{46}$ and $K_{6.8}Si_{46}$ powders by SPS that allow for the investigation of their transport properties over a large temperature range. The formation and evaporation the free alkali metals during the reaction allow us to work only with a small amount (less than 200 mg) of precursors and, in the results, after processing, we obtained 10–20 mg of the materials.

2. RESULTS AND DISCUSSION

As first described in reference [16], clathrates are formed during SPS redox-treatment by oxidation of Si_4^{4-} at the anode while the alkali metal is reduced at the cathode (Figure 2a,b). Clathrate-II formation, according to

$$34\,M_4Si_4 \rightarrow M_{24}Si_{136} + 112\,M$$

can therefore be described by the half-reactions

$$\begin{aligned} 34\,Si_4^{\,4-} - 112\,e^- &\rightarrow Si_{136}^{\,24-}\ (anode) \\ 112\,M^+ + 112\,e^- &\rightarrow 112\,M^0\ (cathode) \end{aligned} \tag{1}$$

For clathrate-I formation according to

$$23\,M_4Si_4 \rightarrow 2\,M_8Si_{46} + 76\,M$$

should be considered by means of the half-reactions

$$\begin{aligned} 23\,Si_4^{\,4-} - 76\,e^- &\rightarrow 2\,Si_{46}^{\,8-}\ (anode) \\ 76\,M^+ + 76\,e^- &\rightarrow 76\,M^0\ (cathode) \end{aligned} \tag{2}$$

Table 1. Spark-plasma synthesis (SPS) conditions for the silicon clathrates.

Precursor	T_R (°C) [a]	t_R (min) [b]	Die/Punches [c]	Products [d]
Na_4Si_4	550	180	C/C	$Na_{24}Si_{136}$ (M) + Na_8Si_{46} (m)
	600	180	C/C	$Na_{24}Si_{136}$ (M)
	700	180	C/C	$Na_{24}Si_{136}$ (M) + Si (m)
K_4Si_4	500	60	BN/SS	$K_{8-x}Si_{46}$ (tr)
	550	60	BN/SS	$K_{8-x}Si_{46}$ (M)
	600	60	BN/SS	$K_{8-x}Si_{46}$ (M) + Si (m)
	650	60	BN/SS	Si (M) + $K_{8-x}Si_{46}$ (m)
Rb_4Si_4	450	60	BN/SS	$Rb_{8-x}Si_{46}$ (M) + $Rb_{24-x}Si_{136}$ (m) + Si (tr)
	500	60	BN/SS	Si (M) + $Rb_{8-x}Si_{46}$ (m) + $Rb_{24-x}Si_{136}$ (m)
Cs_4Si_4	350	60	BN/SS	$Cs_{8-x}Si_{136}$ (M) + Si (tr)

[a] T_R is the reaction temperature; [b] t_R is the treatment time; [c] Materials for the die and punches: C = graphite, BN = *hp*-boron nitride, SS = stainless steel; [d] Excluding unreacted precursor. M = majority phase, m = minority phase, tr = trace.

Reactions (1) and (2) describe the basics of the clathrate formation by SPS, whereby the process is driven by the DC field and the evaporation of the alkali metal at the cathode and its absorption by the graphite die (Figure 2b). In order to qualitatively investigate the influence of electric current on clathrate formation we performed three control experiments with Na_4Si_4 as the precursor. First, the polarity of the DC current was reversed, upon which it was found that the clathrate phase, $Na_{24}Si_{136}$ in this case, consistently forms at the anode (Figure 2c). In a second experiment, equal volumes of Al_2O_3powder were placed above and below the precursor to create an electrically insulating barrier. The electrical current then presumably flowed through the low resistance graphite die such that the current passing through the Na_4Si_4 precursor is minimized. In this case no reaction was observed under the identical conditions used to form the clathrate $Na_{24}Si_{136}$(Figure 2d). In the third experiment, the use of stainless steel punches combined with a die manufactured from boron nitride, a high electrical-resistivity material, presumably forced most of the current to pass through the precursor (as opposed to the surrounding die) (Figure 2e). A more rapid reaction than in the case of using the graphite die was observed. Optical photographs of cross-section of SPS-processed pellets (Figure 2f,h) clearly show that formation of the clathrate phases occurred at the anode during SPS processing. This was also the case for all clathrate compositions prepared in this study.

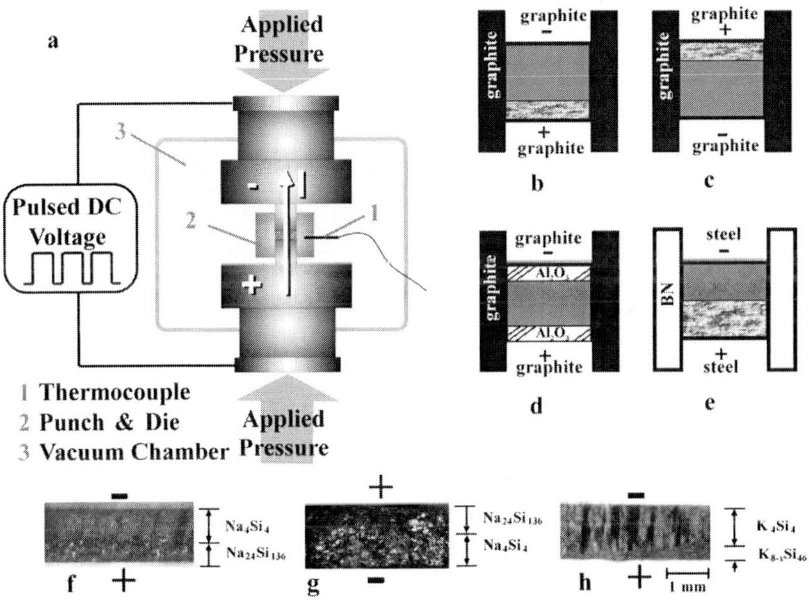

Figure 2. (**a**) SPS setup for the preparation of clathrates; (**b–e**) Location and relative amounts of the reaction products for different polarities and different die and punch materials; (**f–h**) Optical micrographs of the cross-sections from fractured post-reaction pellets illustrating the directional growth of clathrates at the anode.

The crystal structures of the SPS-prepared clathrate phases were investigated by the Rietveld method using powder X-ray diffraction (PXRD) data. Experimental, calculated, and difference diffraction patterns for the clathrate-I and II samples are shown on Figure 3 and

Figure 4, respectively, and the resulting crystallographic data are presented in Table 2 and Table 3, respectively.

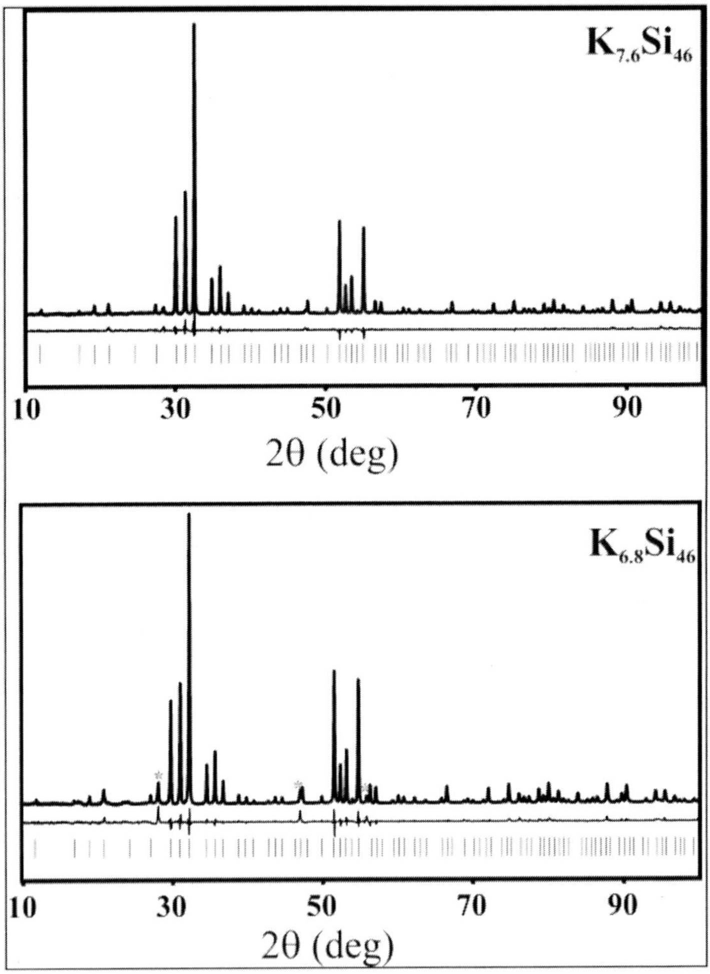

Figure 3. Powder X-ray diffraction (PXRD) patterns of the SPS-prepared clathrates-I $K_{7.6}Si_{46}$ (550 °C) and $K_{6.8}Si_{46}$ (600 °C). Red asterisks mark the reflections of α-Si (<0.2 mass % in $K_{7.6}Si_{46}$ and 1.5 mass % in $K_{6.8}Si_{46}$). The experimental intensities are shown as a solid line. The difference between the experimental and calculated intensities is shown below the experimental data; the tick marks represent the positions of the diffraction reflections of the clathrate-I phase.

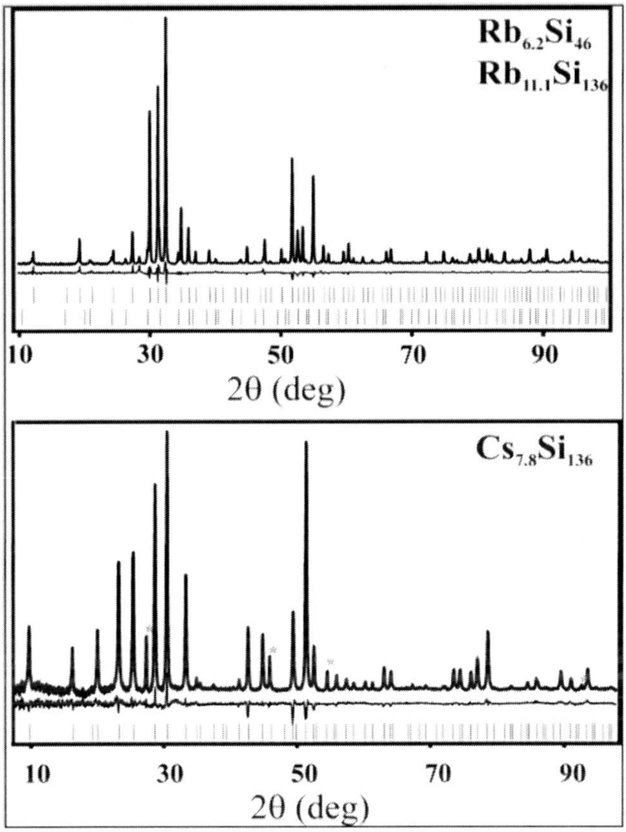

Figure 4. Powder X-ray diffraction (PXRD) patterns of the $Rb_{6.2}Si_{46}$ and $Rb_{11.1}Si_{136}$mixed product and clathrate-II $Cs_{7.8}Si_{136}$. The experimental intensities are shown as a solid line. The differences between the experimental and calculated intensities are shown below the experimental data. The tick marks represent the positions of the diffraction reflections. Red asterisks mark the reflections from α-Si (<0.2 mass % in $Rb_{6.2}Si_{46}$ and $Rb_{11.1}Si_{136}$ and 2.5 mass % in $Cs_{7.8}Si_{136}$).

The lattice parameters for $K_{7.6}Si_{46}$, $Rb_{6.2}Si_{46}$, and $Cs_{7.8}Si_{136}$ show good agreement with those synthesized by other preparation methods [7,16,18,19,24,25,26]. The lattice parameter for $Rb_{11.5}Si_{136}$, as expected, has an intermediate value between that reported for $K_{17.2}Si_{136}$ [19] and Cs_8Si_{136} [6]. In all cases the silicon framework sites were found to be fully occupied (no vacancies), in agreement with prior results for binary silicon clathrates [7,16,18,19,26,27,28]. The stable covalent Si–Si bonds in silicon clathrates allow for a large excess of electrons that fill antibonding states. In a rigid-band model, a high concentration of conduction electrons are provided by charge transfer from the alkali-metal guests, thus inducing metallic behavior, as confirmed experimentally [15,29]. In contrast to SPS-prepared Na_8Si_{46} and $Na_{24}Si_{136}$, which show full occupancy of the polyhedral cages by sodium [19], the SPS-prepared clathrates of potassium, rubidium and caesium exhibit partial occupancy of the alkali-atom positions (the specific refined compositions are given in Table 2 and Table 3).

Table 2. Crystallographic information on the clathrate-I phases prepared by SPS (PXRD data).

Phase	$K_{7.6(1)}Si_{46}$	$K_{6.8(1)}Si_{46}$	$Rb_{6.2(1)}Si_{46}$
Lattice parameter a, Å	10.2776 (1)	10.2747 (1)	10.2848 (1)
Radiation, wavelength λ, Å		Cu $K_{\alpha 1}$, 1.54056	
Maximal diffraction angle 2θ, °		100.30	
Residuals R_I/R_P	0.04/0.10	0.04/0.09	0.02/0.10
$M1$ 2a (0 0 0)	Occ * = 0.783 (1)	Occ = 0.695 (4)	Occ = 0.244 (3)
$M2$ 6d ($\frac{1}{4}$ $\frac{1}{2}$ 0)	Occ = 1.013 (3) **	Occ = 0.918 (3)	Occ = 0.961 (2)
Si1 6c ($\frac{1}{4}$ 0 $\frac{1}{2}$)	–	–	–
Si2 16i (x x x)	x = 0.1845 (1)	x = 0.1846 (1)	x = 0.1840 (1)
Si3 24k (0 y z)	y = 0.3068 (1) z = 0.1183 (1)	y = 0.3060 (1) z = 0.1185 (1)	y = 0.3043 (1) z = 0.1190 (1)

*: Occ—occupancy factor; **: The value 1.013 (3) for the K2 occupancy in $K_{7.6}Si_{46}$ reflects the real error of the refinement (~4 e.s.d.) and helps to understand the reliability of the occupancy values for other clathrates.

Table 3. Crystallographic information on the clathrate-II phases prepared by SPS (PXRD data).

Phase	$Rb_{11.1(1)}Si_{136}$	$Cs_{7.8(1)}Si_{136}$
Lattice parameter a, Å	14.7142 (9)	14.6733 (3)
Radiation, wavelength λ, Å	Cu $K_{\alpha 1}$, 1.54056	
Maximal diffraction angle 2θ, °	100.30	
Residuals R_I/R_P	0.10/0.10	0.05/0.13
$M1$ 8b (⅜ ⅜ ⅜)	Occ * = 0.85 (2)	Occ = 0.968 (2)
$M2$ 16c (0 0 0)	Occ = 0.272 (1)	Occ = 0.0
Si1 8a (⅛ ⅛ ⅛)	–	–
Si2 32e (x x x)	x = 0.2198 (6)	x = 0.2168 (2)
Si3 96g (x x z)	x = 0.1839 (2) z = 0.3698 (6)	x = 0.1822 (1) z = 0.3699 (2)

*: Occ—occupancy factor.

The compositions of clathrates-II prepared in this study together with previously reported in the literature [16,19,29,30] allow for a comparison of the lattice parameter versus guest-atom content in clathrate-II phases. As shown in Figure 5, in all cases the lattice parameter increases with those increasing *M* content in the Si_{20} (crystallographic site 16*c*) cage. In addition, the lattice parameter also increases with increasing alkali-ion radius [31] for a specific *M* concentration. Preferential occupancy of the larger Si_{28} (crystallographic site 8*b*) cages was observed for the case of $Rb_{11.5}Si_{136}$, as well as for for Na_xSi_{136} [30,32].

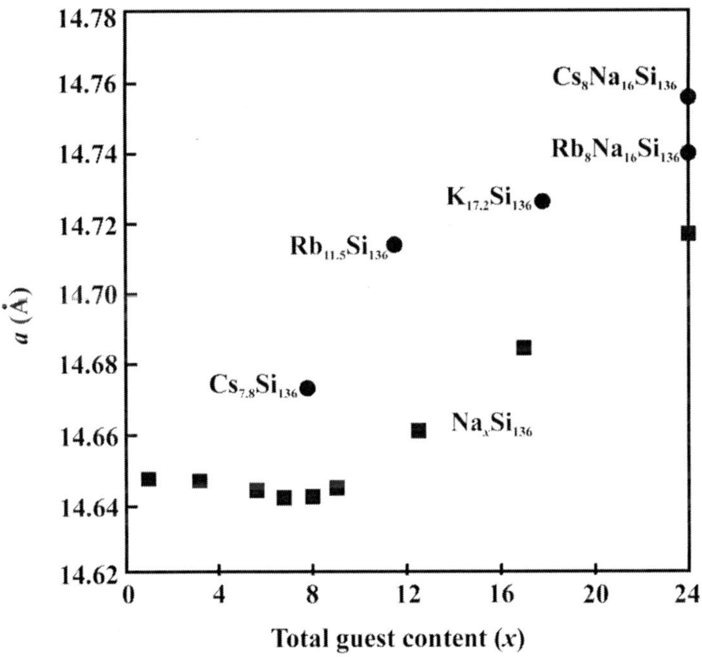

Figure 5. Lattice parameters versus *M* content, *x* for clathrate-I Na_xSi_{136} ($0 < x < 24$) [28], $Na_{24}Si_{136}$ [16], $K_{17.2}Si_{136}$ [19], $Rb_{11.5}Si_{136}$ (this work), and $Cs_{7.8}Si_{136}$ (this work), as well as ternary $Rb_8Na_{16}Si_{136}$ and $Cs_8Na_{16}Si_{136}$ [27].

In order to further characterize the products formed by SPS we measured the electrical and thermal transport properties of $K_{7.6}Si_{46}$ and $K_{6.8}Si_{46}$. The density of the pellets (only $K_{8-x}Si_{46}$) was approximately 90% of the calculated density based on PXRD data. Temperature-dependent electrical resistivity ρ (Figure 6, top), Seebeck coefficient *S* (Figure 6, middle), and thermal conductivity κ (Figure 6, bottom) in the temperature range from 10 to 700 K are shown in Figure 6. The agreement between the high- and low-temperature data, measured with two different setups on two different specimens, is an indication of the homogeneity of the materials prepared by SPS. Both materials $K_{7.6}Si_{46}$ and $K_{6.8}Si_{46}$ show metallic temperature dependence of ρ and S, in agreement with the seminal work by Cros et al. [6]. The fact that these materials show metallic conduction follows from the simple assertion that each K atom donates one electron to the silicon framework resulting in an overall high electron concentration. Measured ρ of cold-pressed $K_{7.6}Si_{46}$, obtained from the thermal decomposition of K_4Si_4, was reported to be much higher: $\rho \sim 140$ mΩ·cm at room temperature with weak temperature dependence [28]. The large discrepancy between those results and the data shown in Figure 6 may be attributed to the microstructure of the specimens and the typically poor inter-granular contact that is achieved using cold pressing [28]. Nevertheless, the ρ values shown in Figure 6 are significantly larger than expected for the presumed high electron concentration. In addition, the residual resistance ratio, $\rho_{300K}/\rho_{10K} = 3.6$, is also relatively low for a good metal and significantly smaller than the values obtained for single crystals of Na_8Si_{46} ($\rho_{300K}/\rho_{10K} = 36$) [33] and $Na_{24}Si_{136}$ ($\rho_{300K}/\rho_{10K} = 14$) [17]. These results reiterate the need for high-density polycrystalline specimens in order to minimize the effects of grain boundary scattering on the transport properties of intermetallic clathrates [34].

3. MATERIALS AND METHODS

3.1. Synthesis of Precursors

Binary precursors M_4Si_4 (M = Na, K, Rb, or Cs) were prepared from alkali metal pieces (ChemPur, Karlsruhe, Germany 99.9 wt %) and coarsely ground silicon (ChemPur, 99.999 wt %), by sealing of the precursor mixture under argon in small, welded tantalum ampoules that were enclosed under vacuum inside silica tubes and heated to 800 °C at an average rate of 3 °C/min. The ampoules were held at 800 °C for 4 h before being slowly cooled to room temperature (furnace cooled). The reaction products were determined to be single phase M_4Si_4 by powder X-ray diffraction (PXRD). All sample manipulations were performed in an argon-filled glove box (H_2O < 1 ppm; O_2 < 1 ppm). **CAUTION**: M_4Si_4 compounds are sensitive to air and moisture (pyrophoric) and must be kept under inert atmosphere at all times.

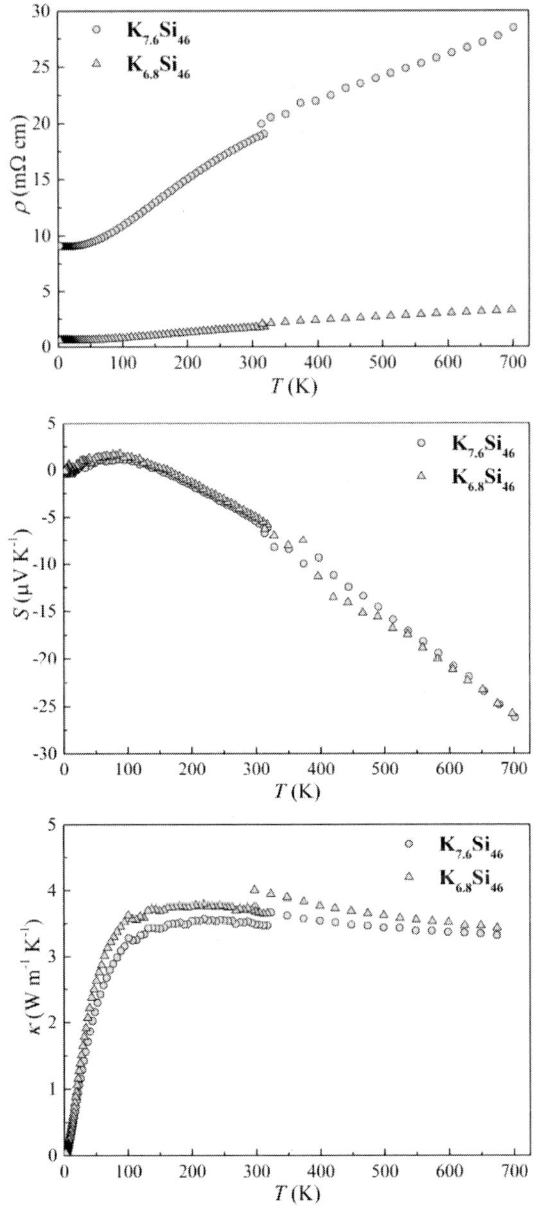

Figure 6. Transport properties of polycrystalline $K_{7.6}Si_{46}$ and $K_{6.8}Si_{46}$. The low temperature, 2 to 300 K, and high temperature, 300 to 700 K, are described in the text. (**Top**) electrical resistivity ρ; (**Middle**) Seebeck coefficient S; (**Bottom**) thermal conductivity κ.

3.2. Spark Plasma Synthesis

SPS experiments were conducted under vacuum (background pressure <10 Pa) using a dedicated SPS setup in a protective argon atmosphere (Figure 2a; SPS: 515 ET, Syntex-Fuji, Japan; glove box: MBraun, Garching, Germany) [24]. Table 1 contains the preparation conditions for the clathrates synthesized within this study. The precursors were ground to fine powder and loaded into graphite or boron nitride (*h*-BN) dies with an inner diameter of 8 mm. Graphite or stainless steel punches were used to apply the uniaxial pressure. Tantalum foil was employed to surround the powder specimen on all sides, isolating the specimen from the die and punches during SPS processing. Graphite dies worked well for Na, however they were observed to react with alkali-metal vapor for precursors of K, Rb, and Cs and often cracked the graphite die during SPS processing. For these alkali metals *h*-BN was found to be most suitable. After SPS the products were isolated from the residual precursor by washing first with ethanol and then distilled water under flowing argon.

3.3. Sample Characterization

Powder X-ray diffraction (PXRD) data were collected using the Guinier technique (Huber G670 camera, Rimsting, Germany, Cu-$K_{\alpha 1}$ radiation, λ = 1.540598 Å, graphite monochromator, $5° \leq 2\theta \leq 100°$, $\Delta 2\theta = 0.005°$). Crystal structure refinements were performed using the software WinCSD [25]. For the transport measurements, the $K_{7.6}Si_{46}$ and $K_{6.8}Si_{46}$powders were first washed then consolidated by SPS using tungsten carbide tool (8 mm inner diameter) at 300 °C and a uniaxial pressure of 600 MPa under vacuum (background pressure of approximately 10 Pa) for 10 min. The density of the pellets was approximately 90% of the calculated density based on PXRD data. The specimens were cut with a diamond wire saw into parallelepiped pieces 1.5 mm × 1.5 mm × 8.0 mm for electrical transport measurements and low-temperature thermal conductivity measurements, and 1.5 mm thick and 8 mm in diameter disks for thermal diffusivity measurements. Electrical resistivity (ρ) and Seebeck coefficient (S) measurements from 22 to 450 °C on SPS densified $K_{7.6}Si_{46}$ and $K_{6.8}Si_{46}$specimens were performed with a commercial ZEM-3 setup (ULVAC-RIKO, Munich, Germany). Specific heat (C_p) measurements were performed on a Netzsch Pegasus DSC (Selb, Germany) while heating to 450 °C at a heating rate of 10 °C/min in an Ar atmosphere. Thermal diffusivity, D, was measured from 22 °C to 450 °C using the laser-flash technique (LFA 457 Micro Flash, Netzsch, Selb, Germany). The temperature dependent thermal conductivity, κ, was calculated using the relation $\kappa = C_p \times D \times d$ where d is the density in g/cm^3. Low temperature ρ, S, and κ data were collected using the TTO-option (Thermal Transport Option) of a PPMS (Physical Property Measurement System) (Quantum Design, San Diego, CA, USA) using a four-point method.

4. CONCLUSIONS

Binary intermetallic clathrates $K_{8-x}Si_{46}$ (x = 0.4; 1.2), $Rb_{6.2}Si_{46}$, $Rb_{11.5}Si_{46}$, and $Cs_{7.8}Si_{136}$ were synthesized by spark plasma synthesis from the respective M_4Si_4 precursors. The formation of polycrystalline products at the anode of the SPS setup was observed for all compounds. The preparation of the new clathrate-II phase $Rb_{11.5}Si_{136}$ is further evidence that SPS is useful for the inorganic synthesis. The role the electric field is in driving the chemical conversion of the precursor into the clathrate phase at the anode during SPS processing is emphasized. Conventional thermal decomposition (CTC) [6,28,35], kinetically controlled thermal decomposition (KCTD) [18,19], chemical oxidation (CO) [7,27], and SPS yield different intermetallic clathrate phases from the same M_4Si_4 precursor (Table 4). While the influence of temperature clearly plays a role in clathrate formation, a definitive understanding of the local formation mechanisms in each of these synthetic approaches [36] is of great interest and is an ongoing task.

Table 4. Clathrate phases obtained by different synthesis techniques from M_4Si_4precursors.

Precursor	Preparation Technique [a]	Clathrate Phases Obtained	References
Na_4Si_4	CTC	Na_8Si_{46} and $Na_{24-x}Si_{136}$	[6,33]
	KCTD	Na_8Si_{46} and $Na_{24-x}Si_{136}$	[18]
	CO	Na_8Si_{46} and $Na_{24-x}Si_{136}$	[7,25]
	SPS	Na_8Si_{46} and $Na_{24-x}Si_{136}$	[16,24], this work
K_4Si_4	CTC	$K_{8-x}Si_{46}$	[6,26]
	KCTD	$K_{8-x}Si_{46}$ and $K_{24-x}Si_{136}$	[19]
	CO	$K_{8-x}Si_{46}$	[7]
	SPS	$K_{8-x}Si_{46}$	this work
Rb_4Si_4	CTC	$Rb_{8-x}Si_{46}$	[6,26]
	SPS	$Rb_{8-x}Si_{46}$ and $Rb_{24-x}Si_{136}$	this work
Cs_4Si_4	CTC	$Cs_{24-x}Si_{136}$	[6]
	SPS	$Cs_{24-x}Si_{136}$	this work

[a] CO = chemical oxidation, KCTD = kinetically controlled thermal decomposition, CTC = conventional thermal decomposition, SPS = spark plasma synthesis.

ACKNOWLEDGMENTS

M.Be. (while at USF) and G.S.N. acknowledge support from the US Department of Energy, Basic Energy Sciences, Division of Materials Science and Engineering, under Award No. DE-FG02-04ER46/45 for SPS synthesis. M.Be. also acknowledges support from Deutscher Akademischer Austausch Dienst. M.Ba. and Y.G. acknowledge support from DFG priority program SPP1415.

AUTHOR CONTRIBUTIONS

Y.G. and G.S.N. developed the concept of this study; I.V. and M.Be. conceived and designed the SPS experiments; I.A. and W.S. performed the physical measurements experiments; I.V., M.Be., and M.Ba. refined the crystal structures from PXRD experiments; all authors contributed to the text of the manuscript.

CONFLICTS OF INTEREST

The authors declare no conflict of interest.

REFERENCES

1. Hohmann, E. Silicide und Germanide der Alkalimetalle. Z. Anorg. Chem. 1948, 257, 113–126.
2. Cros, C.; Pouchard, M.; Hagenmuller, P. Two new phases of the silicon-sodium system. C. R. Hebd. Seances Acad. Sci.1965, 260, 4764–4767.
3. Cros, S.; Pouchard, M. Sur les phases de type clathrate du silicium et des éléments apparentés (C, Ge, Sn): Une approche historique. C. R. Chim. 2009, 12, 1014–1056.
4. Nolas, G.S., Ed.; The Physics and Chemistry of Inorganic Clathrates; Springer: Berlin, Germany, 2014.
5. Kasper, J.S.; Hagenmuller, P.; Pouchard, M.; Cros, C. Clathrate Structure of Silicon Na8Si46 and NaxSi136 (x < 11). Science 1965, 150, 1713–1714. [PubMed]
6. Cros, C.; Pouchard, M.; Hagenmuller, P. Sur une nouvelle famille de clathrates minéraux isotypes des hydrates de gaz et de liquides. Interprétation des résultats obtenus. J. Solid State Chem. 1970, 2, 570–581.
7. Böhme, B.; Guloy, A.; Tang, Z.; Schnelle, W.; Burkhardt, U.; Baitinger, M.; Grin, Y. Oxidation of M4Si4 (M = Na, K) to Clathrates by HCl or H2O. J. Am. Chem. Soc. 2007, 129, 5348–5349.
8. Böhme, B.; Hoffmann, S.; Baitinger, M.; Grin, Y. Application of n-Dodecyltrimethylammonium Chloride for the Oxidation of Intermetallic Phases. Z. Naturforsch. 2011, 66, 230–238.

9. Liang, Y.; Böhme, B.; Reibold, M.; Schnelle, W.; Schwarz, U.; Baitinger, M.; Lichte, H.; Grin, Y. Synthesis of the Clathrate-I Phase Ba8-xSi46 via Redox Reactions. Inorg. Chem. 2011, 50, 4523–4528.
10. Krishna, L.; Chai, P.; Koh, C.A.; Toberer, E.S.; Nolas, G.S. Synthesis and structural properties of type I potassium SiGe alloy clathrates. Mater. Lett. 2015, 149, 123–126.
11. Ammar, A.; Cros, C.; Pouchard, M.; Jaussaud, N.; Bassat, J.-M.; Villeneuve, G.; Duttine, M.; Ménétrier, M.; Reny, E. On the clathrate form of elemental silicon, Si136: Preparation and characterisation of NaxSi136 ($x \rightarrow 0$). Solid State Sci.2004, 6, 393–400.
12. Guloy, A.M.; Ramlau, R.; Tang, Z.; Schnelle, W.; Baitinger, M.; Grin, Y. A guest-free germanium clathrate. Nature 2006, 443, 320–323.
13. Gallmeier, J.; Schäfer, H.; Weiss, A. Cage structure as a common building principle of compounds K8E46 (E = silicon, germanium, tin). Z. Naturforsch. B 1969, 24, 665–667.
14. Bobev, S.; Sevov, S.C. Synthesis and Characterization of Stable Stoichiometric Clathrates of Silicon and Germanium: Cs8Na16Si136 and Cs8Na16Ge136. J. Am. Chem. Soc. 1999, 121, 3795–3796.
15. Nolas, G.S.; Vanderveer, D.G.; Wilkinson, A.P.; Cohn, J.L. Temperature dependent structural and transport properties of the type II clathrates A8Na16E136 (A = Cs or Rb and E = Ge or Si). J. Appl. Phys. 2002, 91, 8970–8973.
16. Beekman, M.; Baitinger, M.; Borrmann, H.; Schnelle, W.; Meier, K.; Nolas, G.S.; Grin, Y. Preparation and Crystal Growth of Na24Si136. J. Am. Chem. Soc. 2009, 131, 9642–9643.
17. Beekman, M.; Schnelle, W.; Borrmann, H.; Baitinger, M.; Grin, Y.; Nolas, G.S. Intrinsic Electrical and Thermal Properties from Single Crystals of Na24Si136. Phys. Rev. Lett. 2010, 104.
18. Stefanoski, S.; Beekman, M.; Wong-Ng, W.; Zavalij, P.; Nolas, G.S. Simple Approach for Selective Crystal Growth of Intermetallic Clathrates. Chem. Mater. 2011, 23, 1491–1495.
19. Stefanoski, S.; Nolas, G.S. Synthesis and Structural Characterization of Single-Crystal K7.5Si46 and K17.8Si136Clathrates. Cryst. Growth Des. 2011, 11, 4533–4537.
20. Baranowski, L.L.; Krishna, L.; Martinez, A.D.; Raharjo, T.; Stevanovic, V.; Tamboli, A.C.; Toberer, E.S. Synthesis and optical band gaps of alloyed Si–Ge type II clathrates. J. Mater. Chem. B 2014, 2, 3231–3237.
21. Dong, Y.; Nolas, G.S. Rapid crystal growth of type-II clathrates A8Na16Si136 (A = K, Rb, Cs) by spark plasma sintering. Cryst. Eng. Commun. 2015, 17, 2242–2244.
22. Dong, Y.; Chai, P.; Beekman, M.; Zeng, X.; Tritt, T.M.; Nolas, G.S. Precursor Routes to Complex Ternary Intermetallics: Single-Crystal and Microcrystalline Preparation of Clathrate-I Na8Al8Si38 from NaSi + NaAlSi. Inorg. Chem. 2015, 54, 5316–5321.
23. Wei, K.; Dong, Y.; Nolas, G.S. Precursor routes to quaternary intermetallics: Synthesis, crystal structure, and physical properties of clathrate-II Cs8Na16Al24Si112. J. Solid State Chem. 2016, 237, 81–85.
24. Reinfried, N.; Höhn, P.; Grin, Y. Spark-Plasma Synthesis an Inert Gas Atmosphere; Max-Planck-Institute for Chemical Physics of Solids: Dresden, Germany, 2006; p. 29.
25. Akselrud, L.; Grin, Y. WinCSD: Software package for crystallographic calculations (Version 4). J. Appl. Crystallogr.2014, 47, 803–805.
26. Stefanoski, S.; Blosser, M.C.; Nolas, G.S. Pressure Effects on the Size of Type-I and Type-II Si-Clathrates Synthesized by Spark Plasma Sintering. Cryst. Growth Des. 2013, 13, 195–197.
27. Blosser, M.; Nolas, G.S. Synthesis of Na8Si46 and Na24Si136 by oxidation of Na4Si4 from ionic liquid decomposition. Mater. Lett. 2013, 99, 161–163.
28. Ramachandran, G.K.; McMillan, P.F.; Dong, J.; Sankey, O.F. K7.62(1)Si46 and Rb6.15(2)Si46: Two Structure I Clathrates with Fully Occupied Framework Sites. J. Solid State Chem. 2000, 154, 626–634.
29. Bobev, S.; Sevov, S.C. Clathrates of Group 14 with Alkali Metals: An Exploration. J. Solid State Chem. 2000, 153, 92–105.
30. Beekman, M.; Nenghabi, E.N.; Biswas, K.; Myles, C.W.; Baitinger, M.; Grin, Y.; Nolas, G.S. Framework Contraction in Na-Stuffed Si(cF136). Inorg. Chem. 2010, 49, 5338–5340.
31. Shannon, R.D. Revised effective ionic radii and systematic studies of interatomic distances in halides and chalcogenides. Acta Crystallogr. A 1976, 32, 751–767.

32. Reny, E.; Gravereau, P.; Cros, C.; Pouchard, M. Structural characterisations of the NaxSi136 and Na8Si46 silicon clathrates using the Rietveld method. J. Mater. Chem. 1998, 8, 2839–2844.
33. Stefanoski, S.; Martin, J.; Nolas, G.S. Low temperature transport properties and heat capacity of single-crystal Na8Si46. J. Phys. Condens. Matter 2010, 22.
34. Beekman, M.; Nolas, G.S. Transport Properties of the Binary Type I Clathrate K8Ge44□2. Int. J. Appl. Ceram. Technol.2007, 4, 332–338.
35. Cros, C.; Benejat, J.-C. Preparation and properties of a clathrate having an extensive domain of existence. Sodium silicide NaxSi136 compounds. Bull. Soc. Chim. Fr. 1972, 5, 1739–1743.
36. Hutchins, P.T.; Leynaud, O.; O'Dell, L.A.; Smith, M.E.; Barnes, P.; McMillan, P.F. Time-Resolved in Situ Synchrotron X-ray Diffraction Studies of Type 1 Silicon Clathrate Formation. Chem. Mater. 2011, 23, 5160–5167.

CHAPTER 9

Synthesis Strategies about 2D Materials

Jianghao Wang, Guangshe Li and Liping Li

ABSTRACT

In recent years, more and more attentions have been paid to two-dimensional (2D) materials due to the excellent electrical, optical, thermal and mechanical properties. To characterize the layer-dependent changes in properties and to provide pathways for their integration into a multitude of applications, it is essential to explore the reliable syntheses of single- and few-layer 2D materials. Therefore, many strategies, such as micromechanical exfoliation, ultrasonic exfoliation, hydrothermal method, topochemical transformation, chemical vapor deposition method and so on, have been developed to synthesize high-quality and ultrathin nanosheets showing their own merits and demerits in preparing 2D nanomaterials. Herein, an overview of the recent progress in the synthetic techniques is presented for 2D materials, in which we would introduce their experimental scheme, advantages and disadvantages, and applications of these synthetic strategies. Eventually, the potential trends and future directions for synthesizing technology for 2D materials are proposed.

Keywords: two-dimensional, nanosheets, top down, bottom up, synthesis strategy

1. INTRODUCTION

The discovery of single-layer graphene in 2004 by Novoselov and Geim has shown that it is highly possible to exfoliate stable, single-atom or single-polyhedral-thick two-dimensional (2D) materials from van der Waals solids, and these 2D materials could exhibit unique and fascinated physical properties, such as ultrahigh carrier mobility at room temperature (~10,000 $cm^2V^{-1}s^{-1}$), quantum hall effect, large theoretical specific surface area (2630 m^2g^{-1}), excellent optical transparency (~97.7%) and so on [1, 2]. This great discovery helps Novoselov and Geim win Nobel Prize in 2010. The success of graphene arouses intensive interests in 2D materials in the world. As a consequence, more and more 2D materials have been synthesized successfully, showing great promise for in many applications.

Ideal 2D materials belong to those with only one atom or several atoms thickness and infinite lateral size. The reliable synthesis of single- and few-layer 2D materials is an essential first step for characterizing the layer-dependent changes in their properties, as well as for providing pathways of their integration into a multitude of applications [1]. For instances, when the thickness of metal Co was reduced to only one or several atoms, the catalytic activity of carbon dioxide reduction would be improved greatly when comparing to bulk metal Co [3]. Single-layer MoS_2 has been synthesized by mechanical exfoliation strategy and exhibits excellent performance in the field of gas sensors and phototransistors. It is worth noting that when bulk MoS_2 is exfoliated into monolayer, the type of band gap would change from indirect type to direct one [1, 4]. Another important example is that single-layered $Co(OH)_2$ can realize an ultrahigh energy density in all-solid-state asymmetric supercapacitor [5].

The emergence of these novel properties is the driving force for the rapid development of research in ultrathin 2D nanosheets, which has been on the forefront of scientific disciplines including chemistry, physics, materials, science, medicine and biology. However, the quest for methods of producing 2D materials with controlled thickness and lateral size has been always a challenging subject. This may be caused by the anisotropic crystal growth and strong chemical bonds in crystal structure. The common classification of crystalline structures according to the type of chemical bonds could be divided into van der Waals solids, layered ionic solids and nonlayered materials [1]. Every synthetic strategy has its own merits and demerits in preparing different kind of materials. Therefore, in this chapter, we concentrate on the different synthetic methods for synthesizing two-dimensional crystals. According to the principle of generating two-dimensional materials, we can divide the synthetic strategies into top-down and bottom-up strategies. The distinction between these two general classifications is based on the processes involved in the creation of the nanometer-sized structures [6]. In the bottom-up approach, nanoscale materials are constructed from atomic or molecular precursors that are allowed to react and grow in size or self-assemble into more complex structures. By contrast, the top-down approach carves nanoscale structures by controlled removal of materials from larger or bulk solids.

2. TOP-DOWN STRATEGY

In top-down nanofabrication, the crafting of ultrathin nanosheets is achieved through physical-based or through chemical-based process. The physical top-down approach employs the use of mechanical force or ultrasonic wave to exfoliate layered van der Waals solids into single- and few-layer 2D materials, whereas the chemical top-down strategy relies essentially on chemical reaction that brought about by ion exchange or by the application of heat and so on. In this section, a brief overview is presented of the top-down synthetic process for 2D nanostructures, including their advantages and limitations.

2.1. Micromechanical exfoliation

Micromechanical exfoliation has been used to exfoliate graphite into single-layer graphene by Novoselov and Geim for the first time [2]. Mechanical exfoliation is a straightforward method to obtain only one- or few-layer nanosheets, which well maintains the crystal structure and properties. Beside this, versatility and low cost of this method make it highly popular for synthesizing 2D materials and extremely convenient for fundamental research. Even so, this method is limited to the laboratory research and seems impossible to scale up for industrial production. Although large size and high quality of ultrathin nanosheets can be obtained by this way, this method is applicable only for layered van der Waals solids. The nanosheets of layered ionic solids and nonlayered materials cannot be obtained by this strategy. Additionally, several other factors (stoichiometry and stacking orders) play the key roles in successful fabrication of monolayer MX_2 nanostructures by mechanical exfoliation. Herein, we take graphene as an example to introduce this method.

Graphene is a monolayer sheet of carbon, showing only one atom thickness but extending indefinitely in two dimensions, which is the typical representative of 2D materials. Many astonishing properties have been discovered for graphene, which include better electrical and thermal conductivity, mechanical strength and optical purity. The procedure of micromechanical exfoliation is very simple. **Figure 1** illustrates the process of mechanical exfoliation [7].

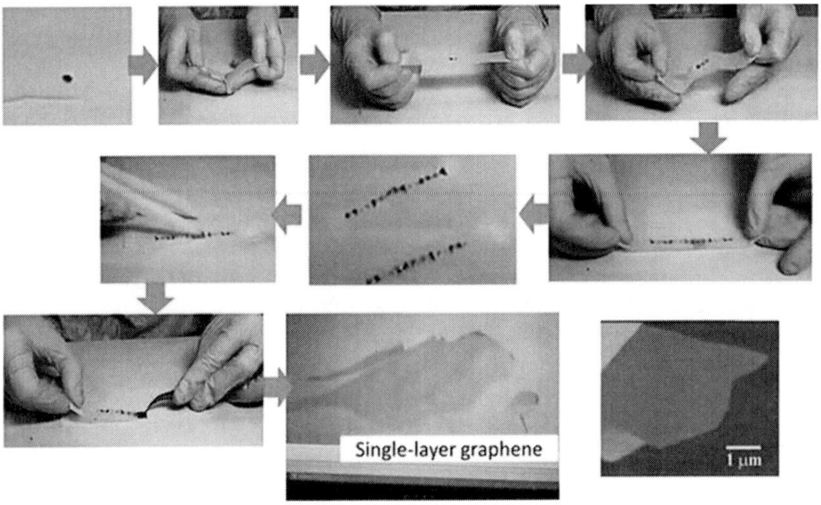

Figure 1. An illustrative procedure of the Scotch-tape–based micromechanical cleavage for graphene [7].

The exfoliation mechanics of this method are utilization of mechanical force to exfoliate graphite from Scotch tape. If one takes great pain to repeat this normal force over and over, the graphitic layer would become thinner and thinner, and eventually one can obtain single-layer graphene.

2.2. Ultrasonic exfoliation

Ultrasonic exfoliation is an effective strategy to delaminate van der Waals solids into single- or few-layer nanosheet. Compared to the mechanical exfoliation, this method is more effective and higher productive. The details for ultrasonic exfoliation process are shown in **Figure 2** [8]. As indicated in **Figure 2a**, sonication time and suitable solvents play the key roles in exfoliation. Suitable solvents are those with appropriate surface energies. In good solvents, the exfoliated nanosheets are stabilized against reaggregation. Otherwise, for "bad" solvents, reaggregation and sedimentation will occur. In recent reports, different types of organic solvents have been used as a dispersing medium for delaminating van der Waals solids (listed in **Table 1**) [8–14]. Although this method has many advantages, it is hard to get high-purity single-layer 2D material, which is a primary need for electronic applications. To make readers understand this method more in-depth, in the following, we will take black phosphorus (BP) as an example to introduce this strategy.

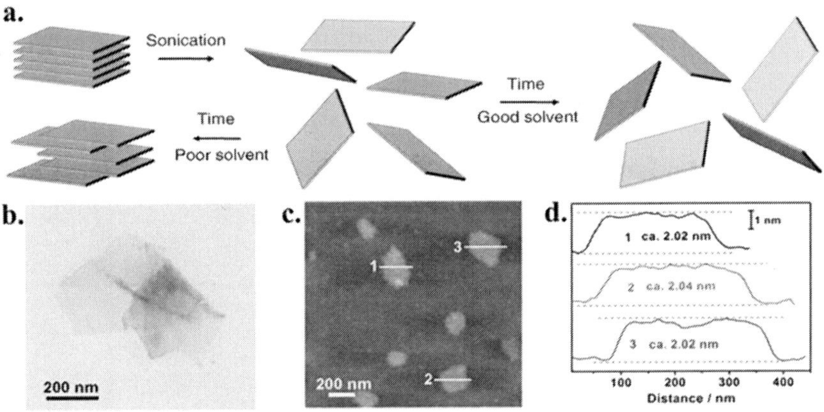

Figure 2. (a) An illustrative procedure of sonication-assisted exfoliation. The layered crystal is sonicated in a solvent, resulting in exfoliation and nanosheet formation [8]. (b) TEM image of ultrathin BP nanosheets. (c–d) AFM image and corresponding height image of ultrathin BP nanosheets [13].

Single-layer BP is a very promising two-dimensional material that can be the substitution of graphene due to its exceptional electronic properties. The direct band gap of BP can be tuned from 0.3 eV in the bulk to 1.5 eV in the monolayer. Recently, Xie's group successfully prepared pristine 2D black phosphorus through direct ultrasonic exfoliation in organic solvent [13]. Briefly, they dispersed 50 mg of bulk black phosphorus in 100 mL of distilled water, which was bubbled with argon to eliminate the dissolved oxygen molecules for avoiding the oxidation. Then, sonicating the mixture solution in ice water for 8 h (Note: keeping the system at a relatively low temperature is important). After ultrasonic treatment, the resultant dispersions were centrifuged at 1500 rpm for 10 min to remove the unexfoliated component and the supernatant was collected for further use.

Ultrathin nanosheet thus obtained show lateral size of about several hundred nanometers (see TEM image in **Figure 2b**), and the ultrathin thickness is indicated by the near transparency of the sheets. AFM image and the corresponding height distribution (**Figure 2c** and **2d**) show that the measured height is about 2.0 nm, consistent with the four individual black phosphorus layers. The black phosphorus nanosheets show excellent photodegradation of organic components such as DPBF and MO.

Table 1. The best 20 solvents for each material [9].

BN		MoS$_2$		WS$_2$	
Solvent	A/l (AU)[1] 300 nm	Solvent	A/l (AU) 670 nm	Solvent	A/l (AU) 630 nm
1 Cyclohexylpyrrolidone (CHP)	100	NVP	100	DMSO	100
2 *N*-dodecyl-pyrrolidone (N12P)	61	N8P	98	NVP	92
3 Benzyl benzoate	44	N12P	97	NMF	90
4 Isopropanol	44	CHP	88	N12P	84
5 *N*-Octyl-pyrrolidone (N8P)	44	NMP	80	DMEU	75
6 *N*-vinylpyrrolidone (NVP)	41	DMEU	73	DMF	73
7 Benzyl ether	40	DMSO	61	Benzyl benzoate	71
8 Dimethyl-imidazolidinone (DMEU)	36	DMF	54	CHP	69
9 Cyclohexanone	29	DMA	54	Cyclohexanone	63
10 Chlorobenzene	28	Benzaldehyde	51	Benzonitrile	59
11 Dimethylsulphoxide (DMSO)	27	Benzonitrile	47	N8P	59
12 Benzonitrile	26	Benzyl benzoate	46	Isopropanol	59
13 Chlorobenzene	25	NMF	41	DMA	57
14 Chloroform	23	Cyclohexanone	38	Benzylether	55
15 Bromobenzene	23	Isopropanol	32	Chlorobenzene	45
16 *N*-Methyl-pyrrolidone(NMP)	23	Qunioline	26	Methanol	45
17 *N*-Methylformamide (NMF)	21	Acetone	24	Formamide	40
18 Dimethylformamide (DMF)	18	Benzylether	23	Bromobenzene	29
19 Dimethylacetamide (DMA)	16	Cyclohexane	22	Quinoline	26
20 Formamide	9	Methanol	21	Acetone	17

[i] - [1]A is absorbance; l is the path length of the beam of light through the material sample. A/l at fixed wavelength was used to estimate the mass remaining in the supernatant.

2.3. Lithium-intercalated and exfoliation

Ultrasonic exfoliation is incapable of peeling off a single layer of 2D nanostructure; herein, lithium (Li) intercalation process is introduced for synthesizing single sheets. The scheme of lithium-intercalated exfoliation strategy is shown in **Figure 3a** [15]. The formation of Li_xXS_2 compound is a key step in lithium intercalation process, and this reaction can be tuned to control the yield of single layers [16, 17]. The yield of this strategy for obtaining single-layer transition metal dichalcogenide is nearly 100%, while some challenges still remain. The first one is that the experiment is carried out at high temperature for long durations. Also, the lithium intercalation must be controlled carefully to obtain single-layer nanosheets, while preventing the formation of metal nanoparticles and precipitation of Li_2S. To make readers understand this method more in-depth, in the following, we would take MoS_2 as an example to introduce this strategy.

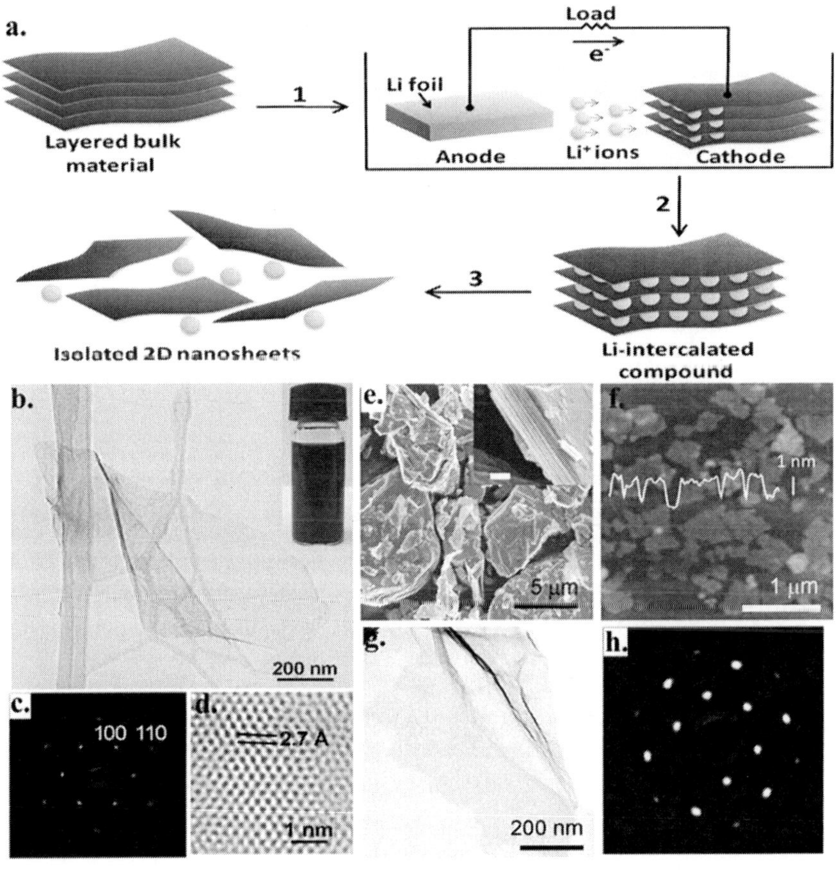

Figure 3. (a) Electrochemical lithiation process for the fabrication of 2D nanosheets from the layered bulk material. (b–d) Morphology characterization of MoS_2 exfoliated by electrochemical lithium-intercalated and exfoliation process [15]. (e–h) Morphology characterization of MoS_2 exfoliated by lithium-intercalated exfoliation process. [18].

2.3.1. Ultrathin nanosheets of mos$_2$ through electrochemical lithium-intercalated exfoliation process

To control the step of the formation of LixMoS_2 compound, Zhang's group developed a simple method—electrochemical lithiation process—to control Li-intercalated and subsequent exfoliation in water or ethanol [15]. Lithium intercalation of MoS_2 is performed in a test cell. The bulk MoS_2 is prepared as cathode. The electrochemical intercalation is performed using galvanostatic discharge at a current density of 0.05 mA. After the discharge process, the product is washed with acetone to remove any residual electrolyte, followed by exfoliation and ultrasonication in distilled water or ethanol. Eventually, the suspension is centrifuged and washed many times with water.

As shown in **Figure 3b–d**, high-quality and ultrathin MoS_2 nanosheets can be obtained by this simple method. The lateral size of nanosheets is about several hundred nanometers with a near transparency. Electron diffraction pattern of a flat area of the nanosheet (**Figure 3c**) and the corresponding high-resolution transmission electron microscope (HRTEM) image (**Figure 3d**) show that the hexagonal lattice structure is formed by this method, showing clearly a lattice spacing of 2.7 Å assigned to the (100) planes. This strategy can also been applied to prepare other single-layer semiconducting nanosheets, such as WS_2, TiS_2, TaS_2 or graphene.

2.3.2. Ultrathin nanosheets of mos$_2$ through lithium-intercalated exfoliation

Recently, Chhowalla's group successfully prepared monolayer MoS_2 through lithium-intercalated and exfoliation [18]. In detail, they immersed 3 g bulk MoS_2 crystals in certain concentration of butyllithium solution in hexane for 2 days in a flask filled with argon gas to obtain lithium intercalation compound. Exfoliation is achieved immediately after this process (within 30 min to avoid deintercalation) by ultra-sonicating Li_xMoS_2 in water for 1 h. The mixture is centrifuged several times to remove excess lithium in the form of LiOH and unexfoliated material.

Commercial MoS_2 powder (**Figure 3e**) was used to prepare highly monodisperse monolayer MoS_2nanosheets through Li intercalation and exfoliation. As shown in **Figure 3f** and **g**, the lateral size of product is 300~800 nm and the average thickness is about 1~1.2 nm, which is larger than the dimension of 0.65~0.7 nm reported for mechanically exfoliated MoS_2 monolayers. This discrepancy may be explained by surface corrugation due to the distortions and the presence of adsorbed or trapped molecules. The absence of any sheets below the thickness values and no evidence of step edges on the nanosheets surface suggest that they consist of monolayers. The selected area electron diffraction (SAED) patterns indicate hexagonal symmetry of the atomic arrangement and that individual sheets consist of a single-crystal domain (**Figure 3h**). All of the results undoubtedly confirm that MoS_2 with monolayer thickness has been successfully synthesized.

2.4. Ion-change exfoliation

Although the exfoliation strategy mentioned previously is convenient to delaminate van der Waals solids into ultrathin nanostructure, it is hard to exfoliate layered ionic solids such as $LiCoO_2$ or LDHs. This is because these ionic solids have strong ionic bonds in the layers. Ion-change exfoliation is a normal method to get this kind of 2D materials [19–23]. The scheme of ion-change exfoliation is shown in **Figure 4a** [19]. In the next section, we will take $LiCoO_2$ and (Co^{2+}-Co^{3+})-LDHs as examples to illuminate this method [23, 24].

2.4.1. LICOO$_2$

$LiCoO_2$ is a kind of cation-exchange layered metal oxides. Seong-Ju's group delaminated $LiCoO_2$ into monolayer recently [23]. The procedure of exfoliating could be described as follows: A proton-exchange reaction for the generation of layered $Li_xH_{1-x}CoO_2$ is carried out in an aqueous solution of HCl (1 M, 1 mg mL^{-1} of $LiCoO_2$) at room temperature for 3 days. During the proton-exchange reaction, HCl solution is replaced with fresh solution every day. Exfoliation of the layered cobalt oxide is achieved by the intercalation of tetrabutylammonium (TBA) cation into the layers of $Li_xH_{1-x}CoO_2$. After the reaction, incompletely exfoliated particles were separated from the colloidal suspension through centrifugation (6000 rpm for 10 min), resulting in a pure colloidal suspension. The powdered samples of completely exfoliated nanosheets are collected from the pure colloidal suspension through high-speed centrifugation at 17,000 rpm for 30 min.

AFM of the as-obtained sample provides direct evidence for the exfoliation of layered cobalt oxide nanosheets (**Figure 4b**). The thickness of $LiCoO_2$ nanosheets is about 1.2 nm, which is slightly thicker than the crystallographic thickness of individual cobalt oxide layers. The observation of Tyndall phenomenon from the pure suspension, characteristic of colloidal suspensions, provides strong evidence for the exfoliation of layered cobalt oxide nanosheets too (**Figure 4c**).

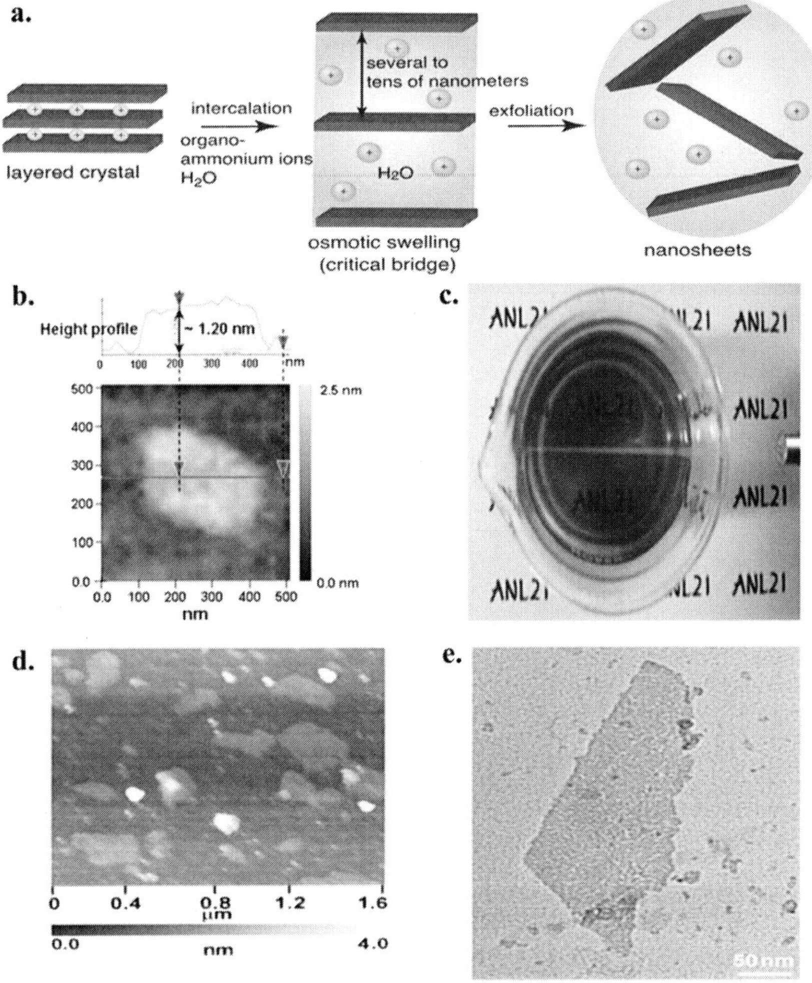

Figure 4. (a) Schematic illustration of the osmotic swelling to exfoliation process [19]. (b) AFM image and height profile of the exfoliated cobalt oxide adsorbed onto PEI-coated mica substrate. Photograph of the colloidal suspension of cobalt oxide. (c) The visible light is illuminated from the side of the beaker to demonstrate the Tyndall scattering effect [23]. Co^{2+}–Co^{3+} LDH nanosheets: (d) AFM image; (e) TEM image [24].

2.4.2. (CO^{2+}-CO^{3+})-LDHS

According to Sasaki's report [24], 0.5 g of the sample is dispersed into an aqueous solution (500 mL) that contains 2.5 M $NaClO_4$ and 2.5 mM HCl. This mixed solution is carefully degassed by purging with nitrogen gas for conversion into ClO−4ClO4−-LDH (addition of HCl is crucial for preventing carbonate, CO2−3CO32− contamination from air). After purging with nitrogen gas, the reaction vessel is tightly capped and shaken for 1 day at room temperature. The sample is filtered, washed with degassed water and air-dried. The obtained ClO−4ClO4− LDH (0.1 g) is mixed with formamide (100 mL) in a conical beaker. After an ultrasonic treatment for 30 min, a translucent colloidal suspension is obtained. The suspension is further centrifuged to remove possible nonexfoliated particles. AFM and TEM images shown in **Figure 4d** and **e** provided direct evidence for successfully exfoliation of the samples.

3. BOTTOM-UP STRATEGY

Top-down strategy is useful in preparing ultrathin and high-quality nanosheets with large lateral size. Even so, it has to be mentioned that all exfoliation methods as mentioned before are only suitable to those materials whose bulk crystals are layered [25]. The mass production by methods mentioned earlier is usually very low. When using bottom-up approach, nanoscale materials are constructed from atomic or molecular precursors that are suitable to react and grow in size, or self-assemble into more complex structures. Then, 2D materials can be enlarged easily [6]. In this section, a brief overview is presented for the bottom-up synthetic process to 2D nanostructures.

3.1. Wet chemical strategy

Wet chemical method is promising for preparing all types of 2D materials owing to its high-yield, low-cost and mass production. Wet chemical strategy is a big kind of synthesis method, which includes hydro/solvothermal synthesis and template synthesis. In this section, we introduce various types of wet-chemical synthesis strategies for preparing 2D nanomaterials.

3.1.1. Hydro/solvothermal synthesis

Hydro/solvothermal method is a common strategy used for the synthesis of inorganic materials. Advantages of hydro/solvothermal method over other types of crystal growth include low temperature (generally in a temperature range of 100–240°C) and convenience of adjusting reaction conditions. Many factors play the key roles in the synthesis of ultrathin nanostructure, which include temperature, reaction time, reactant ratio and so on. Many 2D materials have been prepared by this method, just like MoS_2, TiO_2, ZnO, Co_3O_4, MnO_2 and Rh [26–29]. In the following, we will take Li's work as an example to explain this strategy.

Ultrathin Rh nanosheets were synthesized successfully through solvothermal method by Li et al. recently [29]. For detailed steps, $Rh(acac)_3$ and PVP are dissolved in solvent mixed by benzyl alcohol and formaldehyde. The mixture is stirred vigorously for 1 h and then transferred to a Teflon-lined stainless steel autoclave. The autoclave is sealed and maintained at 180°C for 8 h and then cooled to room temperature. The resulted black product is precipitated with acetone (10 mL), separated in a centrifuge and washed three times with ethanol (10 mL) and finally dried under vacuum.

Detailed structure information for the synthesized atomically thick Rh nanosheets is indicated by TEM, HRTEM and AFM. TEM image clearly shows that ultrathin nanosheet has a dimension lateral size of about 500 nm, while the near transparency of the sheets indicates the ultrathin thickness (**Figure 5a, b**). As shown in **Figure 5c**, Rh nanosheets have a single crystalline nature of Rh nanosheets, which possesses a hexagonal structure, with a lattice parameter of ca. 2.6 Å, closer to the atomic distance of Rh (2.69 Å) in the (111) plane of the bulk Rh phase. AFM image of Rh NS shows that the sheet-like features are consistent with TEM images (see **Figure 5a, b**) Section analysis and height profile (**Figure 5d**) reveal that the height of Rh NS is about 4 Å. This is closer to the diameter of Rh atoms (the radius of Rh atomic is about 1.73 Å). All of these results undoubtedly confirm the successful synthesis of Rh nanosheets with an atomic thickness.

3.1.2. Template synthesis

A typical procedure of template method includes complicating morphology of template by growing crystal confined in specific dimension and then removing template through high temperature or adjusting pH. By this way, nearly all types of nanostructure including quantum dot (0D), nanowire (1D) and nanosheets (2D) can be prepared. The suitable template and condition of removing template are important. Many nonlayered structure 2D nanomaterials can be synthesized using the as-prepared 2D nanomaterials as the template, such as Fe_2O_3, Au, $CuInS_2$ and so on [30–32]. Herein, we take α-Fe_2O_3 as an example to illuminate this method.

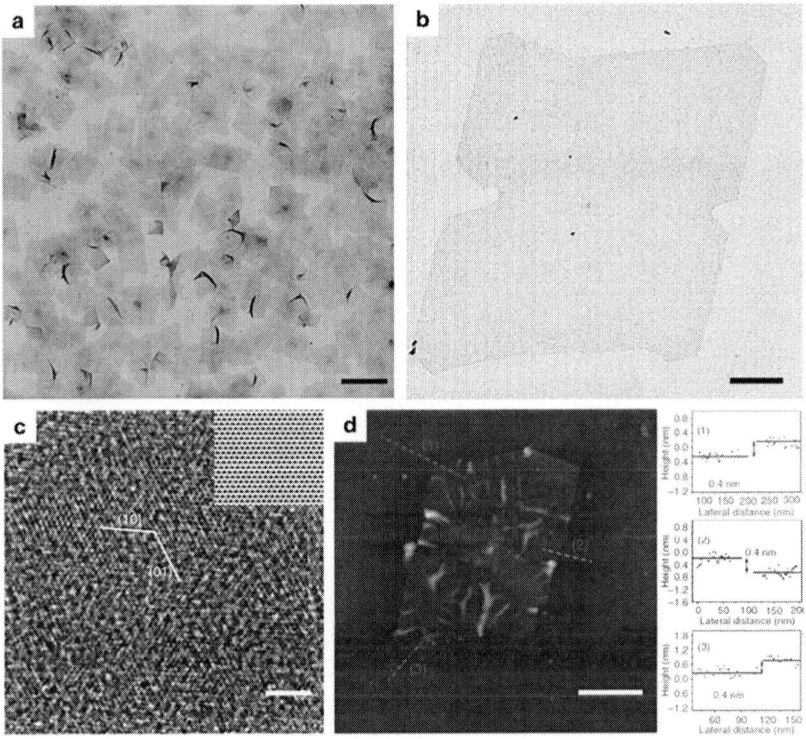

Figure 5. (a) Low-magnification TEM image of PVP-capped Rh nanosheet, (b) high-magnification TEM image of PVP-capped Rh NSs, (c) aberration-corrected microscopy image of PVP-capped Rh NS (inset, the corresponding filtered image using crystallographic average method to improve signal-to-noise ratio), (d) AFM image and the corresponding height profiles of a bare Rh NS [29].

3.1.2.1. Half-unit-cell thickness of alpha-fe_2o_3 semiconductor nanosheets

Recently, Wei's group successfully synthesized free-standing half-unit-cell α-Fe_2O_3 nanosheets using CuO nanoplate as the template [30]. The layered iron hydroxide nanosheets are first prepared on CuO template surface by the mutually promoted slow interfacial reactions of Fe^{2+} hydrolysis and CuO-etching process at a low temperature of 25°C for a whole day, which is beneficial for the synthesis of ultrathin nanosheets. When CuO template is etched out, a heat treatment is carried out for the dehydrogenation of Fe hydroxide nanosheets, which leads to the formation of the stable and free-standing α-Fe_2O_3 nanosheets.

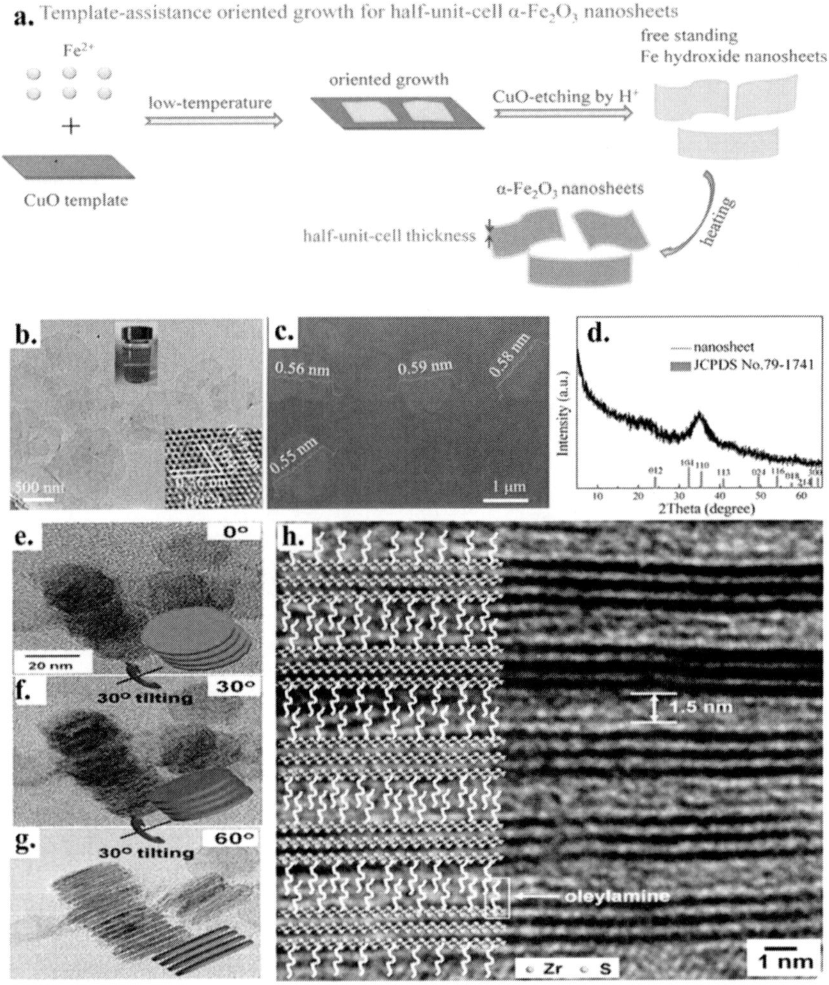

Figure 6. (a) Schematic of synthesis strategy of α-Fe_2O_3 nanosheets. (b) TEM image of the α-Fe_2O_3nanosheets. The insets show the HRTEM image and the Tyndall effect of the α-Fe_2O_3 nanosheets. (c) AFM image and (d) XRD pattern of the α-Fe_2O_3 nanosheets [30]. (e–g) TEM images of the assembled ultrathin ZrS_2 nanodiscs obtained by rotating TEM holder under different angle. (h) HRTEM side-view image of multiply stacked 1.6-nm-thick ZrS_2 nanodiscs [33].

AFM and TEM image provide direct evidence for successful exfoliation of the sample. As shown in **Figure 6b**, **c**, the lateral size of α-Fe_2O_3 nanosheets is up to about 1 μm, and thickness is about 0.55–0.59 nm. The XRD pattern of the α-Fe_2O_3 shown in **Figure 6d** exhibits only a broad and weak diffraction peak corresponding to the (110) orientation plane of α-Fe_2O_3. It is interesting to note that ultrathin α-Fe_2O_3 nanosheet shows an intrinsic ferromagnetism of 0.6 μ_B/atom at 100 K and remains ferromagnetism at room temperature.

3.1.2.2. Ultrathin zirconium disulfide nanodiscs

Except for hard template, soft colloidal templated synthesis is used to prepare two-dimensional2D materials frequently. The long-chain oleylamine and/or oleic acid surfactants are often used as the soft colloidal templates for directing the crystal growth. Cheon et al. used this method to synthesize ultrathin ZrS_2 nanodiscs [36]. Intermediate lamellar complexes composed of 2D arrays of $ZrCl_4$ and alkyl amine are first obtained, in which alkyl amine serves as the soft colloidal template. Then, CS_2 is injected into the mixture aforementioned to form ultrathin ZrS_2 nanodiscs dispersed in solution.

As indicated in **Figure 6e–h**, the resultant ultrathin ZrS_2 nanodiscs possess radius of ~15 nm and thickness of 0.5 nm. The spacing between the discs is ~1.5 nm (**Figure 6h**), which corresponds to the length of the oleylamine surfactant layers. When compared to bulk ZrS_2, ultrathin ZrS_2 discs show the unique nanoscale size effects, enhanced discharge capacity by 230% and greatly improved stability.

3.2. Microwave-assisted method

In recent years, microwave-assisted chemical synthesis strategy has become a well-established technique to promote and enhance chemical reactions. The main advantages of this method are represented by much shorter reaction time (generally in only a few minutes) and higher energy efficiency when comparing to other conventional strategies. Due to these advantages, some 2D nanomaterials can be prepared by this way conveniently, such as SnO_2, α-$Ni(OH)_2$, $K_{0.17}MnO_2$ and CuSe [33–37].

To understand it more clearly, let us take α-$Ni(OH)_2$ as an example (**Figure 7**) [36]: firstly, precursors were prepared by starting materials of $Ni(NO_3)_2 \cdot 6H_2O$, urea, deionized water and ethylene glycol at given proportions. Then, the resulting solution is transferred into a homemade round-bottomed flask and treated under microwave irradiation in a microwave reactor at 700 W for several minutes. Finally, the green powder is obtained by centrifugation and washed several times with distilled water and absolute ethanol. After that, the powder was dried in vacuum at 80°C for 12 h. Detailed structure information for the synthesized $Ni(OH)_2$ nanosheets is unraveled by FESEM and TEM images in **Figure 7b–d**. In comparison with traditional wet-chemical syntheses, the microwave-assisted liquid-phase growth can shorten the reaction time to less than 20 min. Particularly, the ultrathin α-$Ni(OH)_2$nanosheets exhibit a maximum specific capacitance of 4172.5 Fg^{-1} at a current density of 1 Ag^{-1}.

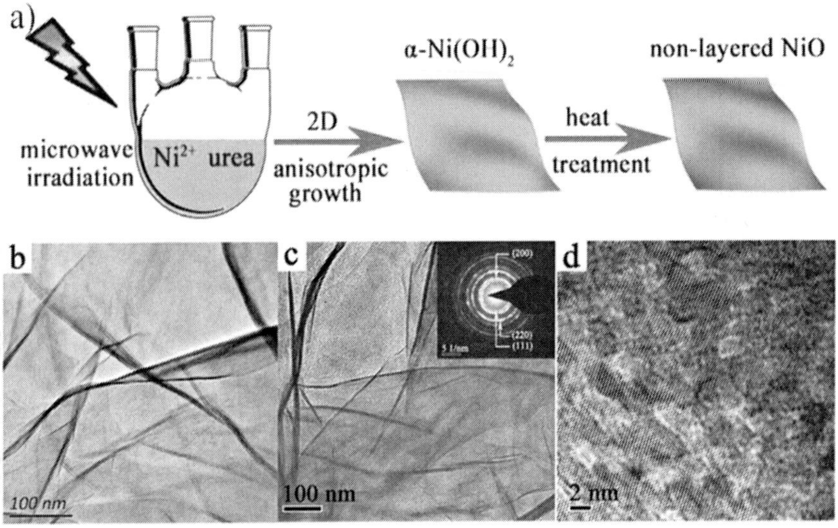

Figure 7. (a) Schematic illustration for the synthesis of nanosheets. (b) High magnification FESEM images of α-$Ni(OH)_2$ nanosheets, (c) TEM image (the inset showing SAED pattern) and (d) a planar HRTEM image [36].

3.3. Topochemical transformation

Topological conversion is a strategy in which the product's morphology is inherited from their precursor through nucleation and growth inside the precursors. The key to success is the degree of lattice match between precursor and their product. From the viewpoint of anisotropy of layered compounds, it is easier to obtain 2D nanostructure of hydroxide rather than oxides. This method is applicable of preparing nonlayer 2D nanostructure oxides, such as Co_3O_4, CeO_2 and δ-FeOOH [38–40]. In the following, we will take Co_3O_4 and Ni as examples to illuminate this method [38, 41].

3.3.1. Atomically thick co_3o_4 nanosheets through topotactic oxidation transformation

The schematic illustration of the preparation of ultrathin Co_3O_4 nanosheets is shown in **Figure 8**. As illustrated in **Figure 8a**, the thickness of (001)-oriented few layered α-$Co(OH)_2$ is closer to the thickness of (111)-oriented

Co_3O_4 nanosheets. More specifically, the discrepancies between α-$Co(OH)_2$ {100} (d=2.76 Å) and Co_3O_4 {220} (d=2.84 Å) are calculated to be below 3%. Moreover, Co^{2+} atomic arrangement in α-$Co(OH)_2$ (001) plane and Co_3O_4 (111) plane is very closer (**Figure 8b**). These features enable atomically thick Co_3O_4 nanosheets to preserve the thickness of the precursor possible.

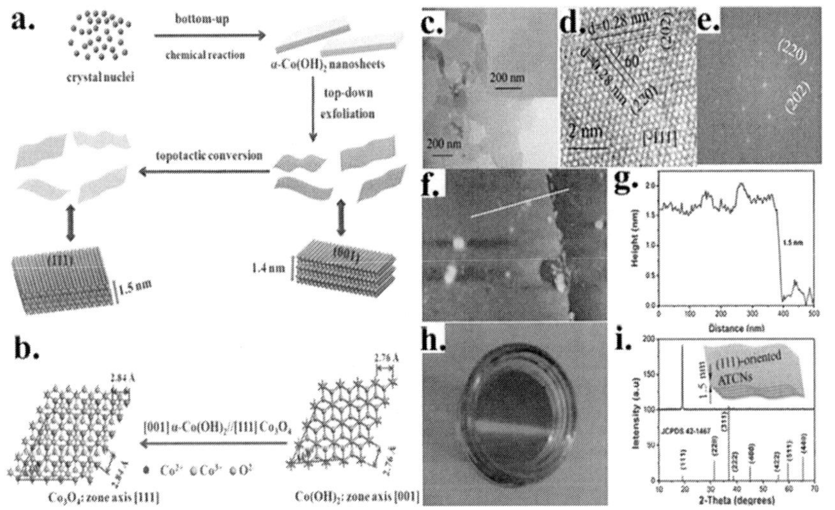

Figure 8. (a) Schematic illustration of the preparation of ATCNs. (b) Schematic model of the top-view illustrating the topochemical transformation from α-$Co(OH)_2$ (001) plane to Co_3O_4 (111) plane. (c–e) TEM image, HRTEM image and the corresponding fast Fourier transform pattern of ATCNs. (f and g) Tapping model AFM image and corresponding height profile of as-obtained nanosheets deposited on a silicon substrate. (h) Photograph of a colloidal suspension of the ATCNs displaying the Tyndall effect. (i) XRD pattern of ATCNs-based film displaying highly preferred [111] orientation [38].

Atomically thick Co_3O_4 nanosheets were prepared by this method, as confirmed by TEM, AMF and XRD. From TEM image, one could clearly see ultrathin nanosheets with lateral size of about 400 nm with the near transparency of the sheets (**Figure 8c**). HRTEM image, and the corresponding fast Fourier transform (FFT) pattern in **Figure 8d** and **e** for the as-obtained product depicts its [111] preferential orientation. The thickness of the as-synthesized Co_3O_4 nanosheets is also evaluated by tapping-mode atomic force microscopy (AFM). AFM image and the corresponding height distribution (**Figure 8f, g**) show that the measured height is about 1.5 nm. Microscopically, the colloidal suspension of the product displays the Tyndall phenomenon (**Figure 8h**), so that the formation of homogeneous ultrathin nanosheets in ethanol can be inferred. As shown in **Figure 8i**, XRD pattern of the atomically thick nanosheet-based films can be readily indexed to be pure Co_3O_4 (JCPDS no. 42-1467) and also displays the highly preferred [111] orientation, which is corresponding to the result of HRTEM. All of the results undoubtedly confirm that Co_3O_4 nanosheets with ultrathin thickness have been successfully synthesized. The oriented growth of ATCNs is attributed to α-$Co(OH)_2$-to-Co_3O_4topochemical conversion with the relationship of [001] α-$Co(OH)_2$ and [111] Co_3O_4.

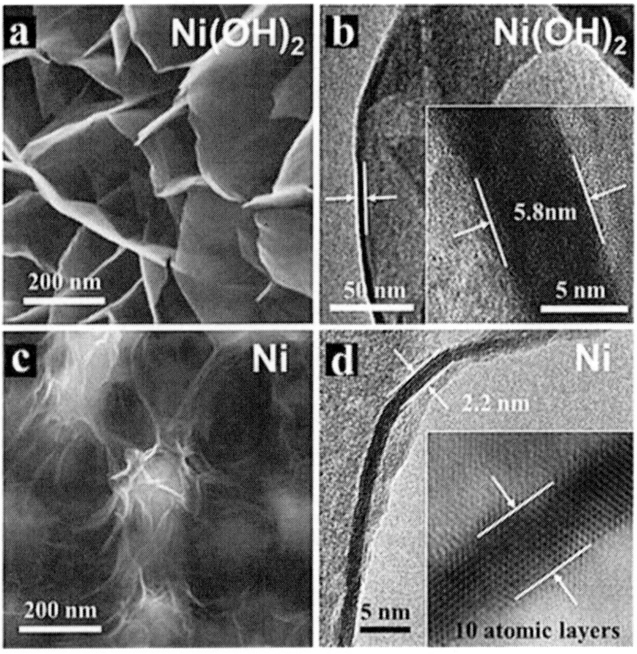

Figure 9. (a and c) SEM images of the as-synthesized $Ni(OH)_2$ nanowall array and Ni-NSA. (b and d) HRTEM image of a scrolled $Ni(OH)_2$ nanowall and a vertically laid Ni nanosheet. The inset shows the thickness of $Ni(OH)_2$ nanowall and Ni nanosheet [41].

3.3.2. Ultrathin nickel nanosheets through topotactic reduction transformation

Except for topotactic oxidation, toporeduction is also frequently used to prepare nonlayered materials. Sun [41] used this method and synthesized ultrathin nickel nanosheets array recently. $Ni(OH)_2$nanosheet array on a metal substrate is prepared first. And then, ethylene glycol is used as the reduction agent to get Ni nanosheets. The slow conversion kinetics keeps the ultrathin nanosheet morphology and atomic thickness.

Figure 9 reveals the morphology inherence between Ni and $Ni(OH)_2$ nanosheet array. **Figure 9a, b** shows SEM and HRTEM of $Ni(OH)_2$ sheet array. It is seen that the thickness of $Ni(OH)_2$ is about 5.8 nm (10–12 layers), whereas the thickness of Ni nanosheet shown in **Figure 9c, d** is about 2.2 nm (10 atomic layers). This provides strong evidence that Ni nanosheets are obtained by toporeduction transformation.

3.4. Chemical vapor deposition (cvd)

Recently, CVD technique has shown promise to generate high-quality TMD layers with a scalable size, controllable thickness and excellent electronic properties. CVD is a high-temperature chemical synthesis process by which a desired material is deposited on substrates. CVD processes have been extensively studied for synthesizing thin film coatings of a wide range of materials, including metals, semiconductors, and insulators. The scheme of CVD method is shown in **Figure 10a** [1]. The precursor vapor is introduced from outside or generated inside the tube furnace. The main advantages of CVD synthesis process are represented by the accesses to high quality, high purity 2D nanomaterials with controlled properties, which allow one to control the morphology, crystallinity and defects of 2D nanostructures by tuning the process parameters [42]. Due to these advantages, some 2D nanomaterals have been prepared by this way conveniently, such as graphene, MoS_2, h-BN [43–49]. CVD method is also another annealing strategy for the growth of high-quality single-crystalline 2D sheet on substrates. Even so, CVD suffers from the requirement of high temperature, high vacuum and specific substrates. In the following, we will take h-BN and MoS_2 as an example to elaborate this strategy.

Hexagonal BN is called "white graphite" due to its similar lamellar structure to graphite. h-BN possesses many unique properties such as chemical stability, strong mechanical strength, high-thermal conductivity and low dielectric constant. What is more, the band gap of h-BN can be tuned with thickness. To produce BN precursor, the first heating zone is ramped up to T_1=60~90°C with a heating belt. Turn on the on-off valve between the first and second heating zone only during CVD process, one can prevent the undesired growth. The second heating zone is heated up to T_2 = 1000°C for typically 10~90 min under 10 sccm hydrogen atmosphere at a pressure of 350 mTorr after mounting the copper foil. Copper foil (25 μm, Alfa Aesar) is used as the metal catalyst substrate. Before the growth of h-BN, copper foil is annealed at 1000°C for 20 min under 10 sccm H_2 to grow the copper grain and to obtain a smooth surface. AFM of as-obtained sample provides direct evidence for single-layer h-BN synthesized by chemical vapor deposition method (**Figure 10b**). AFM image and the corresponding height distribution inset in **Figure 10b** show that the measured height is about 0.42 nm, consistent with monolayer thickness (c-axis spacing for h-BN is ~0.32 nm). In addition, TEM image of monolayer h-BN clearly demonstrates that the number of layer of h-BN is one (**Figure 10c**). Moreover, for a complete h-BN layer (**Figure 10d**), the wrinkles of h-BN film (indicated by a yellow arrow) under scanning electron microscopy (SEM) can be clearly seen. These wrinkles are characteristic of h-BN and graphene films due to the negative thermal expansion coefficients of h-BN and graphene. All of the results undoubtedly confirm that h-BN with monolayer thickness have been successfully synthesized.

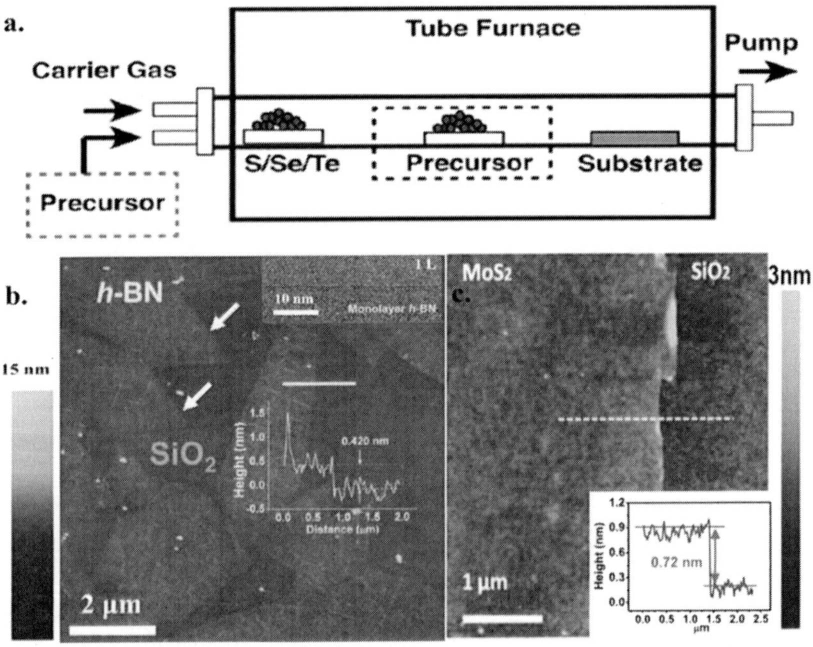

Figure 10. (a) Schematic illustration of experimental setup for the 2D materials synthesis by a noncatalytic vapor deposition process. The precursor vapor can be introduced from outside or generated inside the tube furnace [1]. (b) AFM images of the initial flakes of h-BN. White arrows point to the sharp edges of these flakes. Inset: TEM image of the folded edge of a monolayer h-BN [43]. (c) AFM image of a monolayer MoS_2 film on a SiO_2/Si substrate. Inset: corresponding height profile of as-obtained nanosheets [46].

Synthesis of transition metal dichalcogenides (TMDs) using CVD is the cutting edge research area in recent years. Lee [46] synthesized single-layer MoS_2 by this method recently. As shown in **Figure 10c**, ultrathin MoS_2 nanosheets with smooth surface are observed with AFM. The cross-sectional height in inset of **Figure 10c** reveals that the thickness of MoS_2 film is ~0.72 nm, which corresponds to a monolayer MoS_2.

4. PERSPECTIVE AND CONCLUSIONS

With the advantages of very high surface-to-volume ratios and special electronic properties in two-dimensional materials, ultrathin nanosheets are very promising in catalysis, supercapacitors, photoconductive materials, batteries and magneto-optical components. Therefore, it is imperative to exploit more technologically advanced synthesis strategies to obtain high quality, large size, ultrathin two-dimensional materials. At present, thinner, larger size, high productive and quality nanosheets are still the chief purpose for scientists. Usually, the thinner nanosheets give a higher catalytic performance and other special properties. Besides these, controllable thickness, architectures assembled by nanosheets and composite nanostructure, such as quantum-dot/nanosheets, sandwich structure and modification on two-dimensional nanosheets, also provide new opportunities for material science.

In summary, synthetic strategies, such as top-down and bottom-up method, have been employed for synthesizing two-dimensional crystals. The top-down strategy includes micromechanical exfoliation, ultrasonic exfoliation, lithium-intercalated and exfoliation and ion-change exfoliation. Using top-down strategies, one is able to obtain high-quality and large size two-dimensional crystal conveniently. Even so, this method is only appropriate for those materials whose bulk crystals are layered. Comparatively, bottom-up strategy could overcome this shortage. Bottom-up strategy is composed of wet chemical method, microwave-assisted chemical, topological conversion strategy and chemical vapor deposition method. All these strategies may present different features and questions for preparing two-dimensional materials. In the future, scientists should take more and more efforts to develop new synthetic technologies for high-quality 2D material and the relevant applications.

REFERENCES

1. Joshua EG. Progress, challenges, and opportunities in two-dimensional materials beyond graphene. ACS Nano. 2014;7(4): 2898–2962. DOI: 10.1021/nn400280c
2. Novoselov KS. Electric field effect in atomically electric field effect in atomically. Science. 2004;306(5696): 666–669. DOI: 10.1126/science.1102896
3. Xie Y. Partially oxidized atomic cobalt layers for carbon dioxide electroreduction to liquid fuel. Nature. 2016;529(7484): 68–71. DOI: 10.1038/nature16455
4. Zhang H. Single-layer MoS2 phototransistors. ACS Nano. 2012;6(1): 74–80. DOI: 10.1021/nn2024557
5. Gao S. Ultrahigh energy density realized by a single-layer beta-Co(OH)2 all-solid-state asymmetric supercapacitor. Angew Chem Int Ed Engl. 2014;53(47): 12789–12793. DOI: 10.1002/anie.201407836
6. Han MY. Chemical routes to top-down nanofabrication. Chem Soc Rev. 2013; 42(14):6006–6018. DOI: 10.1039/c3cs60113g
7. Shen ZG. A review on mechanical exfoliation for the scalable production of graphene. J. Mater. Chem. A. 2015;3(22):11700–11715. DOI: 10.1039/c5ta00252d
8. Coleman JN. Liquid exfoliation of layered materials. Science. 2013;334(6052): 72–75. DOI: 10.1126/science.1208891
9. Coleman JN. Two-dimensional nanosheets produced by liquid exfoliation of layered materials. Science. 2011;331(6017): 568–571. DOI: 10.1126/science.1194975
10. Coleman JN. Large-scale exfoliation of inorganic layered compounds in aqueous surfactant solutions. Adv Mater. 2011;23(34): 3944–3948. DOI: 10.1002/ adma. 201102584
11. Zhang HL. A Mixed-solvent strategy for efficient exfoliation of inorganic graphene analogues. Angew Chem Int Ed Engl. 2011;50(46): 10839–10842. DOI: 10.1002/ anie.201105364
12. Coleman JN. Preparation of high concentration dispersions of exfoliated MoS2 with increased flake size. Chem Mater. 2012;24(12): 2414–2421. DOI: 10.1021/Cm301515z
13. Xie Y. Ultrathin black phosphorus nanosheets for efficient singlet oxygen generation. J Am Chem Soc. 2015;137(35): 11376–11382. DOI: 10.1021/jacs.5b06025
14. Hersam MC. Solvent exfoliation of electronic-grade, two-dimensional black phosphorus. ACS Nano. 2015;9(4):3596–3604. DOI: 10.1021/acsnano.5b01143
15. Zhang H. Single-layer semiconducting nanosheets: high-yield preparation and device fabrication. Angewandte Chemie. 2011;50(47): 11093–11097. DOI: 10.1002/ anie.201106004
16. Zhang H. The chemistry of two-dimensional layered transition metal dichalcogenide nanosheets. Nat Chem. 2013;5(4): 263–275. DOI: 10.1038/nchem.1589
17. Chen M. Chemically exfoliated ReS2 nanosheets. Nanoscale. 2014;6(21): 12458–12462. DOI: 10.1039/c4nr03740e
18. Chhowalla M. Photoluminescence from chemically exfoliated MoS2. Nano Lett. 2011;11(12): 5111–5116. DOI: 10.1021/nl201874w
19. Sasaki T. Two-dimensional oxide and hydroxide nanosheets: controllable high-quality exfoliation, molecular assembly, and exploration of functionality. Acc Chem Res. 2014;48(1): 136–143. DOI: 10.1021/ar500311w
20. Hwang SJ. Composition-tailored 2 D Mn(1-x)Ru(x)O(2) nanosheets and their reassembled nanocomposites: improvement of electrode performance upon Ru substitution. Chem Eur J. 2014;20(17): 5132–5140. DOI: 10.1002/chem.201304009
21. Sasaki T. Exfoliated oxide nanosheets: new solution to nanoelectronics. J Mater Chem. 2009;19(17): 2503. DOI: 10.1039/b820160a
22. Sasaki T. Osmotic swelling to exfoliation. Exceptionally high degrees of hydration of a layered titanate. J Am Chem Soc. 1998;120(19): 4682–4689. DOI: 10.1021/Ja974262l

23. Hwang SJ. Soft-chemical exfoliation route to layered cobalt oxide monolayers and its application for film deposition and nanoparticle synthesis. Chem Eur J. 2009;15(41): 10752–10761. DOI: 10.1002/chem.200901590
24. Sasaki T. Topochemical synthesis of monometallic (Co2+-Co3+) layered double hydroxide and its exfoliation into positively charged Co(OH)2 nanosheets. Angew Chem Int Ed Engl. 2008;47(1): 86–89. DOI: 10.1002/anie.200703941
25. Zhang H. Wet-chemical synthesis and applications of non-layer structured two-dimensional nanomaterials. Nat Commun. 2015;6:7873. DOI: 10.1038/ncomms8873
26. Xue JM. Ultrasmall Fe3O4 nanoparticle/MoS2 nanosheet composites with superior performances for lithium ion batteries. Small. 2014;10(8): 1536–1543. DOI: 10.1002/smll.201302879
27. Bai H. Large-scale, ultrathin and (001) facet exposed TiO2 nanosheet superstructures and their applications in photocatalysis. J Mater Chem A. 2014;2(7): 2040. DOI: 10.1039/c3ta14343k
28. Dou SX. Generalized self-assembly of scalable two-dimensional transition metal oxide nanosheets. Nat Commun. 2014;5: 3813. DOI: 10.1038/ncomms4813
29. Li J. Ultrathin rhodium nanosheets. Nat Commun. 2014;5: 3039. DOI: 10.1038/ncomms4093
30. Wei S. Half-unit-cell alpha-Fe2O3 semiconductor nanosheets with intrinsic and robust ferromagnetism. J Am Chem Soc. 2014;136(29): 10393–10398. DOI: 10.1021/ja504088n
31. Zhang H. Synthesis of hexagonal close-packed gold nanostructures. Nat Commun. 2011;2: 292. DOI: 10.1038/ncomms1291
32. Xie Y. CuInSe2 ultrathin nanoplatelets: novel self-sacrificial template-directed synthesis and application for flexible photodetectors. Chem Commun. 2012;48(73): 9162–9164. DOI: 10.1039/c2cc34727j
33. Xie Y. Ultra-rapid microwave-assisted synthesis of layered ultrathin birnessite K0.17MnO2 nanosheets for efficient energy storage. J Mater Chem A. 2013;1(28): 8154–8159. DOI: 10.1039/C3ta11194f
34. Cheon J. Ultrathin zirconium disulfide nanodiscs. J Am Chem Soc. 2011;133(20): 7636–7639. DOI: 10.1021/ja200400n
35. Cao C. Microwave-assisted and gram-scale synthesis of ultrathin SnO2 nanosheets with enhanced lithium storage properties. ACS Appl Mater Interfaces. 2015;7(4):2745–2753. DOI: 10.1021/am507826d
36. Li YD. Ultrathin nickel hydroxide and oxide nanosheets: synthesis, characterizations and excellent supercapacitor performances. Sci Rep. 2014;4:5787. DOI: 10.1038/srep05787
37. Pan GB. Facile microwave-assisted synthesis of Klockmannite CuSe nanosheets and their exceptional electrical properties. Sci Rep. 2014;4:5998. DOI: 10.1038/srep05998
38. Xie Y. Topochemical transformation route to atomically thick Co3O4 nanosheets realizing enhanced lithium storage performance. Nanoscale. 2013;5(12):5241–5246. DOI: 10.1039/c3nr01178j
39. Sun Y. Pits confined in ultrathin cerium (IV) oxide for studying catalytic centers in carbon monoxide oxidation. Nat Commun. 2013;4:2899. DOI: 10.1038/ncomms3899
40. Chen PZ. Ultrathin nanosheets of feroxyhyte: a new two-dimensional material with robust ferromagnetic behavior. Chem Sci. 2014;5(6):2251. DOI: 10.1039/c3sc53303d
41. Sun X. Single-crystalline ultrathin nickel nanosheets array from in situ topotactic reduction for active and stable electrocatalysis. Angew Chem Int Ed Engl. 2016;55(2):693–697. DOI: 10.1002/anie.201509616
42. Li LJ. Recent advances in controlled synthesis of two-dimensional transition metal dichalcogenides via vapour deposition techniques. Chem Soc Rev. 2014;DOI: 10.1039/c4cs00256c
43. Kong J. Synthesis of monolayer hexagonal boron nitride on Cu foil using chemical vapor deposition. Nano Lett. 2012;12(1):161–166. DOI: 10.1021/nl203249a
44. Rodney SR. Large-area synthesis of high-quality and uniform graphene films on copper foils. Science. 2009;324(5932):1312–1314. DOI: 10.1126/science.117124
45. Kong J. Role of the seeding promoter in MoS2 growth by chemical vapor deposition. ACS Nano. 2014;14(2):464–472. DOI: 10.1021/nl4033704

46. Lee YH. Synthesis of large-area MoS2 atomic layers with chemical vapor deposition. Adv Mater. 2012;24(17):2320–2325. DOI: 10.1002/adma.201104798
47. Nayak PK. Robust room temperature valley polarization in monolayer and bilayer WS2. Nanoscale. 2016;8:6035–6042. DOI: 10.1039/C5NR08395H
48. Shaw JC. Chemical vapor deposition growth of monolayer MoSe2 nanosheets. Nano Res. 2014;7(4):511–517. DOI: 10.1007/s12274-014-0417-z
49. Huang JK. Large-area synthesis of highly crystalline WSe2 monolayers and device applications. ACS Nano. 2014;8(1):923–930. DOI: 10.1021/nn405719x

CHAPTER 10

Radiation Effects in Optical Materials and Photonic Devices

Dan Sporea[1*] and Adelina Sporea[1]

ABSTRACT

The chapter continues previous reviews on radiation effects in optical fibers and on the use of optical fibers/optical fiber sensors in radiation monitoring, published by InTech in 2010 and 2012, by referring to radiation effects in optical materials, with an emphasis on those operating from visible to mid-IR, and on some photonic devices such as optical fibers for amplifiers, fiber Bragg gratings and long period gratings. The focus is on optical materials and fiber-based devices designed for both terrestrial and spaceborne applications. For the presented subjects, an overview of available data on X-rays or gamma rays, electron beams, alpha particles, neutrons, and protons effects is provided. In addition, comments on dose rate, dose, and/or temperature effects on materials and devices degradation under irradiation are mentioned, where appropriate. The optical materials and photonic devices reliability under ionizing radiation exposure is discussed as well, as the opportunities to use them in developing radiation sensors or dosimeters. The chapter includes an extensive bibliography and references to last published results in the field. Novel proposed applications of photonic devices in charged particle beam diagnostics, quasi-distributed radiation field mapping and the evaluation of radiation effects in materials for mid-IR spectroscopy are briefly introduced to the reader.

Keywords: radiation effects, optical materials, optical fibers, fiber Bragg grating, long period grating

1. INTRODUCTION

The goal of this chapter is to continue previous reviews on radiation effects in optical fibers [1] and on the use of optical fibers and optical fiber sensors in radiation related measurements [2], by referring to radiation effects in optical materials and some photonic devices. According to this vision, the chapter is organized to cover the interaction of ionizing radiation with some optical materials and optical fibers, followed by a reference to radiation effects on some photonic devices based on optical fibers. The discussion addresses radiation effects produced by both energetic photons and charged particles, as appropriate [3]. In this context, an overview of some recently published results in the field is included, with a focus on original authors' contributions.

The terrestrial radiation environments where optical and photonics components can be found include, but are not limited to, high energy physics experiments, nuclear power plants [3], fusion installations as the International Thermonuclear Experimental Reactor – ITER, or the Laser Mégajoule – LMJ [4–8], high power laser installations [9], nuclear waste repositories [10], high energy physics [11, 12], medical equipments for diagnostics or treatment [13]. On the other side, applications of optical components or photonic devices can be found in spaceborne instrumentation [14–16]. These environments involve various types of ionizing radiations, depending on the application considered: X-rays or gamma rays, electron beams, alpha particles, neutrons, protons, and Bremsstrahlung [3, 14, 17].

2. RADIATION EFFECTS IN OPTICAL MATERIALS

2.1. Optical materials

Extensive research was involved in the elucidation of defects formation in glasses, as investigations were performed in glasses with various compositions under ionizing and non-ionizing radiation exposure. The studies were focused either on the materials degradation upon irradiation or on the possible use of such materials in radiation dosimetry [18, 19]. The radiation induced changes depend on the glass composition, total dose, dose rate, temperature and humidity during exposure, and post irradiation heating of the sample [19]. The operation of a glass-based dosimeter can be decided as function of radiation sensitivity, linearity of the response, stability of the radiation produced effect, and possibility to re-use the material.

Besides glass-based optical materials, radiation hardening tests were performed on various other optical materials. More than 20 years ago radiation induced defects were studied in BaF_2 crystals by exposing them to gamma rays (from 10 Gy to 47 kGy) and observing the optical attenuation recovery (between 300 and 700 nm) under UV radiation and the scintillating signal [20]. Samples from different manufacturers exhibited radiation induced attenuation (RIA) saturation starting from 10^2 Gy. Crystal impurities and defects are the primary source of the optical attenuation increase in the 190–250 nm and 500–600 nm spectral bands induced by gamma rays [21]. BaF_2 and LaF_3 were subjected to Ne and U ions (at energies from 1.4 to 13.3 MeV/u) bombardment, and their degradation was investigated by scanning force microscopy (SFM), optical spectroscopy and surface profilometry. RIA for BaF_2 shown an increase at λ = 240, 420, 550 and 750 nm, while LiF_3 crystals remained almost unchanged spectrally. Surface topography studies indicated the presence of hillock in the irradiated zone [22].

Neutron irradiation was done on $Y_3Al_5O_{12}$, CaF_2 and LiF and RIA was monitored for UV-visible spectra. For $Y_3Al_5O_{12}$ samples an increase of the optical attenuation was present for wavelength lower than 350–400 nm. CaF_2 and LiF single crystals degrade their optical transmission after neutron exposure mostly in the 400–500 nm region. When heated after the irradiation RIA for the three crystals recovers according to different patterns [23].

Gamma irradiation (dose rate 110 Gy/h, total doses of 500 Gy, 2 kGy, 8 kGy, 20 kGy) was conducted on CaF_2, Fused Silica and Clearceram in order to evaluate their qualification for space applications. RIA modification was measured over the 350–800 nm spectral interval at normal incidence. The optical investigations were completed by ellipsometry tests before and after the irradiation, from 200 nm to 1 μm. In the case of Clearceram, for example, three absorption bands located at 3.20, 2.20, and 1.81 eV are present. Over a quite long period these peaks decrease exponentially [24].

The scintillation properties of different optical materials were studied for their possible use in the development of radiation detectors. Two radiation induced luminescence (RIL) bands were observed in polycrystalline BaF_2 irradiated by X-rays at 295 K, for the wavelengths intervals 380–600 nm and around 659 nm [25]. Under X-ray irradiation, ZnSe crystals present a degradation of the optical transmission at 475–575 nm, and four luminescence spectra at λ = 460 nm, 610 nm, 645 nm and 970 nm [26]. An

X-ray induced RIL peak was reported in CaF_2 crystals at λ = 420 nm, accompanied by a small one at about 350 nm [27].

Sapphire is one of the most intensively studied optical material under different ionizing radiation: X-ray (40 kV, 15 mA) and β from a ^{90}Sr source at 1.5 Gy/min [28]; 8 MeV proton (flux 2×10^{12} p cm^{-2} s^{-1} and 150 MeV argon (4.3×10^{8} ion cm^{-2} s^{-1}) and 253 MeV krypton (4.3×10^{8} ion cm^{-2} s^{-1}) [29]; gamma rays (dose rate 6.7 Gy/s, total dose 10^{8} Gy) combined with neutrons (energy 2.4 MeV and fluences of 10^{17}–10^{20} n cm^{-2}) [30]; fast neutrons (energy 1.2 MeV and fluence of 1.4×10^{18} n cm^{-2}) [31], neutron irradiation followed by heating (energy 10 MeV, flux of 6.6×10^{12} n cm^{-2} $s^{-1}$and fluence up to 10^{19} n cm^{-2}), maximum temperature 1000°C) [32]. Sapphire proved to be radiation hardened under gamma exposure up to high doses (10^{8} Gy), but is more susceptible to optical transmission degradation under gamma-neutron irradiation as the optical attenuation increases with the fluence for wavelengths below 600 nm. RIL spectra change under gamma irradiation in relation to RIA modification [30]. For fast neutron irradiation, absorption bands develop at λ = 203, 255, 300, 357 and 450 nm, while post irradiation excitation produce photoluminescence signals at λ = 320, 377 and 551 nm [31]. High purity Al_2O_3 crystals present absorption peaks at λ = 206, 230, 258, 305, 358 and 452 nm, when irradiated by fast neutrons. Post irradiation annealing (up to 1000°C heating) contributes to partial recover of RIA. Excitation at λ = 302 nm induces luminescence at λ = 325, 482, 543 nm [32]. The RIL associated to the F- and F^+- bands was found to be dependent on the proton dose [29].

The invention of quantum cascade laser (QCL) [33] and subsequent research in the field made possible the development of compact, very accurate, and portable spectroscopic instruments for the mid-IR spectral range of interest for organic compounds identification (3–12 μm). Tunable QCLs or arrays of QCLs operating at different wavelengths proved to be affordable substitutes for Fourier transform IR (FTIR) systems to be used in astronomy, astrophysics, astrochemistry, and space missions [34, 35]. In order to operate reliably under extreme conditions as components to be included in spaceborne equipment, the composing parts of such equipment have to be tested under specific radiation exposure. Within this context, a program to evaluate for the Romanian Space Agency passive and active mid-IR components is under way. In this chapter, reference will be made to tests carried out on mid-IR windows materials (CaF_2, BaF_2, ZnSe, and sapphire—Al_2O_3). The subjects of these investigations are COTS (components-on-the-shelf) products, manufactured in Europe and China. The windows have 10–25 mm diameter and a thickness of 2–3 mm. Considering the complex space radiation environments which can be encountered during extra planetary missions, irradiation tests were run under various irradiation conditions: gamma rays (dose rate of 5.7 kGy/h +/−1.8%, four irradiation steps at total doses of 0.1 kGy; 1 kGy; 10 kGy; 20

kGy), alpha particles (doses from 1.46 × 10^6 kGy to 5.79 × 10^6 kGy, depending on the window material, at three beam currents of 100, 200 and 300 μC, beam diameter 3–4 mm), protons (at 10^{15} p cm^{-2}, 10^{16} p cm^{-2}, 10^{17} p cm^{-2} fluences), electron beam (dose rate of 4 kGy/min, total dose – several kGy). Gamma and electron beam irradiation were performed in air at room temperature, while proton and alpha particle irradiation were done in vacuum [36–38]. Prior and after each irradiation step the samples were measured in relation to their optical transmittance and reflectance over the spectral range from 250 nm to 18 μm. **Figure 1** illustrates the setup for spectral reflectance evaluation, which was carried out with the Gooch and Housego OL Series 750 Automated Spectroradiometric Measurement System and the accessories (tungsten lamp and IR glower, diffracting gratings, integrating spheres, reflectance standards, optical detectors - Si & Ge, PbS, InSb and HgCdTe) appropriate to each spectral interval over which the measurements were done. Optical transmittance of the irradiated samples was used to assess the color centers generation (RIA) in mid-IR optical materials under various irradiation conditions (type of radiation, dose, and dose rate). In the mean time, during alpha particle irradiation the RIL was monitored on-line in order to associate this signal with the presence of some dopants/impurities. The peak wavelengths of the detected radioluminescence are: λ = 285 nm (Caf_2), λ = 405 nm (BaF_2), λ = 465 nm (ZnSe), and λ = 700 nm (Al_2O_3). Significant decrease of the optical transmission was obtained for CaF_2 and BaF_2, below 1 μm, while a drop of the optical transmission over the investigated spectral range (0.3–16 μm) was noticed for all the materials except for sapphire, case when the transmission change was smaller as compared to other mid-IR materials. For the same purpose, RBS (Rutherford backscattering spectrometry) measurements were done on the investigated samples.

Spectral reflectance measurements were associated to optical microscopy tests as charged particles impinging on windows surface affect its quality (**Figure 2**).

For the first time, THz (Terahertz) spectroscopy and imaging were introduced (**Figure 3**) in the analysis of irradiated mid-IR optical materials.

2.2. Optical fibers

Optical fibers were suggested to be possible candidates for radiation dosimetry through RIA, RIL or thermoluminescence monitoring [2]. Such applications were considered in the case of glass optical fibers and plastic optical fibers as well. The radiation induced effects depends strongly on the optical fiber composition of the irradiated optical fibers, a detailed discussion on radiation induced defects in Pure Silica Core (PSC) and doped optical fibers can be found in [39, 40].

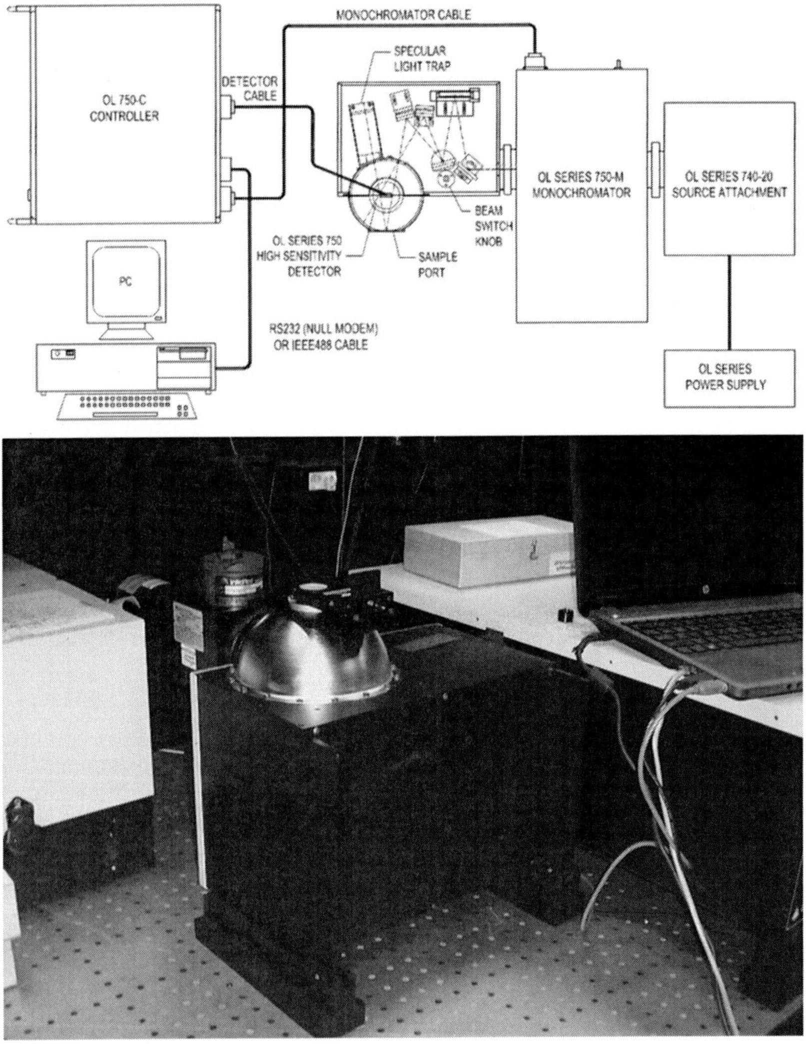

Figure 1. The sketch (a) and the picture (b) of the setup for spectral reflectance measurements over the UV to mid-IR (Reproduced with permission and courtesy of Gooch and Housego).

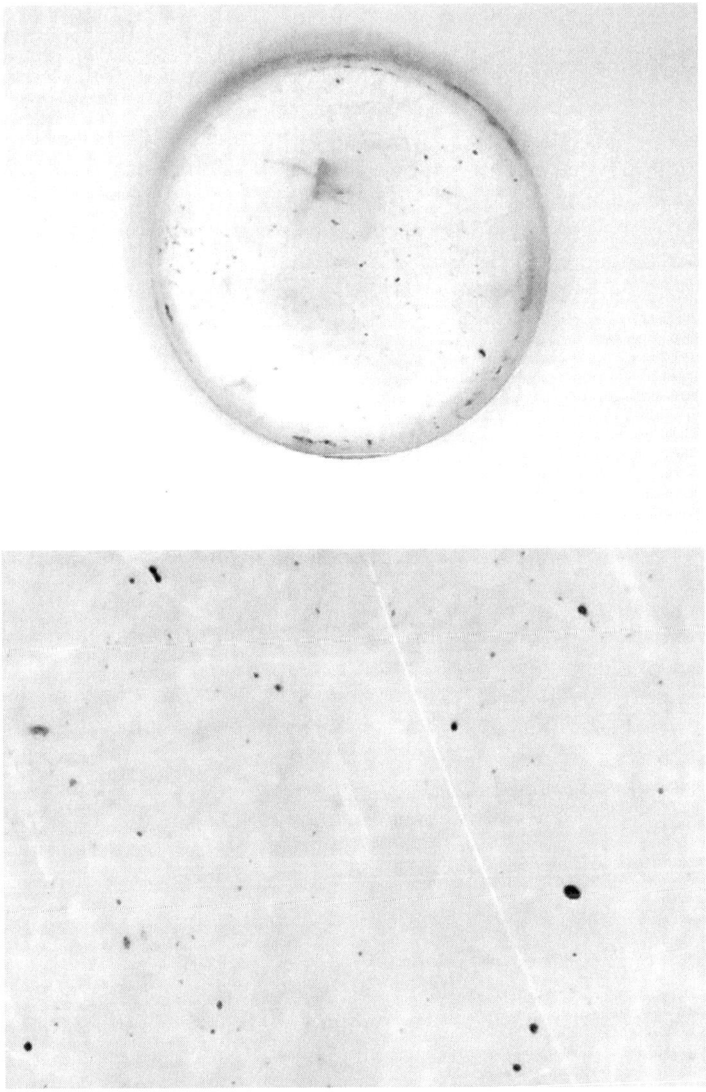

Figure 2. The degradation of windows surface quality after alpha particle irradiation: (a) BaF2; (b) ZnSe (Courtesy of Laura Mihai).

Optical fibers based on different dopants were proposed for radiation dosimetry using the optical attenuation change under gamma irradiation: TiO_2 and GeO_2 + TiO_2 in the silica core SM optical fibers [41]; Ge/Al co-doped SM fibers [42]; P_2O_5 doped step-index MM [43]. Depending on the experiment, low doses (dose rates of 0.01–1 Gy/h, for total doses up to 1

Gy) [43], moderate (dose rates 5–10 Gy/h, total doses of 10–100 Gy) [41] or high doses (dose rates of 2, 4 or 6 Gy/min, and the maximum total dose of 13 kGy) [42] were used. RIA was measured either at λ = 980, 1310 and 1530 nm [41, 42] or at λ = 502, 540 and 560 nm [43]. The recovery of the irradiation induced attenuation was studied at room temperature, through photobleaching (Ar laser radiation at λ = 514 nm) or by heating the samples (1000°C in O_2 atmosphere at the pressure of 10 psi). Some of the studied samples (depending of the dopants concentration, wavelength considered, dose used, and radiation sensitivity) indicate a linear dependence of RIA with the dose, its independence with the dose rate and low recovery, which provide the ground for applications in radiation dosimetry.

Step index multimode, 16.0 mole% P_2O_5 co-doped (with and without 6.0 mole% GeO_2 in the core) optical fibers were tested under gamma irradiation and shown a linear response at 505 nm for dose rates of 0.1 to 1 Gy/h, and a radiation sensitivity of 0.69–0.97 dB m^{-1} Gy, up to a maximum dose 1 Gy [44]. As the samples present an independence of the radiation response on the dose rate and a low recovery they seem to be suitable for medical dosimetry.

An investigation of gamma rays on polarization maintaining (PM) optical fibers to be used in interferometric fiber optic gyroscope (IFOG) for spaceborne assembles was carried out on three types of fibers: (i) pure-silica-core, (ii) P-doped (1 mol%), and germanium Ge-doped (15 mol%), as they were exposed to total doses of 100 Gy and 1 kGy. RIA was monitored over the spectral range 1100–1700 nm, with a focus on 1310 and 1550 nm. All the three fibers are more sensitive to radiation at 1550 nm than at 1310 nm, and the most degradation as it concerns RIA was observed in the case of the P-doped optical fiber [45].

On-line evaluation of gamma radiation produced RIA was investigated in 40 cm long Co/Fe co-doped alumino-silicate optical fibers fabricated by modified chemical vapor deposition (MCVD), and exposed to dose rates of 6.7, 18.4, 37.0, and 78.3 Gy/min, for 30 min to total doses of 201, 551, 1110, and 2348 Gy [46]. The measurements were done at 1310 nm with a Yokogawa AQ6370C optical spectrum analyzer. Following the irradiation, the attenuation in the doped fiber increases 258 times as compared to a reference optical fiber (SMF-28™), and a small decay of the RIA signal was noticed after the irradiation ended. An almost linear relationship was present between RIA and the dose up to a total dose of 2.4 kGy at 78.3 Gy/min, which recommend this optical fiber for radiation dosimetry, as RIA dose rate independence was observed.

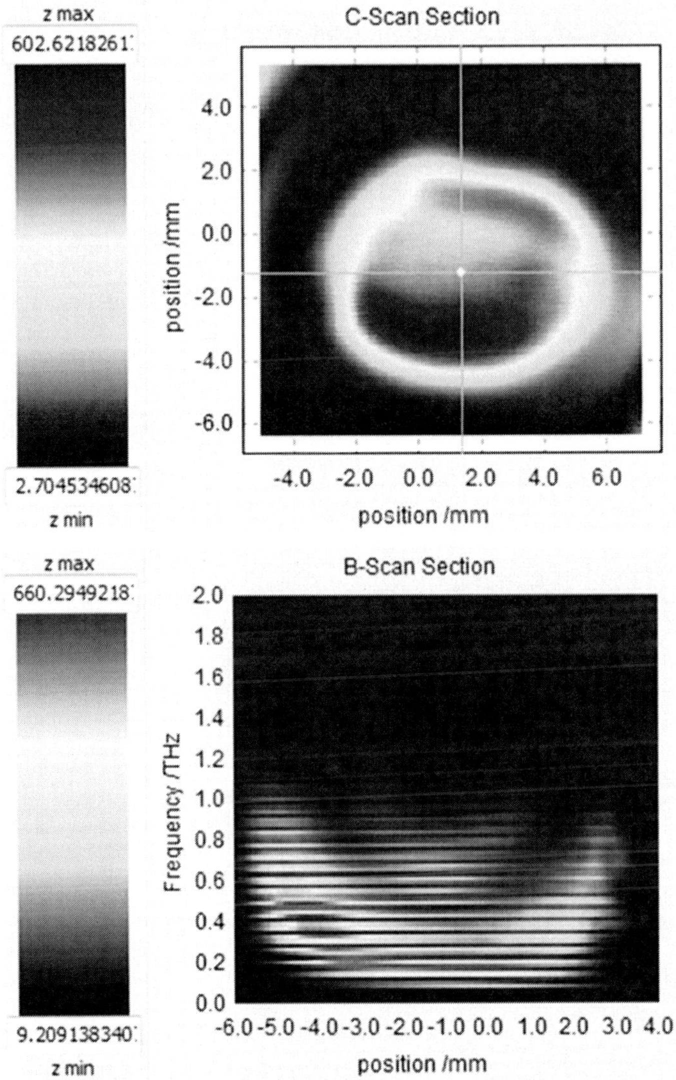

Figure 3. 2D plot of the THz reflected signal for BaF_2 window as the sample was scanned in the XY plane, for 0–2 THz (a); the frequency signal along the X axis (b), at a resolution of 50 μm (Courtesy of Laura Mihai).

The performances of Ce-doped alumino-phospho-silicate optical fibers without Au (5.0 wt.% Al_2O_3, 0.15 wt.% P_2O_5, 0.3 wt.% CeO_2) and with Au co-doping (5.1 wt.% Al_2O_3, 0.15 wt.% P_2O_5, 0.27 wt.% CeO_2, and 0.2 wt.% Au_2O_3), produced by MCVD, were evaluated under electron beam irradiation at energy of ~6 MeV up to the total doses of 10^{12} e/cm^2, 5 × 10^{12} e/cm^2, 10^{13} e/cm^2, 5 × 10^{13} e/cm^2, 10^{14} e/cm^2, and 2.5 × 10^{15} e/cm^2. RIA was monitored between 400 nm and 1100 nm, and post irradiation photobleaching under He-Ne laser radiation laser (λ = 543 nm) and UV. The results of the study indicated that the Ce/Au-doped optical fibers are more suitable for fiber-based dosimetry [47].

In the last years, the research focused also on the X-ray irradiation (10 keV energy) effects on various optical fibers: PSC, Ge-doped, P-doped, Fluorine-doped/ co-doped, in some cases the irradiation being combined with sample heating to 300°C [48, 49].

The irradiation outcomes in changing the optical fiber characteristics were investigated by on-line RIA measurements, confocal micro-luminescence (CML) and electron paramagnetic resonance (EPR), for total doses up to 3 MGy (dose rate of 50 Gy/s). RIA modifications in the UV-visible spectral range for Ge/F co-doped optical fibers produced under different conditions (draw speed and tension) were monitored with a mini optical fiber spectrometer (Ocean Optics HR4000) and deuterium-halogen DH2000 light source. Almost complete recovery of the irradiation induced attenuation was observed at room temperature for λ = 360 nm, corresponding to Ge(1) defects, at room temperature during the day following the exposure [49]. For various production conditions and different chemical composition the RIA vs. total irradiation dose graph indicates a linear dependency up to the dose of 300 kGy.

An interesting aspect in the research of optical fiber behaviour exposed to radiation is represented by the evaluation of radiation induced changes as heating is applied to the investigated sample. Different types of multimode UV optical fibers (deep UV enhanced, high OH step-index; solarization resistant, high OH step-index; UV enhanced extended spectral response, H_2-loaded, step-index) were tested under gamma irradiation. In some circumstances, the samples were heated to 573 K while irradiated, as the variation of the optical attenuation at specific wavelengths in the UV spectral range (λ = 248 nm, λ = 265 nm, λ = 320 nm and λ = 330 nm) was monitored. For some wavelengths (λ = 248 nm and λ = 265 nm) H_2-loading improves radiation resistance, while heating during the irradiation of the same optical fiber increases their radiation sensitivity [50].

A complex research was carried out to evaluate the combined effect of radiation (1keV X-ray, 50 Gy/s dose rate and total dose from 1.5 KGy to 1 MGy) and heating up to 300°C, for different types of optical fibers [48]. The results of this investigation indicated that no unitary behavior with temperature during radiation exposure can be predicted in the case of PSC, Ge-doped, P-doped, Fluorine-doped optical fibers, as in some situations (fiber compositing, temperature range, total dose) the sample heating has no effect in compensating the defects generation, while in some others RIA increases or decreases with the temperature increase. In any case, some of the study conclusions refer (i) to the possible use of such fibers in radiation dosimetry, especially at low doses, and (ii) to the limited length of fiber which can be incorporated into distributed radiation sensing systems.

A recent contribution to the use of optical fibers in radiation dosimetry refers to some studies carried out in cooperation with a team working at the European Synchrotron Radiation Facility, in Grenoble on UV multimode optical fibers subjected to synchrotron radiation [51]. In this experiment were tested five types of commercially available UV optical fibers and off-line measurements concerning the degradation of the optical fibers transmission in the UV-visible spectral range were done, as color centers developed upon exposure to synchrotron radiation. The measurements were performed for different doses: 5, 10, 30, 60, 200, 400, 1000 and 2000 Gy. The radiation induced optical attenuation was monitored in relation to recovery phenomenon both at room temperature and after samples heating to 560 K. The change of optical transmission and the increase of the attenuation at specific wavelengths (λ = 215 nm; λ = 229 nm; λ = 248 nm; λ = 265 nm; λ = 330 nm) were performed using the setup presented in **Figure 4**. The measurements were done with a dedicated software control developed under LabVIEW, by using a broadband UV-visible light source, a sensitive optical fiber spectrometer, and an optical fiber multiplexer.

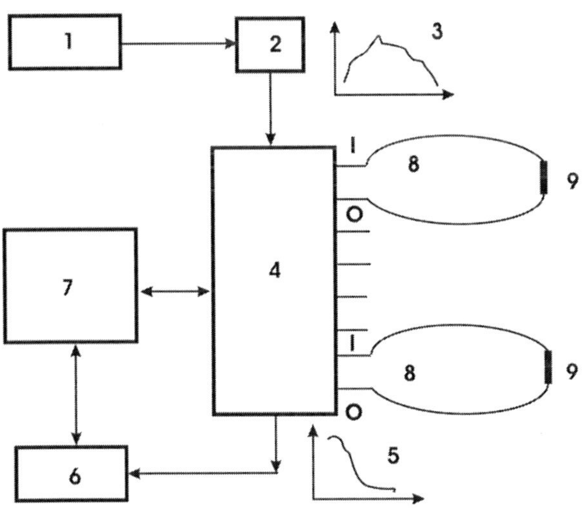

Figure 4. The sketch of the measuring setup: (1) light source; (2) optical fiber attenuator; (3) spectrum of the light source; (4) optical fiber multiplexer; (5) optical fiber sample absorption spectrum; (6) optical fiber mini spectrometer; (7) laptop; (8) connecting optical fibers; (9) irradiated optical fiber samples. I, multiplexer input; O, multiplexer output.

The graphical user's interface is illustrated in **Figure 5**. The multiplexer is employed to connect different tested samples to the light source and the spectrometer and help for the determination of the optical absorption from 215 nm to 600 nm. The operator can preset the number of averaging cycles, the box car value and can select the wavelengths of interest to be monitored. Of interest for these tests was to monitor specific wavelength associated to the presence of color centers in glass fibers. By observing the dependency of the optical absorption on the total irradiation dose at the specified wavelengths (**Figure 6**) can be estimated the linearity of this dependency and the spectral interval over which a saturation effect occurs. This can help to evaluate the possible use of these optical fibers in radiation dosimetry, at specified dose rates.

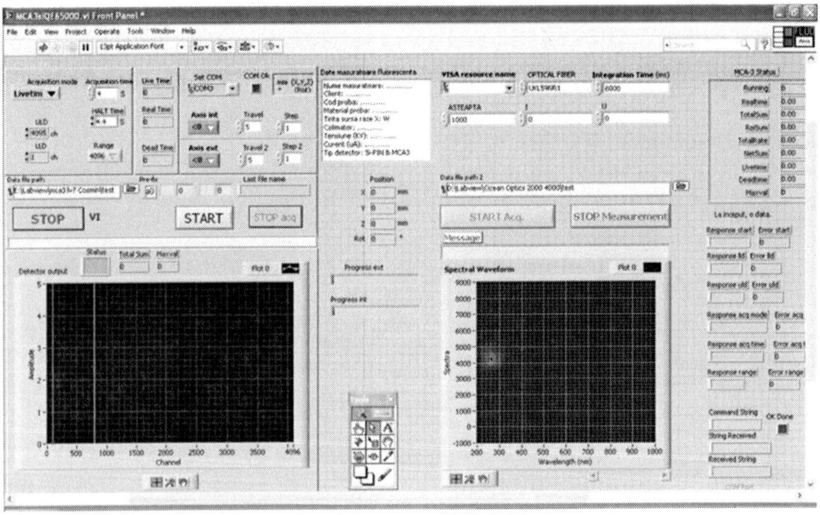

Figure 5. The LabVIEW graphical user's interface for the experimental setup (Courtesy of Laura Mihai).

As a premiere, the use of THz spectroscopy and imaging to assess the synchrotron radiation induced changes in the core and coating of the investigated optical fibers (**Figure 7**) was reported. This approach makes possible the "visualization" of the dielectric constants change of the irradiated materials.

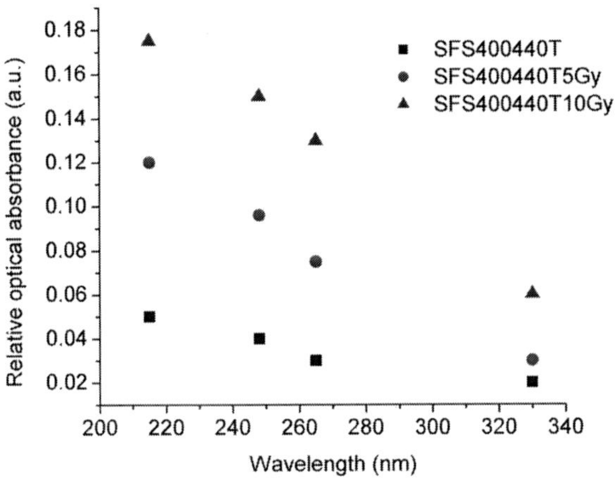
Relative optical absorbance (a.u.)
0.18
0.16
0.14
0.12
0.10
0.08
0.06
0.04
0.02
200
220
240
260
280
300
320
340
Wavelength (nm)
SFS400440T
SFS400440T5Gy
SFS400440T10Gy

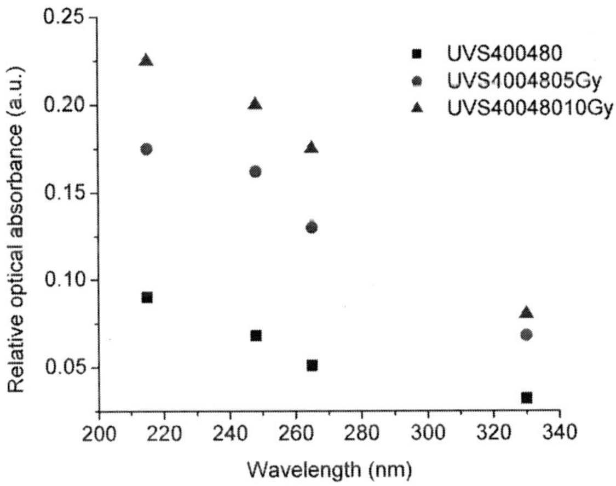
Relative optical absorbance (a.u.)
0.25
0.20
0.15
0.10
0.05
200
220
240
260
280
300
320
340
Wavelength (nm)
UVS400480
UVS4004805Gy
UVS40048010Gy

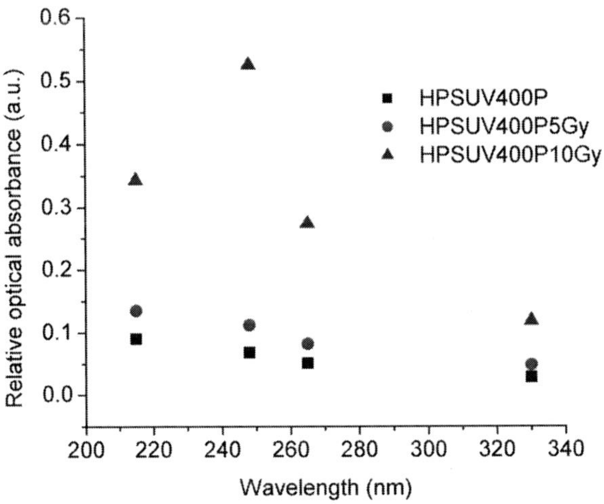
Relative optical absorbance (a.u.)
0.6
0.5
0.4
0.3
0.2
0.1
0.0
200
220
240
260
280
300
320
340
Wavelength (nm)
HPSUV400P
HPSUV400P5Gy
HPSUV400P10Gy

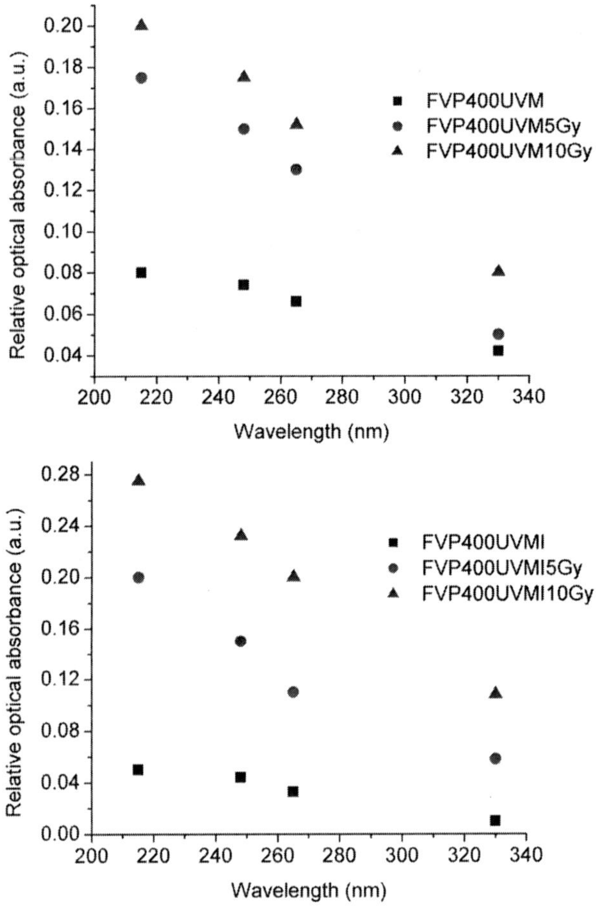

Figure 6. The dynamics of the color center formation as function of the dose (i.e. 5 Gy and 2000 Gy in this case) for the five tested optical fibers, at λ = 215 nm, λ = 248 nm, λ = 265 nm, λ = 330 nm: (a) SFS400/440T (Fiberguide Industries Inc.); (b) UVS400/480 (Fiberguide Industries Inc.); (c) HPSUV400P (Oxford Electronics); (d) FVP400/UVM (Polymicro Technologies); (e) FVP400/UVMI (Polymicro Technologies).

The potential to employ P-doped core multimode optical fibers as X-ray (1 keV) radiation detectors for doses up to 3 kGy, under different dose rates (1, 10, 50 Gy/s) was studied by monitoring in real time the RIA in the spectral range from 200 to 900 nm. The results indicated a sub linear dependence of the irradiation induced attenuation with the dose and dose rate. The temperature did not change RIA during the irradiation for values between 5 and 50°C. The highest radiation sensitivity was observed at 300 nm, ~0.60 dB/m^{-1} Gy^{-1} [52].

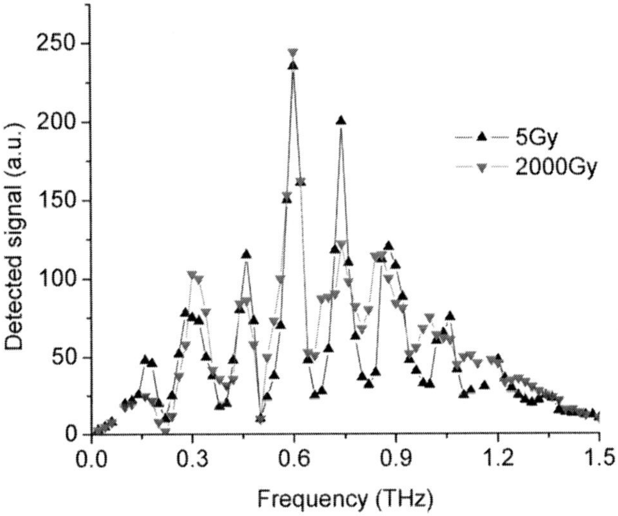

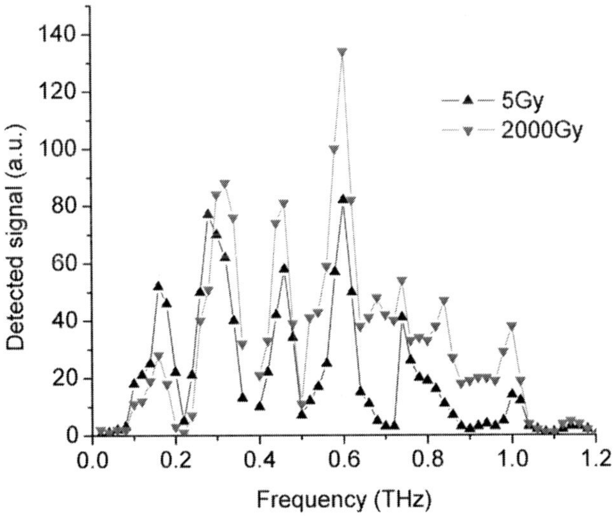

Figure 7. The THz spectral signal of the core (a) and coating (b) of an irradiated UV optical fiber, for two synchrotron radiation doses: 5 Gy and 2000 Gy (Courtesy of Laura Mihai).

For the case RIA values during the irradiation is dependent on the radiation dose and exhibit a recovery after the exposure is interrupted, a differential scheme can be applied when the measurand is represented by the difference of RIA at two wavelengths, λ = 413 nm and λ = 470 nm, respectively [53].

Another novelty recently introduced refers to the test of gamma radiation on perfluorinated polymer optical fibers (pPOF). These results complement the investigations previously reported in literature on the spectral characteristics of gamma irradiated PMMA optical fibers. In this case, commercially available MM perfluorinated fiber GigaPOF-50SR from Chromis Fiberoptics were exposed to gamma ray at a dose rate of 5.7 kGy/h for total doses of 1, 5, 20, 50 and 100 kGy. The irradiation took place at maximum temperature of 32°C, and the samples were measured before and after the irradiation using a ν-OTDR from Luciol Instruments, operating at 1310 and 1550 nm [54]. In addition, strain tests were also performed to evaluate the mechanical degradation of the polymer fiber under gamma irradiation. For doses up to 5 kGy the attenuation vs. the dose presents a linear dependency with a radiation sensitivity of 0.096 ± 0.006 dB m-1 kGy^{-1} at 1310 nm, and 0.25 ± 0.05 dB m-1 kGy^{-1} at 1550 nm. The obtained results suggest the possible use of these optical fibers in radiation dosimetry. The dynamics of the color centers formation under gamma irradiation was investigated by on-line measurements and by monitoring RIA change at several wavelengths (λ = 420, 525, and 750 nm). The highest radiation sensitivity was noticed at λ = 420 nm (up to 0.5 kGy) dose, while the best linearity response of RIA vs. dose was obtained at λ = 750 nm for a dose reaching 2.5 kGy [55].

The last decade recorded as a breakthrough the use of sapphire optical fibers in radiation environments, as sapphire supports very high operating temperatures and presents minor optical transmission changes when exposed to ionizing radiation over a wide spectral range [56]. Recently, sapphire optical fibers were subjected to neutron flux of 1.5×10^{12} n cm^{-2} s^{-1} and a gamma dose rate of 84 kGy/hr (dose in sapphire) for a total neutron fluence of 1.1×10^{17} n cm^{-2} and total gamma dose of 1.8 MGy [57]. The optical attenuation in the sapphire fiber was measured on-line over the 500–2200 nm spectral range. The major RIA occurs below the 500 nm measuring limit. A change of RIA was noticed as the sample was heated to 1000°C, under gamma irradiation [58].

The use of optical fibers for distributed measurements in ionizing radiation environments became a hot topic in the last years as several tests were run and some possible applications were suggested: Raman and Brillouin strain/ temperature sensing [10, 59]; Rayleigh temperature sensing [60]. Tests were performed on standard communication, highly doped GeO_2, F-doped optical fibers [10, 59] and (PSC), Ge-doped, radiation hardened and Al-doped optical fibers [60], at dose rates of 700 Gy/h [60], 1.5 kGy/h [10], 28 kGy/h [59], and total doses of 110 kGy [60], 618 kGy [10], 10 MGy [59].

RIA proved to be the major challenge in using optical fibers distributed Brillouin temperature sensors, as the operating distance is reduced to several hundreds of meters due to irradiation [59]. Limitations in temperature monitoring are present also in the case of Rayleigh-based setup [60]. Temperature measurements using Raman scattering encounters some difficulties as the temperature evaluation uncertainty is quite high upon irradiation [10].

The complementary aspect of interest is represented by the possible use of radiation sensitive optical fibers for distributed dosimetry. Two approaches were proposed to be used: (i) the optical time domain reflectometry – OTDR [61], (ii) the optical frequency domain reflectometery – OFDR [62]. In both experiments the fiber samples were subjected to gamma irradiation: P-doped optical fibers [61] and standard communication, Al-doped, and P-doped optical fibers [62], at dose rates 590 Gy/h at 30°C and 605 Gy/h at 80°C, total dose of about 71 kGy [62], and dose rate of 23 mGy/s, for a total dose of 300 Gy [61]. The OTDR approach exhibits a limited spatial resolution for distributed dosimetry. Depending on the optical fiber type, the interrogating instrument dynamic range and the dose, the usable length of the dosimetric fiber is limited to several centimeters [62].

2.3. Rare earth-doped optical fibers

Generally, radiation hardening tests of rare earth-doped optical fibers were performed in relation to their possible use in space applications such as inter-satellite communications and fiber optic gyroscopes (FOGs) applications [63–65]. The investigations focused on the radiation induced absorption in the optical fiber and on changes of the emission efficiency. Tests were carried out under gamma ray [63, 65, 66], neutrons [65], protons [65–67], or X-ray [66] exposure. The reported studies indicate that:

- the irradiation results are primarily dependent on the dose and not on the type of irradiation involved [65],
- the device degradation is function more on the co-dopant concentration than on the rare-earth elements [65],
- annealing and photobleaching can contribute to partial recovery of the irradiation induced effects [65, 67, 68].

Tests on photobleaching of two Erbium-doped fibers (EDFs) subjected to 1.3 and 2.24 kGy gamma irradiation were carried out at two laser wavelengths (532 and 976 nm). A higher efficiency was achieved with the shorter wavelength excitation. The annealing effect induced by the 976 nm laser is comparable to that present in the case of thermal annealing at 500 K [69].

In another experiment, an EDF was tested under electron irradiation (beam energy 1.2 MeV, total dose 10 kGy), and the output power, noise figure and the central wavelength were measured after the irradiation, with the pump at 980 nm and detection at 1550 nm. The central wavelength did not change after irradiation, while the output power and noise figure were deteriorated (i.e. the output power dropped from 10 to –60 dBm for the same pump level). Within 2 weeks, the two parameters partially recovered [70].

Single mode and multi-mode Yb-doped optical fibers, acting as amplifiers at 1064 nm, were subjected to gamma rays and mixed gamma-neutron irradiation, up to a total dose of 1 Gy(Si) [71]. It was noticed a linear decrease of the output power with the irradiation dose. A slight recovery was observed at room temperature, after 20 hours.

In his PhD thesis, Fox [72] concluded that EDFs are more radiation sensitive than Yb^{3+}-doped fibers, and the most radiation hardened to gamma ray are optical fibers of Er^{3+}/Yb^{3+} co-doped type. In this investigation, the radiation induced degradation increases with the increase of the dose rate.

Optical fiber preforms of Yb-free aluminosilicate core, Yb-doped Al-free silicate core, Yb-doped alumino-silicate cores and Yb-free germanosilicate core produced by MCVD and solution-doping techniques were irradiated by 45 kV X-rays, at ~2.5 Gy(SiO_2)/ min, for doses up to 0.3–0.5 kGy [73]. The effects of the irradiation were studied by thermally stimulated luminescence (TSL) and optical absorption measurements.

Investigations were carried out to evaluate the influence of H_2-loading of EDFs, under pumping at 980 and 1480 nm, as the samples were subjected to gamma irradiation, having doses from 0.1 to 10 kGy. One such sample was a H_2-loaded EDF, while the other was H_2-free carbon-coated [74]. At 980 nm pumping a photobleaching effect was observed which increases the efficiency of the process especially in the H_2-loaded optical fiber. No such effect was present with the pump radiation of 1480 nm. H_2-loading and the use of a hermetical coating of the optical fiber, which prevents H_2 diffusion, produce a radiation resistant EDF [75].

One approach to enhance the radiation resistance of the optical amplifier consists in the use of Er-doped-nanoparticles optical fibers [76]. Another proposed method is based on the doping of the optical fiber core with Ce, reducing in this way the radiation sensitivity [77].

Tests on a thulium doped optical fiber amplifier were performed under neutron irradiation and a 17.1 dB was observed for a dose of 720 Gy, which is close to the computed values [78].

The development of Bi co-doped silica optical fibers [79] having an extensive emission operation range (from 1100 to 1800 nm) called researchers attention on their investigation under ionizing radiation. Tests run under gamma irradiation, dose rate – 5.5 kGy/h and maximum total dose – 50 kGy, on Bismuth/Erbium/Ytterbium co-Doped Fiber – BEYDF ([Bi] ~ 0.07, [Er] < 0.1, [Al] ~ 0.35, [Ge] ~ 1.18, [Yb] ~0.02 atom%, respectively, indicated the formation of bismuth related active center (BAC), as the absorption at 830nm is increased significantly, while the absorption at longer wavelength diminishes [80].

In the case of Bi/Al-co-doped silica optical fibers the fluorescence peak increases with the increase of the gamma dose (dose rate, 800 Gy/h; maximum total dose, 3 kGy), when the sample is pumped at 980 nm. The preform of the investigated materials had the following concentrations: Si—28 mol/% (core), Ge— 6 mol/% (core), O—66 mol/% (core and inner layer), Bi—0.016 mol/% (core and inner layer), and Al—0.04 mol/% (core and inner layer). It is presumed that this enhancement of the photoluminescence signal upon irradiation is due to creation of subvalence Bi ions [81].

Another investigation was focused on the effect of gamma irradiation (dose rate – 5.5 kGY/h, total doses of 1, 5, 15, 30 and 50 kGy) on samples of Bismuth/Erbium co-doped fiber (BEDF). The irradiation has significantly increased unsaturable absorption by about 8 dB. This corresponds to a very significant unsaturable attenuation coefficient change of ~32 dB/cm. The results indicate that the saturable absorption of the BEDF at 830 nm is also increased by about 4 dB. This corresponds to a significant saturable attenuation coefficient change of 16 dB/cm. The increase in saturable pump absorption implies that the irradiation has significantly increased the number of bismuth related active centers (BAC-Si). In this case the overall emission at 1430 nm (which is related to BAC-Si) is only slightly decreased while the unsaturable absorption is significantly increased. A good radiation survivability of the BEDF for emission or amplification was noticed [82].

3. RADIATION EFFECTS IN PHOTONIC DEVICES

3.1. Fiber bragg gratings

Fiber Bragg gratings were excessively studied under various type of ionizing radiation: gamma ray [12, 83, 84], neutron [85], mixed gamma-neutron [86–88] or 13.5 MeV protons [89]. In most cases, the changes of the FBG Bragg wavelength are relatively low. Tests were carried out using Type I, Type II, Type IA, Type IIA, chemical composition or fs laser engraved gratings, and were focused on the effect of fiber composition, dose rate, total dose, heating during the irradiation or possible photobleaching during the exposure to ionizing radiation [90]. For example, Type I gratings produced in H_2-loaded standard communication optical fibers (SMF28™) or B/Ge co-doped optical fibers present an increase of the central wavelength followed by a plateau, with an overall change between 10 pm to 34 pm, under 0.54 MGy exposure. In the case of Type II gratings written in B/Ge co-doped optical fibers exhibit a decrease of the central wavelength between -30 and -60 pm. Degradation (30–50%) of gratings reflectivity was produced after irradiation, while no recovery was observed in the studied sensors [91]. A quite significant impact of gamma radiation on Ge-doped core optical fibers was observed as such uncoated fibers are immersed in water during gamma exposure [83], for a total dose of up to 5 MGy. Generally, all the investigations carried out on radiation effects on FBGs focused on the possible use of such devices in radiation environments for temperature or humidity [12] monitoring. Under some design circumstances (optical fiber type, technology used to write the grating), these modifications are significant, making possible the use of FBGs as radiation detectors [92, 93].

More recently, the use of FBGs developed in sapphire optical fiber was proposed for monitoring very high temperature in nuclear reactors [94].

Through these studies, the paradigm was changed as [95]:

- FBGs were produced in radiation hardened optical fibers by two-beam interferometer and deep ultraviolet fs laser radiation;
- for the first time FBGs written in both standard commercial optical fibers and radiation hardener optical fibers were tested, exposed to the electron beam from a linear accelerator;
- a mash of FBGs was used for beam diagnostics of charged particle beams, as a novelty.

This research indicated a linear dependence of the Bragg wavelength shift with the irradiation dose for commercially available FBGs and grating produced in radiation hardened optical fibers, and this shift was monitored by measuring simultaneously with a thermocouple the temperature change in the irradiation plane. This aspect is important because no saturation or permanent modification of FBG wavelength was observed; hence these sensors can be used for on-line measurements. The principle of the charged particle beam diagnostics is illustrated in **Figure 8**, for the case of an electron beam having a diameter of about 100 mm. The spatial resolution is limited by the number of FBGs engraved in an optical fiber, their individual length and the distance between two adjacent FBGs. In the described proof-of-concept design, gratings having 12 mm and 4 mm length were employed. As compared to other solutions for charge particle beam monitoring (arrays of Faraday cups; ionization chambers; micro strip metal detector; pepperpot device, slit-grid or rotating slits; scintillating screen or gas detector; flat panel detectors, arrays of p-i-n diodes; moving wire or vibrating wire scanners) there are several advantage of this approach: its immunity to electro-magnetic noise, remote monitoring capability and the possibility to multiplex the acquired signals by using few connecting lines.

The operation of the proposed instrument is similar to that employed for laser beam diagnostics. In the evaluation of a laser beam quality an image sensor is used to acquire the transversal distribution of the laser beam's intensity. Periodically, the electric charge generated inside the image sensors is removed upon reading and new acquisition starts. In this implementation the instrument operation is based on the proved reliability of the tested FBGs under electron beam exposure and on the linear shift of FBGs Bragg wavelength with the temperature in the detection plan, and hence with the deposed energy by the charge particle beam. Prior to the use, the FBGs were calibrated as it concerns their wavelength change vs. temperature.

When the charge particle beam (1) is propagating from the linear accelerator output (2) its diameter increases (4) in the detection plan (5) due to its divergence (**Figure 8a**). In the detecting plan (item 1 in **Figure 8b**) a mesh composed of FBGs (3) is placed, the sensors being embedded into a thermally insulating material to prevent the lateral dissipation of the heat. As the individual gratings are exposed to different beam energies their temperature increases according to the energy deposed on each detecting site. In this way, in time, a map of the transferred energy at each location is obtained.

The detector operates as an energy integrator (dose related measurements). Permanently, the central wavelength of all sensors is acquired by a Micron Optics optical fiber interrogator, which makes possible the real time mapping of the electron beam cross section energy distribution. In order to avoid saturation, periodically the FBGs signal has to be "reset" by blocking the electron beam and forced cooling the FBGs array. For this purpose a thick (20 mm) Al shutter restricts under the software control (**Figure 8a** and **8d**) the exposure of the FBGs by interrupting the electron beam. During the interval the beam is blocked, a cooler pushes air over the sensors' matrix. The Al shutter goes back and forth as the array has to be exposed or cooled. The flow chart for the instrument operation is presented in **Figure 9**.

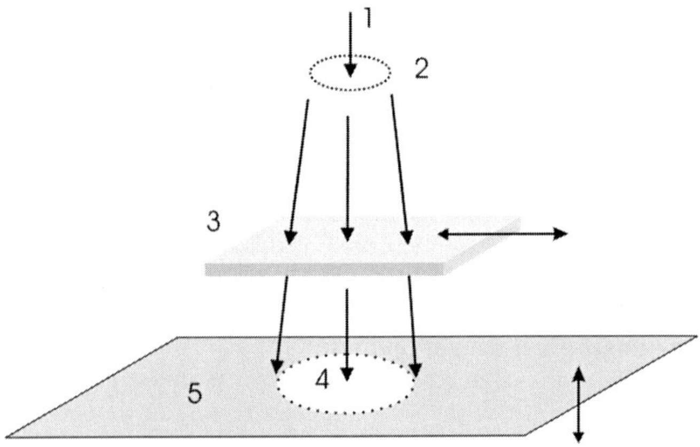
1
2
3
4
5

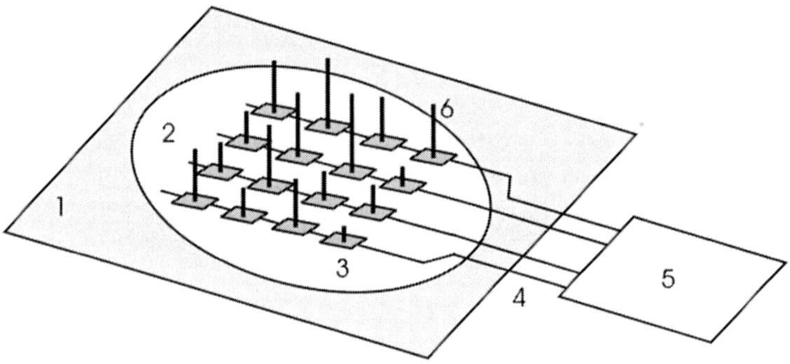
1
2
3
4
5
6

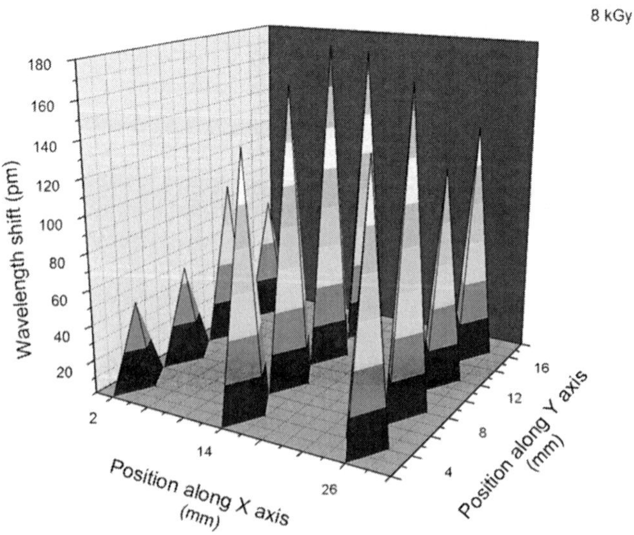
8 kGy
Wavelength shift (pm)
180
160
140
120
100
80
60
40
20
2
14
26
Position along X axis (mm)
4
8
12
16
Position along Y axis (mm)

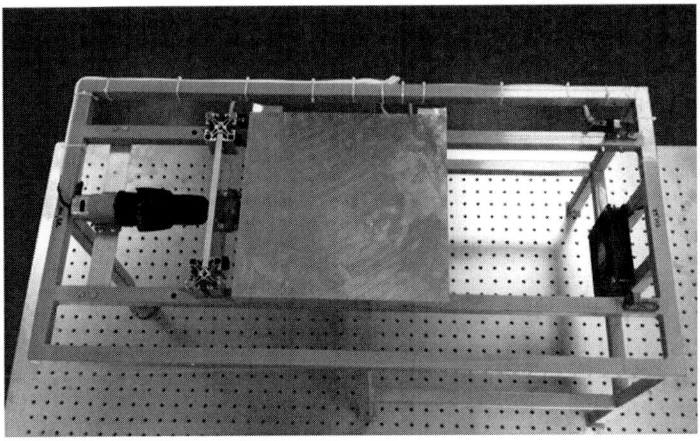

Figure 8. The operation principle of the electron beam analyzer: (a) the moving shutter used to interrupt FBGs exposure to the electron beam: (1) the incident electron beam; (2) beam diameter at the exit of the linear accelerator focusing system; (3) moving shutter; (4) the electron beam diameter at its incidence on the detection plan; (5) detecting plan. (b) The position of the FBGs mash: (1) the detecting plan; (2) electron beam pattern on the detection plan; (3) FBGs; (4) connecting optical fibers; (5) optical fiber interrogator; (6) bars symbolizing the integral energy deposited on a particular FBG at a specified moment. (c) Example of the data acquired for the 8 kGy dose [95]. (d) Top view of the shutter and cooling system.

During the "Initialization" step, the shutter is closed, the FBGs are cooled, and the functioning parameters set by the operator are introduced in the system. If a STOP command was issued by the operator the system stops and displays in 3D format the integrated energy to which each sensor was subjected, measured as Bragg wavelength shift in response to the local temperature increase. If no STOP command was given the system periodically acquires on the central wavelength shift for each sensor and checks if the maximum set temperature was reached by any of the FBGs. If this temperature was reached the system stops and displays the results as a 3D representation. If no such signal was received the software commands during the "Data acquisition" step the closing of the shutter and the cooling of the FBGs array. The periodicity of the cooling cycles, and the upper limit of the temperature to which any of the sensors can be subjected are set by the operator. The Bragg wavelength shift with the dose increase is given in **Figure 10**. In this case, during the irradiation pause, the grating's cooling was obtained by convection in air and not by forced cooling. The proposed instrument can be used during the adjusting process of the charged particle accelerator or to check the stability in time of the output beam.

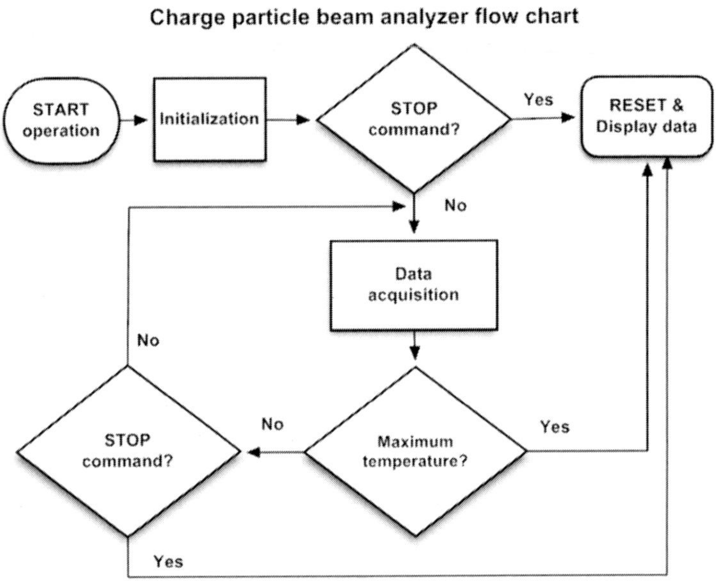

Figure 9. The flow chart of the electron beam analyzer.

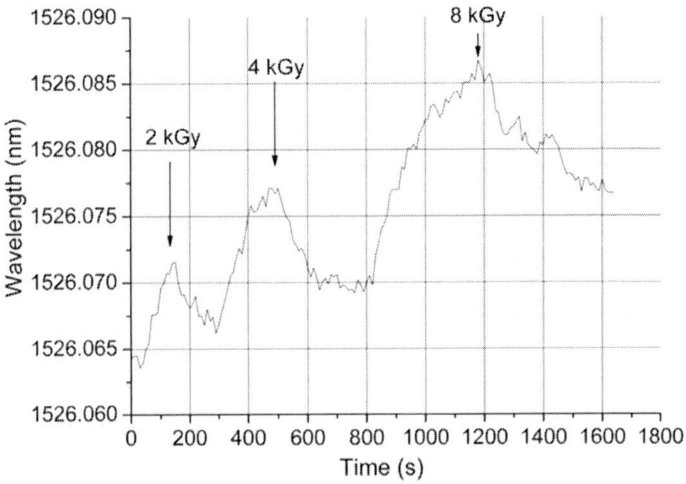

Figure 10. The change of the Bragg wavelength with the dose increase. Periodically, between two exposures the grating was cooled by natural convection [95].

A novelty in the field can be considered the investigation of FBGs performances written in polymer fibers under fast neutrons irradiation. In the paper it is suggested that neutrons produce a degradation of the fiber structure, which in turn causes a shift of the Bragg wavelength up to 14 pm. The wavelength change with neutron dose can be exploited in radiation dosimetry [96].

3.2. Long period gratings

Long period gratings (LPGs) were tested only under gamma ray exposure [90]. Most of the measurements were performed off-line [97–99], with some exceptions when the device behavior was monitored during the irradiation [100, 101]. The LPGs were obtained by different techniques: CO laser or UV engraving [99], electric arc-discharge (EAD) technique [101], CO_2 laser point-by-point writing [97], as chiral gratings [100], having a turnaround point (TAP) design produced by a CO_2laser [98]. Various optical fibers were employed: N-doped and Ge-doped optical fibers [99], pure-silica-core/ F-doped silica cladding fibers [101], SMF-28™ [97], core doped with P, Ge, F, rare earth elements [100], photosensitive fibers (B/Ge co-doped) [98]. The irradiation conditions are modified from 5 kGy [97] up to 1.47 MGy [99]. Depending on the optical fiber type, the writing technique, the dose rate/total dose used, the reported results led to different conclusions: (i) no effect was observed on the LPG transmission spectrum for doses lower than 5 kGy [97], lower than 500 kGy [101] or as high as 1.47 MGy [99], (ii) a 10 nm shift of the transmission deep was noticed for doses of 100 kGy [100], (iii) very high shift of the dual resonance deep (35 nm) is present for a dose of 6 kGy [98]. This variety of investigations outcomes suggest that, by an appropriate selection of the above mentioned parameters, LPGs can be produced either to be radiation hardened or with pronounced radiation sensitivity, appropriate to be used in radiation dosimetry. No change in the temperature sensitivity of LPGs upon irradiation was observed [101].

A new proposed approach in studying gamma irradiation effects on LGPs Includes several novelties [102, 103]:

- the use of an OFDR model LUNA OBR 4600, operating in the transmission mode or, for comparative purpose, of an optical fiber interrogator model sm125;
- LPGs on-line monitoring when the irradiation is on and off, which makes possible the observation of room temperature recovery;
- the comparative evaluation of the radiation induced LPG changes for grating produced in standard communication and specially designed radiation hardened optical fibers.

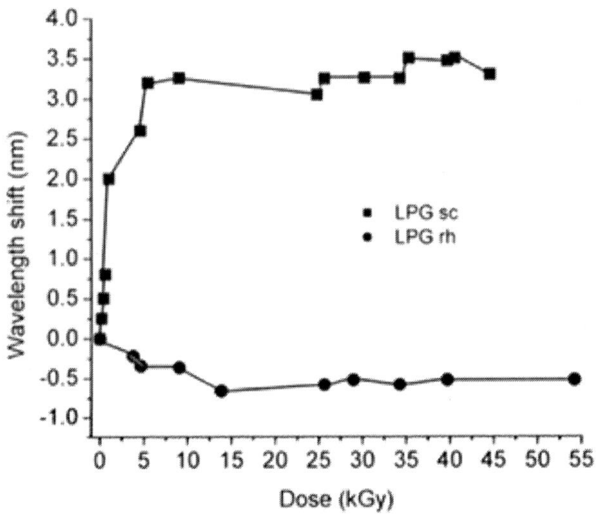

Figure 11. The change of the wavelength deep for two LPGs written in standard communication optical fibers LPGsc and in F-doped core optical fiber LPGrh.

The use of the OFDR improved drastically the detection S/N as compared to classical reading with an optical spectrum analyzer. During the irradiation the sensors were encapsulated into ceramic radiation transparent cases to avoid any strain induced changes in the LPG spectrum, and in the mean time, the gratings were placed into a thermally insulated box. The temperature was permanently monitored both inside this box and in the irradiation chamber and temperature related corrections were applied to the LPG characteristics. Based on the referred papers, **Figure 11** illustrates comparatively the behaviour of gamma irradiation on the LPG developed into a standard communication optical fiber (LPGsc) and one written in a radiation hardened optical fiber (LPGrh). The wavelength deep of the two samples move in opposite directions, the wavelength of the grating produced in the standard communication fiber increases with the dose increase, while for the other one the wavelength decreases with the dose increase. Besides that, it can be noticed the magnitude of the two changes, LPGsc being by far more sensitive. The fluctuations of the wavelength deep values during gamma exposure as well as the partial saturation trend are associated to the on-off operation of the irradiation facility. In this way, for the first time the recovery effect during the irradiation pauses was observed.

Because for the LPGrh the wavelength change is by far less significant than that in the case of LPGsc, it can be expected that the first one can be used in radiation environments as a sensor for temperature/humidity, while the second LPG can be employed within specific dose range as an optical fiber-based dosimeter.

4. CONCLUSIONS

The last four to five years brought additional data concerning radiation effects in optical fibers and optical fiber-based devices as new materials, technologies and possible applications in radiation environments emerged. In the frame of this chapter, the major trends were presented and some novel techniques and applications promoted recently were addressed, in relation to different types of irradiations (i.e. electron beam, synchrotron radiation). In this way, the data reviewed in previous publications are updated and the extensive possibilities offered by new materials and experimental setups in using radiation effects on optical fibers were demonstrated. By focusing such type of research on newly developed fibers (plastic, micro-structured, multi-core) and by using novel photonic devices based on optical fibers, new classes of equipment will emerge for terrestrial applications (dosimetry, medicine, remote and distributed monitoring) and for space missions.

ACKNOWLEDGEMENTS

The authors acknowledge the financial support of the Romanian Executive Agency for Higher Education, Research, Development and Innovation Funding (UEFISCDI), under grant 8/2012, project "Sensor Systems for Secure Operation of Critical Installations - SOCI" and of the Romanian Space Agency (ROSA) through the project "Evaluation of Components for Space Applications - ECSA", grant 67/ 2013.

REFERENCE

1. Sporea D, Agnello S, Gelardi FM. Irradiation effects in optical fibers. In: Pal B, editor. Frontiers in Guided Wave Optics and Optoelectronics. InTech; 2010. p. 46–66.
2. Sporea D, Sporea A, O'Keeffe S, McCarthy D, Lewis E, Optical fibers and optical fiber sensors used in radiation monitoring. In: Yasin M, Harun SW, Arof H, editors. Selected Topics on Optical Fiber Technology. InTech; 2012. p. 607-652.
3. Berghmans F, Brichard B, Fernandez FA, Gusarov M, Van Uffelen M, Girard S. An introduction to radiation effects on optical components and fiber optic sensors. In: Bock WJ, Gannot I, Tanev S, editors. Optical Waveguide Sensing and Imaging, NATO Science for Peace and Security Series B: Physics and Biophysics. Springer; 2008. p. 127-165.

4. Bourgade JL, Allouche A, Baggio J, Bayer C, Bonneau F, Chollet C, et al. New constraints for plasma diagnostics development due to the harsh environment of MJ class lasers. Rev. Sci. Instrum. 2004;75(10):4204-4212.
5. Bourgade JL, Costley A.E, Reichle R, Hodgson ER, Hsing W, Glebov V, et al. Diagnostic components in harsh radiation environments: Possible overlap in R&D requirements of inertial confinement and magnetic fusion systems. Rev. Sci. Instrum. 2008;79:10F304.
6. ITER. Available from: https://www.iter.org/mach [Accessed: January 29, 2016]
7. Girard S, Baggio J, Leray J-L, Meunier J-P, Boukenter A, Ouerdane Y. Vulnerability analysis of optical fibers for Laser Megajoule Facility: Preliminary studies. IEEE Trans. Nucl. Sci. 2005;52(5):1497-1503.
8. Yamamoto S, Shikama T, Belyakov V, Farnum E, Hodgson E, Nishitani T,et al. Impact of irradiation effects on design solutions for ITER diagnostics. J. Nucl. Mater. 2000;283–287(1):60–69.
9. ELI-Nuclear Physics Working Groups. The White Book of ELI Nuclear Physics Bucharest-Magurele, Romania [Internet]. Available from: http://www.eli-np.ro/documents/ELI-NP-WhiteBook.pdf [Accessed: January 29, 2016]
10. Delepine-Lesoille S, Pheron X, Bertrand J, Pilorget G, Hermand G, Farhoud R, et al. Industrial qualification process for optical fibers distributed strain and temperature sensing in nuclear waste repositories. J. Sensors. 2012;369375. DOI: 10.1155/2012/369375
11. Wijnands TJ, De Jonge LK, Kuhnhenn J, Hoeffgen SK, Weinand U. Optical absorption in commercial single mode optical fibres for the LHC machine. Proceedings of the Topical Workshop on Electronics for Particle Physics [Internet], Prague, 3–7 September 2007, pp. 121-124. Available from: http://cds.cern.ch/record/1089249/files/p121.pdf [Accessed: January 29, 2016]
12. Makovec A, Berruti G, Consales M, Giordano M, Petagna P, Buontempo S, et al. Radiation hard polyimide-coated FBG optical sensors for relative humidity monitoring in the CMS experiment at CERN. J. Instrum. 2014;9(03):C03040.
13. O'Keeffe S, McCarthy D, Woulfe P, Grattan MW, Hounsell AR, Sporea D, et al. A review of recent advances in optical fibre sensors for in-vivo dosimetry during radiotherapy. Br. J. Radiol. 2015;88:1050.
14. Barth JL, Dyer CS, Stassinopoulos E.G. Space, atmospheric, and terrestrial radiation environments. IEEE Trans. Nucl. Sci. 2003;50(3):466-482.
15. Barth JL, LaBel KA, Poivey, C. Radiation assurance for the space environment. In: International Conference on Integrated Circuit Design and Technology ICICDT '04; Austin. 2004. p. 323-333.
16. ECSS Secretariat. Space environment, ECSS-E-10-04A, ESA Publications Division [Internet], 2000. Available from: www.spacewx.com/Docs/ECSS-E-ST-10-04C_15Nov 2008.pdf [Accessed: January 29, 2016]
17. Schwank JR, Shaneyfelt MR, Dodd PE. Radiation hardness assurance testing of microelectronic devices and integrated circuits: Radiation environments, physical mechanisms, and foundations for hardness assurance. IEEE Trans. Nucl. Sci. 2013;60(3):2074-2100.
18. El-Kheshen AA. Glass as radiation sensor. In: Nenoi M, editors. Current Topics in Ionizing Radiation Research. InTech; 2012. p. 579-602.
19. Farah K, Mejri A, Hosni F, Hamzaoui AH, Boizot B. Formation and decay of colour centres in a silicate glasses exposed to gamma radiation: Application to high-dose dosimetry. In: Nenoi M., editors. Current Topics in Ionizing Radiation Research. InTech; 2012. p. 603-624.
20. Woody CL, Kierstead JA, Levy PW, Stoll S. Radiation damage in BaF2 crystals. IEEE Trans. Nucl. Sci. 1992;39:515-523.
21. Li PJ, Xie YY, Zhao YL, Yin ZW. Radiation damage barium fluoride in large crystals [Internet], 1992. Available from: http://lss.fnal.gov/archive/other/ssc/ssc-gem-tn-92-000077.pdf [Accessed: January 29, 2016]

22. El-Said AS, Cranney M, Ishikawa N, Iwase A, Neumann R, Schwartz K, et al. Study of heavy-ion induced modifications in BaF2 and LaF3 single crystals. Nucl. Instrum. Methods B. 2004;218:492–497.
23. Izerrouken M, Meftah A, Nekka M. Color centers in neutron-irradiated Y3Al5O12, CAF2 and LiF single crystals. J. Lumin. 2007;127:696–702.
24. Fernández-Rodríguez M, Alvarado CG, Núñez A, Álvarez-Herrero A. Analysis of optical properties behaviour of Clearceram, fused silica and CaF2 glasses exposed to simulated space conditions. In: Proceedings of the International Conference on Space Optics (ICSO 2010), 4–8 October 2010. Available from: http://www.congrexprojects.com/custom/icso/Papers/Session%2011a/FCXNL-10A02-2018721-1-FERNANDEZ-RODRIGUEZ_ICSO_PAPER.pdf
25. Brovetto B, Maxia V, Salis M. On the radioluminescent emission of BaF2. Il Nuovo Cimento D. 1996;18(9):1107-1110.
26. Dafinei I, Fasoli M, Ferroni F, Mihokova E, Orio F, Pirro S, et al. Low temperature scintillation in ZnSe crystals. IEEE Trans. Nucl.Sci. 2010;57(3):1470-1474.
27. Yanagida T, Fujimoto Y, Koshimizu M, Fukuda K. Scintillation properties of CdF2 crystal. J. Lumin. 2015;157:293–296.
28. Kristianpoller N, Rehavi A, Shmilevich A, Weiss D, Chen R.Radiation effects in pure and doped Al2O3 crystals. Nucl. Instrum. Meth B. 1998;141:343-346.
29. Plaksin OA, Stepanov VA, Demenkov PV, Stepanov PA, Skuratov VA, Kishimoto N. Radioluminescence of alumina during proton and heavy ion irradiation. Nucl. Instrum. Methods B. 2003;206:1083–1087.
30. Islamov AKh, Ibragimova EM, Nuritdinov I. Radiation-optical characteristics of quartz glass and sapphire. J. Nucl. Mater. 2007;362:222–226.
31. Izerrouken M, Benyahia T. Absorption and photoluminescence study of Al2O3 single crystal irradiated with fast neutrons. Nucl. Instrum. Methods B. 2010;268:2987-2990.
32. Zhang MF, Zhang HL, Han JC, Guo HX, Xu CH, Ying GB. Effects of neutron irradiation and subsequent annealing on the optical characteristics of sapphire. Effects of neutron irradiation and subsequent annealing on the optical characteristics of sapphire. Physica B. 2011;406:494–497.
33. Faist J, Capasso F, Sivco DL, Sirtori C, Hutchinson AL, Cho AY. Quantum cascade laser. Science. 1994;264(5158):553-556.
34. Wilson CF, Chassefière E, Hinglais E, Baines KH, Balint TS, Berthelier J-J, et al. The 2010 European Venus Explorer (EVE) mission proposal. Exp. Astron. 2012;33(2):305-335.
35. Forouhar S, Borgentun C, Frez C, Briggs R, Bagheri M, Webster CR, et al. Mid-IR semiconductor lasers for space. In: Mid-Infrared Optoelectronics: Materials and Devices (MIOMD-XII); 5–9 October 2014; Montpellier, France.
36. Mihai L, Sporea D, Sporea A. Vâţă I. Testing of ZnSe, CaF2, BaF2, and sapphire windows under alpha particles irradiation. In: 3rd International Conference and Exhibition on Lasers, Optics and Photonics; 1–3 September 2015; Valencia, Spain.
37. Mihai L, Sporea D, Sporea A, Ighigeanu D, Neguţ D. Optical and THz evaluation of components for gas sensing spectroscopy in hazardous environments. In: NATO ARW on THz Diagnostics of CBRN effects and Detection of Explosives & CBRN; 3–6 November 2015; Izmir, Turkey.
38. Mihai L, Sporea D, Sporea A, Craciun G, Manaila E. Electron beam irradiation of materials and components to be used in mid-IR spectroscopy. In: Photonics Europe 2016; 3–7 April, 2016; Brussels, Belgium. paper EPE117-144.
39. Girard S, Kuhnhenn J, Gusarov A, Brichard B, Van Uffelen M, Ouerdane Y, et al. Radiation effects on silica-based optical fibers: Recent advances and future challenges. IEEE Trans. Nucl. Sci. 2013;60(3):2015-2036.
40. Griscom DL. A minireview of the natures of radiation-induced point defects in pure and doped silica glasses and their visible/near-IR absorption bands, with emphasis on self-trapped holes and how they can be controlled. Phys Res Int. 2013;ID 379041.
41. Paul MC, Sen R, Bhadra SK, Pal M, Giri PP, Dasgupta K, et al. Gamma ray radiation induced absorption in Ti doped single mode optical fibres at low dose levels. Opt. Mater. 2007;29:738–745.

42. Paul MC, Sen R, Bhadra SK, Dasgupta K. Radiation response behaviour of Al codoped germano-silicate SM fiber at high radiation dose. Opt. Commun.. 2009;282:872–878.
43. Paul MC, Bohra D, Dhar A, Sen R, Bhatnagar PK, Dasgupta K. Radiation response behavior of high phosphorous doped step-index multimode optical fibers under low dose gamma irradiation. J. Non-Cryst. Solids. 2009;355:1496-1507.
44. Ghosh S, Das S, Paul MC, Dasgupta K, Bohra D, Chaudhary HS,et al. Evaluation of the performance of high phosphorous with germanium codoped multimode optical fiber for use as a radiation sensor at low dose rates. Appl. Opt. 2011;50(25):E80-E85.
45. Jin J, Liu J, Xu J. Effect of spectral dependencies of radiation-induced attenuation in polarization maintaining fibers on interferometric fiber optic gyroscope at near-infrared wavelengths. Optik. 2013;124:5679-5682.
46. Kim Y, Ju S, Jeong S, Kim JY, Lee NH, Jung HK, et al. Gamma-ray irradiation-induced optical attenuation in Co/Fe co-doped alumino-silicate optical fiber for dosimeter application. J. Lightwave Technol. . 2014;32(22):3791-3797.
47. Kir'yanov AV, Ghosh S, Paul MC, Barmenkov YO, Aboites V, Kozlova NS. Ce-doped and Ce/Au-codoped alumino-phosphosilicate fibers: Spectral attenuation trends at high-energy electron irradiation and posterior low-power optical bleaching. Opt. Mater. Express. 2014;4(3):434-448.
48. Girard S, Marcandella C, Morana A, Perisse J, Di Francesca D, Paillet P, et al. Combined high dose and temperature radiation effects on multimode silica-based optical fibers. IEEE Trans. Nucl. Sci. 2013;6(6):4305-4313.
49. Di Francesca D, Boukenter A, Agnello S, Girard S Alessi A, Paillet P, et al. X-ray irradiation effects on fluorine-doped germanosilicate optical fibers. Opt. Mater. Express. 2014; 4(8):1683-1695.
50. Sporea D, Sporea A, Oproiu C. Effects of hydrogen loading on optical attenuation of gamma-irradiated UV fibers. J. Nucl. Mater. 2012;423(1-3):142-148.
51. Sporea D, Mihai L, Sporea A, Lixandru A, Bräuer-Krisch E. Investigation of UV optical fibers under synchrotron irradiation. Opt. Express. 2014;22(25):31473-31485.
52. Girard S, Ouerdane Y, Marcandella C, Boukenter A, Quenard S, Authier N. Feasibility of radiation dosimetry with phosphorus-doped optical fibers in the ultraviolet and visible domain. J. Non-Cryst. Solids. 2011;357:1871-1874.
53. Tomashuk AL, Grekov MV, Vasiliev SA, Svetukhin VV. Fiber-optic dosimeter based on radiation-induced attenuation in P-doped fiber: suppression of post-irradiation fading by using two working wavelengths in visible range. Opt Express. 2014;22(14):16779-16783.
54. Stajanca P, Sporea D, Mihai L, Negut D, Schukar M, Krebber K. Gamma radiation induced effects on perfluorinated polymer optical fibers for sensing applications. In: Bunge C-A, Kruglov R, editors. Perfluorinated Polymer Optical Fibers for Sensing Applications. BoD - Books on Demand: Norderstedt; 2015. ISBN 978-3-7392-1499-3.
55. Mihai L, Sporea D, Negut D, Stajanca P, Krebber K. On-line monitoring of gamma irradiated perfluorinated polymer optical fiber. In: Photonics Europe 2016; 3–7 April, 2016; paper EPE104-58.
56. Sporea D, Sporea A. Radiation effects in sapphire optical fibers. Physica Status Solidi (c). 2007;4(3):1356–1359.
57. Petrie ChM, Windl W, Blue ThE. In-situ reactor radiation-induced attenuation in sapphire optical fibers. J. Am. Ceram. Soc. 2014;97:3883-3889.
58. Petrie ChM, Wilson B, Blue ThE. In situ gamma radiation-induced attenuation in sapphire optical fibers heated to 1000°C. J. Am. Ceram. Soc. 2014;97:3150-3156.
59. Phéron X, Girard S, Boukenter A, Brichard B, Delepine-Lesoille S, Bertrand J,et al. High γ-ray dose radiation effects on the performances of Brillouin scattering based optical fiber sensors. Opt. Express. 2012;20(24):26978-26985.
60. Faustov A, Gusarov A, Wuilpart M, Fotiadi AA, Liokumovich LB, Kotov OI, et al. Distributed optical fibre temperature measurements in a low dose rate radiation environment based on Rayleigh backscattering. In: Berghmans F, Mignani AG, De Moor P, editors. Proceedings of Optical Sensing and Detection II; SPIE 8439: 84390C; 2012.
61. Toccafondo I, Brugger M, Di Pasquale F, Guillermain E, Kuhnhenn J. First steps towards a distributed optical fiber radiation sensing system. International Conference on Space

Optics, Tenerife, Canary Islands, Spain [Internet], 7–10 October 2014. Available from: http://congrexprojects.com/Custom/ICSO/2014/Papers/5.%20Posters/Poster%20Session%202/2.10.66978_Toccafondo.pdf [Accessed: January 29, 2016]

62. Faustov AV, Gusarova A, Wuilpart M, Fotiadi A, Liokumovichc LB, Zolotovskiyf IO, at al. Remote distributed optical fibre dose measuring of high gamma-irradiation with highly sensitive Al- and P-doped fibres. In: Baldini F, Homola J, Lieberman RA, editors. Proceedings of Optical Sensors 2013; SIPE 8774:877404; 2013.
63. Brichard B, Fernandez, AF, Ooms H, Berghmans F. Gamma dose rate effect in erbium-doped fibers for space gyroscopes. In: Proceedings of the 16th International Conference on Optical Fiber Sensors OFS16, Nara, 13–17 October 2003, pp. 336-339. Available from: https://www.researchgate.net/profile/Francis_Berghmans/publication/228759091_Gamma_Dose_Rate_Effect_in_Erbium-Doped_Fibers_for_Space_Gyroscopes/links/ 09e4150b 62 5689e300000000.pdf [Accessed: January 29, 2016]
64. Brichard B, Fernandez AF, Ooms H, van Uffelen M, Berghmans F. Study of the radiation-induced optical sensitivity in erbium and aluminium doped fibres. In: Proceedings of the 7th European Conference on Radiation and Its Effects on Components and Systems RADECS2003, Noordwijk, 15–19 September 2003, pp. 35-38. Available from: http://citeseerx.ist.psu.edu/viewdoc/download?doi=10.1.1.59.2175&rep=rep1&type=pdf [Accessed: January 29, 2016]
65. Lezius M, Predehl K, Stower W, Turler A, Greiter M, Hoeschen C, et al. Radiation induced absorption in rare earth doped optical fibers. IEEE Trans. Nucl. Sci. 2012;59(2):425-433.
66. Girard S, Ouerdane Y, Tortech B, Marcandella C, Robin T, Cadier B, et al. Radiation effects on Ytterbium- and Ytterbium/Erbium-doped double-clad optical fibers. IEEE Trans. Nucl. Sci. 2009;56(6):3293-3399.
67. Rose TS, Gunn D, Valley GC. Gamma and proton radiation effects in Erbium-doped fiber amplifiers: active and passive measurements. J. Lightwave Technol. 2001;19(12):1918-1923.
68. Chang SH, Liu R-Y, Lin C-E, Chou F-I, Tai C-Y, Chen CC. Photo-annealing effect of gamma-irradiated erbium-doped fibre by femtosecond pulsed laser. J. Phys. D: Appl. Phys. 2013;46:495113-495119.
69. Peng T-S, Huang Y-W, Wang LA, Liu R-Y, Chou F-I. Photo-annealing effects for Erbium doped fiber sources after gamma irradiation tests by using 532 nm and 976 nm lasers. In: Jones J, Culshaw B, Ecke W, López-Higuera JM, Willsch R, editors. Proceedings of the 20th International Conference on Optical Fibre Sensors; SPIE 7503: 750375; 2009.
70. Ma J, Li M, Tan L-Y, Zhou Y-P, Yu S-Y, Che C. Space radiation effect on EDFA for inter-satellite optical communication. Optik - Int. J. Light Electron Opt. 2010;121(6):535-538.
71. Singleton BJ. Radiation Effects on Ytterbium-doped Optical Fibers. No. AFIT-ENP-DS-14-J-15, Air Force Institute of Technology Wright-Patterson AFB OH Graduate School of Engineering and Management [Internet], 2014. Available from: http://www.dtic.mil/cgi-bin/GetTRDoc?Location=U2&doc=GetTRDoc.pdf&AD=ADA602 890 [Accessed: January 29, 2016]
72. Fox BP. [thesis]. Investigation of ionizing-radiation-induced photodarkening in rare-earth-doped optical fiber amplifier materials, The University of Arizona; 2013.
73. Mady F, Benabdesselam M, Mebrouk Y, Dussardier B. Radiation effects in ytterbium-doped silica optical fibers: traps and color centers related to the radiation-induced optical losses. In: Proceedings of RADECS 2010, Paper LN2; 2010. Available from: https://hal.archives-ouvertes.fr/hal-00559422/file/Paper_LN2.pdf [Accessed: January 29, 2016]
74. Zotov KV, Likhachev ME, Tomashuk AL, Kosolapov AF, Bubnov MM, Yashkov MV, et al. Radiation resistant Er-doped fibers: optimization of pump wavelength. IEEE Photonics Technol. Lett. 2008;20(17):1476–1478.
75. Zotov KV, Likhachev ME, Tomashuk AL, Bubnov MM, Yashkov MV, Guryanov AN, et al. Radiation-resistant erbium-doped fiber for spacecraft applications. IEEE Trans. Nucl. Sci. 2008;55(4):2213-2215.

76. Thomas J, Myara M, Troussellier L, Burov E, Pastouret A, Boivin D, et al. Radiation-resistant erbium-doped-nanoparticles optical fiber for space applications. Opt. Express. 2012;20(3):2435-2444.
77. Girard S, Mescia L, Vivona M, Laurent A, Ouerdane Y, Marcandella C, et al. Design of radiation-hardened rare-earth doped amplifiers through a coupled experiment/simulation approach. J. Lightwave Technol. 2013;31(8):1247-1254.
78. Hamdalla, TA, Nafee SS. Radiation effects on the gain of thulium doped fiber amplifier: Experiment and modeling. Opt. Laser Technol. 2014;55:46-49.
79. Luo Y, Wen J, Zhang J, Canning J, Peng G-D. Bismuth and erbium codoped optical fiber with ultrabroadband luminescence across O-, E-, S-, C-, and L-bands. Opt. Lett. 2012;37(16):3447-3449.
80. Yan B, Luo Y, Sporea D, Mihai L, Neguţ D, Sang X, et al. Gamma radiation-induced formation of bismuth related active centre in Bi/Er/Yb co-doped fibre. In: Asia Communications and Photonics Conference (ACP); 19–23 November 2015; Hong Kong.
81. Wen J, Liu W, Dong Y, Luo Y, Peng G-D, Chen N, et al. Radiation-induced photoluminescence enhancement of Bi/Al-codoped silica optical fibers via atomic layer deposition. Opt. Express. 2015;23(22):29004-29013.
82. Sporea D, Mihai L, Negut D, Luo Y, Yan B, Ding M, et al. Effects of gamma rays on rare Bi/Er co-doped optical fibers. In: ROMOPTO2015 Conference; 1–4 September 2015; Bucharest, Romania.
83. Butov OV, Golant KM, Shevtsov IA, Fedorov AN. Fiber Bragg gratings in the radiation environment: Change under the influence of radiolytic hydrogen. J. Appl. Phys. 2015;118:074502.
84. Fernandez AF, Brichard B, Berghmans F, Decréton M. Dose-rate dependencies in gamma-irradiated fiber Bragg grating filters. IEEE Trans. Nucl.Sci. 2002;49(6):2874-2878.
85. Fielder RS, Klemer D, Stinson-Bagby KL. High neutron fluence survivability testing of advanced fiber Bragg grating sensors. In: AIP Conference Proceedings on Thermophysics in Microgravity; Commercial/Civil Next Generation Space Transportation; 21st Symposium Space Nuclear Power & Propulsion; Human Space Explorer; Space Colonization; New Frontiers & Future Concepts; 2004, Vol. 699, pp. 650-657.
86. Fernandez AF, Brichard B, Berghmans F, Rabii H, Fokine M, Popov M. Chemical composition fiber gratings in a high mixed gamma neutron radiation field. IEEE Trans. Nucl. Sci. 2006;53(3):1607-1613.
87. Gusarov A, Fernandez AF, Vasiliev S, Medvedkov O, Blondel M, Berghmans F. Effect of gamma–neutron nuclear reactor radiation on the properties of Bragg gratings written in photosensitive Ge-doped optical fiber. Nucl. Instrum. Methods B. 2002;187(1):79-86.
88. Perry M, Niewczas P, Johnston M. Effects of neutron-gamma radiation on fiber Bragg grating sensors: a review. IEEE Sensors. 2012;12(11):3248-3257.
89. Curras E, Virto AL, Moya D, Vila I, Carrión JG, Frövel M, Palomo FR. Influence of the fiber coating type on the strain response of proton-irradiated fiber Bragg gratings. IEEE Trans. Nucl.Sci. 2012;59(4):937-942.
90. Gusarov A, Hoeffgen SK. Radiation effects on fiber gratings. IEEE Trans. Nucl. Sci. 2013;60(3):2037-2053.
91. Maier RRJ, MacPherson WN, Barton JS, Jones JDC, McCulloch S, Fernandez AF, et al. Fibre Bragg gratings of type I in SMF-28 and B/Ge fibre and type IIA B/Ge fibre under gamma radiation up to 0.54 MGy. In: Proceedings of SPIE 17th International Conference on Optical Fibre Sensors, 2005, Vol. 5855, pp. 511-514.
92. Faustov A, Saffari P, Koutsides C, Gusarov A, Wuilpart M, Mégret P, et al. Highly radiation sensitive type IA FBGs for future dosimetry applications. IEEE Trans. Nucl. Sci. 2012;59(4):1180-1185.
93. Krebber K, Henschel H, Weinand U. Fibre Bragg gratings as high dose radiation sensors? Meas. Sci. Technol. 2006;17:1095-1102.
94. de Villiers GJ, Treurnicht J, Dobson RT. In-core high temperature measurement using fiber-Bragg gratings for nuclear reactors. Appl. Therm. Eng. 2012;38:143-150.
95. Sporea D, Stancalie A, Becherescu N, Becker M, Rothhardt M. An electron beam profile instrument based on FBGs. Sensors 2014. 2014;14(9):15786-15801.

96. Hamdalla TA, Nafee SS. Bragg wavelength shift for irradiated polymer fiber Bragg grating. Opt. Laser Technol. 2015;74:167–172.
97. Chaubey S, Joshi P, Kumar M, Arya R, Nath AK, Kher S. Design and development of long-period grating sensors for temperature monitoring. Sādhanā. 32(5):513-519.
98. Kher S, Chaubey S, Kashyap R, Oak SM. Turnaround-Point Long period fiber gratings (TAP-LPGS) as high radiation dose sensors. IEEE Photonics Technol. Lett. 2012;24(9):742-744.
99. Vasiliev SA, Dianov EM, Golant KM, Medvedkov OI, Tomashuk AL, Karpov VI,et al. Performance of Bragg and long-period gratings written in N- and Ge-doped silica fibers under y-radiation. IEEE Trans. Nucl. Sci. 1998;45(3):1580-1583.
100. Henschel H, Hoeffgen SK, Kuhnhenn J, Weinand U. High radiation sensitivity of chiral long period gratings. IEEE Trans. Nucl. Sci. 2010;57(5):2915-2922.
101. Rego G, Fernandez AF, Gusarov G, Brichard B, Berghmans F, Santos JL. Effect of ionizing radiation on the properties of arc-induced long-period fiber gratings. App. Opt. 2005;44(29):6258-6263.
102. Sporea D, Stancălie A, Negut D, Pilorget G, Delepine-Lesoille S, Lablonde L. On-line tests of an optical fiber long-period grating subjected to gamma irradiation. IEEE Photonics J. 2014;6(6):1-9.
103. Sporea D, Stăncalie A, Neguţ D, Pilorget G, Delepine-Lesoilled S, Lablonde L. Comparative study of long period and fiber Bragg gratings under gamma irradiation. Sensor Actuator A-Phys. 2015;233:295-301.

CHAPTER 11

Electrochemical Synthesis of Nanocomposites

Randa Abdel-Karim[1]

[1] *Faculty of Engineering, Cairo University, Egypt*

ABSTRACT

This chapter presents an overview of research efforts focused on both fabrication and properties of nanocomposites prepared by electrodeposition. The nanoparticles can improve the base material in terms of wear resistance, damping properties, and mechanical strength as well as electrical properties. Different kinds of matrix, such as metals, polymers, and ceramic matrix, have been employed for the production of composites reinforced by nano-ceramic particles such as carbides, nitrides, and oxides as well as carbon nanotubes. Theoretical aspects and mechanisms related to the electrodeposition process of nanocomposite films, from aqueous solutions, are discussed.

Keywords: Nano-structured coatings, Electrocodeposition, Theoretical models, Characterization techniques, Applications

1. INTRODUCTION

Materials are considered nanosized when one of the components dimensions are in the nanometer scale, with typical dimensions smaller than 100 nm. Nanocomposites are composites in which one or more of the phases has dimensions in the nanoscale [1].

Nanocomposites are commonly known as materials consisting of two or more dissimilar materials with well-defined interfaces. Generally, one material forms a continuous matrix while the other provides the reinforcement [2].

They are reported to be the materials of 21st century in the view of possessing design uniqueness and property combinations that are not found in conventional composites. The general understanding of these properties is yet to be reached, even though the first inference on them was reported as early as 1992 [3]. Various techniques have been considered to prepare nanocomposite materials including thermal, plasma spraying and physical and chemical vapour deposition. Among these methods that is widely used, electrodeposition (ECD)), which offers several advantages when compared with other techniques [4]. Electrodeposition offers a lot of advantages over other surface modification techniques:

- Capability of manufacturing nanostructured multi-component films.
- High purity of deposited materials.
- Applicable to substrates of complex shape.
- Can be used for deposition of ceramics, glasses, polymers, composites.
- Rigid control of the composition and microstructure of deposit.
- Low cost of equipment and materials.
- Easy to be scaled up to industry level.
- This process avoids the problems associated with high temperature and high pressure processing.
- Reduction of waste often encountered in dipping or spraying techniques [5, 6].

Research into the preparation of nanocomposite coatings, by electrochemically co- deposition of fine particles with metal from electrolytic solutions, has been investigated by numerous authors [7-12].

As in the case of micro-composites, nanocomposite materials can be classified, according to their matrix materials, in three different categories;

- Metal Matrix Nanocomposites (MMNC) such as Cr/Al_2O_3 [13], Ni/Al_2O_3 [14], $Co\text{-}TiO_2$ [15], Zn-Ni/ TiO_2 [16], Al/CNT [17], Mg/CNT [18].
- Ceramic Matrix Nanocomposites (CMNC); such as Al_2O_3/ZrO_2 [19], ceramic/CNT [20],
- Polymer Matrix Nanocomposites (PMNC), such as Thermoplastic/ thermoset polymer/layered silicates [21], polyester/ Fe_2O_3 [22], polyester/TiO_2 [23].

Nanocomposite materials have been extensively investigated in bulk and thin film forms because of their wide range of applications, starting from traditional industries, such as general mechanics and automobiles, paper mills, textiles, and food industries, to high- technology industries, such as microelectronics and magnetoelectronics [24]. In addition, the applications of nanocomposite coatings include wear and abrasion-resistant surfaces, lubrication, high hardness tools, dispersion-strengthened alloys, and protection against oxidation and hot corrosion. It has been also used to produce high surface area cathodes that have been used as electro catalysts for hydrogen electrodes in industrial water electrolysis [25, 26].

2. THEORETICAL MODELS OF COMPOSITES ELECTRODEPOSITION

2.1. Models neglecting the hydrodynamics of fluids

In 1972 Guglielmi [27] proposed the first mechanism on electrocodeposition of inert particles in the metal matrix, and later this mechanism has been adopted by various authors. The model proposed by Guglielmi does not consider mass transfer.

According to this mechanism the process involves a two-step mechanism, as follows;

- In the in the first step, when the particles approach the cathode they become weakly adsorbed at the cathode surface by Van der Waals forces.
- In the second step, particles are adsorbed strongly on the cathode surface by Coulomb forces and consequently are incorporated into the growing metal matrix.

The loose adsorption coverage, a ratio of the area covered by loosely adsorbed particles to the total electrode area, was expressed in terms of the concentration of suspended particles using the classical Langmuir adsorption isotherm. For the strong adsorption rate, the volume of particles strongly adsorbed was given by a Tafel-type exponential relationship at high over potentials that depended on kinetic constants. The volume of metal electrodeposited was obtained by Faraday's law. Thus, the volume fraction of incorporated particles was then formulated as a function of the bulk concentration and the electrode over potential. From this model the volume fraction of incorporated particles, α', can be mathematically expressed by:

$$\frac{\alpha'}{1-\alpha'} = \frac{zF\rho_m V_0}{M_m i_0} e^{(B-A)} \frac{KC_{p,b}}{1+KC_{p,b}} \tag{1}$$

where M_m is the atomic weight, ρ_m is and the density of electrodeposited metal respectively, i_0 the exchanging current density, z the valence of the electrodeposited metal, F the Faraday constant, η the electrode reaction over potential, $C_{p,b}$ the particle concentration in the bulk electrolyte. k is the Langmuir isotherm constant, which is mainly determined by the intensity of interaction between particles and cathode. The parameters V_0 and B are related to particle deposition, and they play a symmetrical role with the parameters. In addition, i_0 and A are related to metal deposition.

In 1987, in order to overcome the shortcoming in the Guglielmi's model Celis et al. [28] proposed another model. The model consists of five consecutive steps:

- The particles are surrounded by ionic clouds;
- Convection towards the cathode surface;
- Diffusion through the hydrodynamic boundary layer;
- Diffusion through the concentration boundary layer;
- Adsorption of particles at the cathode and particles are entrapped within the metal deposit by the reduction of the ionic cloud.

In Celis model, an equation relating weight of particle (W_p), with the weight increase per unit time and surface area (ΔW_p) due to particle incorporation. The weight fraction of particles embedded in percentage is;

$$\frac{\Delta W_p}{\Delta W_m + \Delta W_p} x100 \tag{2}$$

One of the limitations of this model is that the data needed for probability coefficient are not available [29].

In 1995, a theoretical model was proposed by Fransaer et al. [30] to describe the variations in the flow of current to disk electrodes caused by a finite number of spherical and probate particles. They developed a boundary collocation method to calculate the resistance variations and the influence on current in the presence of particles and in their absence. The increase in resistance to the flow of particles was measured as a function of the diameter of the particle. The resistance is increased by larger particles to a greater degree than smaller particles. The primary current distribution around a particle was plotted as a function of the aspect ratio of the particle.

The position of the particle was also determined, from its influence on the current as it flows past the electrode.

2.2. Models including the hydrodynamics of fluids

In 1974 Foster and Kariapper et al. [31] performed adsorption studies on composite coatings. They proposed that if the particles in suspension acquire a positive surface charge, they could be incorporated into the metal film by electrostatic attraction. Their model is based on the following expression;

$$\frac{dVp}{dt} = \frac{N^{*} h C_v}{1 + h C_v} \tag{3}$$

where Vp is the volume fraction of the particles in the deposit, N* is the number of collisions of particles suitable for the co-deposition per second and C_v is volume percent of particles in the plating bath (%).

The parameter h was related to several parameters given by the following expression;

$$h = h^{*}\left(q\Delta E + L i^{2} - ab\right) \tag{4}$$

where h* is a constant, q is the charge density on a particle, ΔE ΔE is the potential field at the cathode, I is the current density, L is the bond strength of the metal/particle per surface area, a is related to the shape and b depends on agitation.

In 1987, Valdes et al [32] reported a model for co-deposition on a rotating ring disk electrode (RRDE). The equation of continuity based on differential mass balance was chosen as a starting point where different mass transport processes (Brownian diffusion and convection) for particles have been considered. According to this model, the electrochemical rate of particle deposition can be written as;

$$r_p = K^0 (C_S^P) n \left[\exp\left(\frac{\alpha ZF}{RT}\right)\eta_a - \exp\left(\frac{1-\alpha ZF}{RT}\right)\eta_a \right] \tag{5}$$

Where K^0 is an electrochemical rate constant which depends on CPSCSP, the concentration of electroactive species (metal cations) adsorbed on the surface of the particels, n is valance and ηaηa is the activation overpotential.

In 1992, Fransaer et al. [33] proposed a trajectory model suggested for particle co-deposition on a rotating disk electrode (RDE). This particle is based on particles larger than 1 micrometer in size. In this model, Navier-Stocks equation for RDE was resolved by using Taylor expansion. The velocity of the particle that can reach towards the cathode is;

$$\frac{dz_p}{dt} = a_p^2 v_{stag} \frac{F'_{stagn}}{F'} - \frac{F_{ext}}{6\pi a_p F'} \tag{6}$$

Where FextFext is the external force at any instant along the particle, F′stagnFstagn' is the force propelling the particle towards the electrode, F′F' is the resistance force felt by the particle while it is moving perpendicular towards electrode, a_p is the particle radius.

In 2000, according to the model of Vereecken et al. [34], the particles kinetics and residence time at the electrode surface have been considered. Convective-diffusion controls the transport of particles to the surface. The influence of particle gravitational force and hydrodynamics is accounted for various current densities. It is valid only when the particle size is smaller than the diffusion layer thickness. In their model they used Fick's first law and diffusion layer thickness. By combining both they got;

$$J_p = -1.554 v^{-\frac{1}{6}} D_p^{213} \left(C_{P,b} - C_{P,s}\right) w^{\frac{1}{2}} \tag{7}$$

where CP, sCP, s, is the particle concentration at the surface, CP,bCP,b is particle concentration in the bulk, $D_p$is the diffusion coefficient, ν is the kinematic viscosity of the solution and ω is the rotation speed of the electrode.

The ratio of the number of moles of particles to the number of moles of metal atoms equals the ratio of their fluxes J_p / J_m.

J_p can be written as;

$$J_p = \frac{3V_{m,M}}{4\pi r^3 zFN_A} \frac{x_y}{1-x_y} i \tag{8}$$

Where Vm,MVm,M is the molar volume of metal film, NANA is the Avagadros number, xyxy is the volume fraction of the particles in the film, r is the radius of the particles, F is Faradays constant, z is the charge of metal ions, i is the current density.

$$Jm = \frac{i}{zF} \tag{9}$$

Combining equations 7, 8, 9;

$$\frac{x_y}{1-x_y} = \frac{4\pi r^3 zFN_A}{3V_{m,M}}\left(1.554 v^{-\frac{1}{6}} D_p^{213}\right)\left(C_{P,b} - C_{P,s}\right)\frac{w^{\frac{1}{2}}}{i} \tag{10}$$

Where ν_p is the volume fraction of particles, ω is the rotation rate.

As indicated in Figure 1, the model of nanoparticles co-deposition suggested by Timoshkov et al. [35], is based on the following stages:

- Coagulation of ultra-fine particles in electrolytic bath,
- Formation of quasi-stable aggregates,
- Transport of the aggregates to the cathode surface by convection, migration and
- diffusion,
- Disintegration of the aggregates in the near-cathode surface,
- Weak adsorption of ultra-fine particles and aggregate fragments onto the cathode surface,
- Strong adsorption of dispersion fraction and embedment.

3. PARTICLE INTERACTIONS AND SUSPENSION STABILITY

3.1. The dlvo theory

The state of dispersion of particles in suspension can be controlled by careful manipulation of the inter-particle forces and their interactions. A quantitative description of the relationship between stability of suspension and energies of interactions between colloidal particles and other surfaces in a liquid has been given by the classical DLVO (Derjaguin-Landau-Verwey-Overbeek) theory. According to this theory, the stability of a colloidal system is determined by the total pair interaction between colloidal particles, which consists of coulombic double-layer repulsion and van der Waals' attraction. The total energy V_T of interaction of two

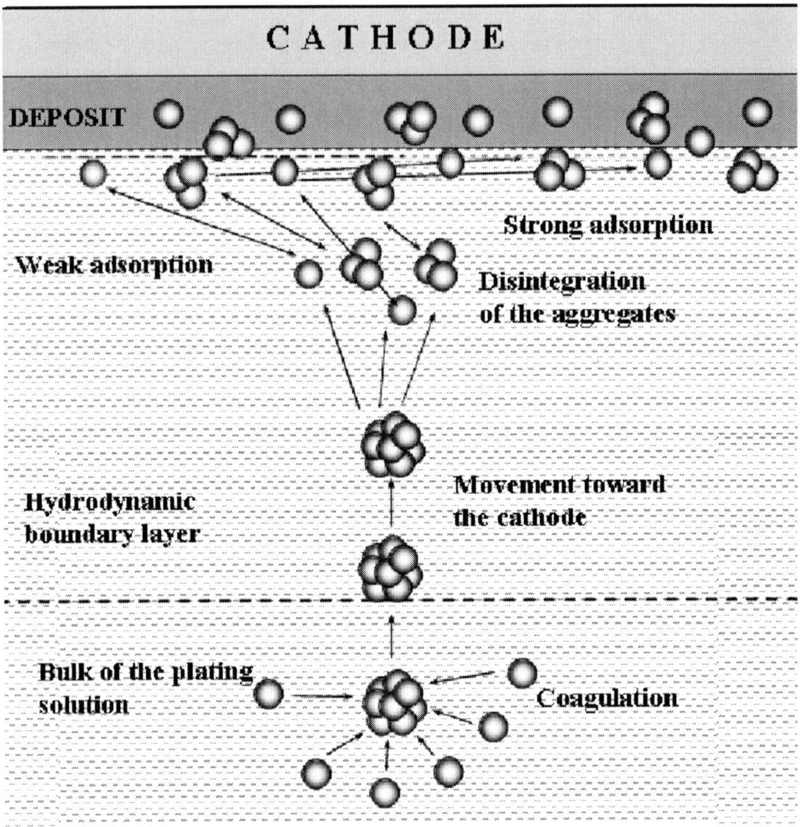

Figure 1. Model of nanoparticles co-deposition process [6, 35]

$$V_T = V_A + V_R \tag{11}$$

The attractive energy V_A of the London-van der Waals' interaction between two spherical particles can be expressed by:

$$V_A = -\frac{A}{6}\left(\frac{2}{S^2-4}+\frac{2}{S^2}\right) \tag{12}$$

where A is the Hamaker constant and S = 2 + H/a, with H the shortest distance between the two spheres and a the particle radius. If H << a, Equation (2) can be simplified to:

$$V_A = -A\frac{a}{12H} \tag{13}$$

The repulsive energy V R is:

$$V_R = 2\pi\varepsilon\varepsilon_0\, a\Psi lin\left[1+e^{-kH}\right] \tag{14}$$

where εε is the dielectric constant of the solvent, ε0ε0 is the vacuum dielectric permittivity, ΨΨ is the surface potential, l/k is the Debye length:

$$k = \left(\frac{e^2\sum n_i z_i}{\varepsilon\varepsilon_0\, kT}\right)^{\frac{1}{2}} \tag{15}$$

where e is the electron charge, k is the Boltzmann constant, T is the absolute temperature, n_i is the concentration of ions with valence z_i [6].

3.2. Particle charging

The charge on the particles can be developed according to the following mechanisms:

a) Ions are selectively adsorbed onto the solid particle from the electrolyte.

b) Ions are dissociated from the solid phase into the electrolyte.

c) Dipolar molecules are adsorbed or orientated at the particles surface.

d) Electrons are transferred between the solid and liquid phase due to differences in work function.

Charged particle in a suspension is surrounded by oppositely charged ions. In the so-called boundary layer, the concentration of these ions is higher than their concentration in the bulk electrolyte. These ions and the particle should move in opposite directions when an electric field is applied. At the same time, the ions are also attracted by the particle, and as a result, a fraction of the ions surrounding the particle will not move in the opposite direction but move along with the particle. Accordingly, the speed of a particle is not determined by the surface charge but by the net charge enclosed in the liquid sphere, which moves along with the particle [6, 29].

3.3. Particle transport

The particle transfer toward the cathode surface occurs by four mechanisms, namely convection, migration, diffusion, and Brownian movement.

Convection: It includes thermal and stirring effects, which can be increased extensively by applying vibration, shock, and other stirring types and temperature gradients.

Migration: It is the movement of positive ions and negative ions, or charged particles, through the electrolyte under the effect of applied potential between the electrodes immersed in that electrolyte. The migration process occurs only for charged particles.

Diffusion: Electrode reaction decreases the concentration of oxidant or reductant at the electrode surface, producing a concentration gradient. Thus, the species movement from the higher to the lower concentration is enhanced. The diffusion process occurs for both charged and uncharged particles.

Brownian's movement: It is dependent on the particle size and may be ignored for particle size larger than 1 μm [24].

3.4. Other interparticle forces

During the electrodeposition process of nanocomposites, the following forces are taken into consideration:

- Mechanical forces, resulting from interaction with the fluid flow and other particles, gravity and buoyancy.
- Electrical forces, due to the electric field presented in the electrolyte.
- Molecular forces working on the particle in the vicinity of the cathode electrodes surface [24].

4. EFFECT OF DEPOSITION PARAMETERS ON ECD

The electrodeposition process and thus the resulted microstructure of composite coating can be affected by many parameters, such as:

- The electrolyte condition: chemical composition of the electrolyte bath, presence of additives, pH, temperature, and electrolyte stirring.

- Current conditions: current density and type of applied current (pulse or DC).
- The properties of the reinforcing particles: particle size, surface properties, concentration, and type of dispersion in the electrolyte [23, 36].

Current Density

Current density plays an important role in controlling the deposition rate which will in turn affect the concentration, composition and morphology of incorporated particles in the coatings. It also influences the thickness of the composite films, such that as the current density increases the thickness of the coatings increases. Low current density produces films with large surface irregularities. When the current density is increased, the amount of particle incorporation obtained has been found to increase for the Ni-TiO_2 system with a relatively slow agitation, decrease for natural or synthetic diamond in Ni and for Cr particles co-deposited in Ni and to be unaffected when co-depositing alumina in Ni [24].

Particles reinforcement in the composite coatings varies with current density. At first, incorporation increases sharply at the beginning with increase in current density till it reaches maximum value followed by sharp decrease. Therefore, hardness of composite coatings mainly increases due to the combined effect of both grain refining as well as of dispersive strengthening. When electroplating at lower current densities, metal ions dissolved from anode are transported at low rate and hence there is insufficient time for these ions to absorb on particles resulting in weak Coulomb force between anions adsorbed on particles leading to lower concentration of electrodeposited particles in the composite coatings. On the other hand, at higher current densities, metal ions dissolved from anode are transported faster than particles by the mechanical agitation which causes a decrease in co-deposition of particles as well as hardness of composite coatings. Therefore, selection of optimum current density is important to enhance the concentration of particles in the composite coatings [37]. The maximum current density for preparing nanocomposite coatings is limited by the limiting current density.

The DC electrodeposition methods are often associated with slower deposition rates and coating defects such as surface roughness, porosity, poor adhesion, undesirable microstructure, etc. Recently, pulse current (PC) and pulse reverse current (PRC). ECD methods have attracted significant attention to improve deposition rates and microstructure of the coatings for improved mechanical and corrosion properties [24, 25].

Bath Temperature

According to Akarapu [25], two contrary behaviors were observed regarding the effect of temperature on the obtained crystallization size. This discrepancy was due to the two contradictory effects of temperature increase on the thermodynamic and kinetic driving force of nucleation process. As the electrolyte's temperature increases, the thermodynamic driving force of crystallization decreases and the critical size of the nucleus will increase. This will lead to lower nucleus densities and formation of coarse grain. On the other hand, the increase in temperature leads to enhancement of the kinetic driving force. This results in an increase in the rate of nucleation and thus fine grains formation [25].

Bath Agitation

Agitation of the plating solution is important in determining particle incorporation. There are various methods of agitation employed include circulation by pumping, purging, of air, ultrasonic agitation, and the plate pumper technique. In general, if the agitation is too slow (laminar flow), the particles in the bath may not disperse completely, except when their density is low. If the agitation is too high (turbulent), particles will not have sufficient time to get attached to the surface, and this results in poor particles incorporation [25].

Increase in the bath agitation in the parallel plate electrode setup has been found to increase the amount of particles co-deposited within the electroplated film for the Ni-Al_2O_3 and Ni-TiO_2 systems. When the agitation is increases, a greater number of particles arrive at the electrode surface and the amount of particle incorporation in the metal film increases. However, if the agitation is too intense, the residence time for the particles at the electrode surface is insufficient and the particles are swept away before they can be incorporated into the growing metal film [25]. The amount of co-deposition has also been observed to decrease in the Cu-SiC and Cu-CrB_2 systems with increasing agitation [24].

Particle Characteristics

Particles can be characterized by their composition and crystallographic phase, as well as by their size, density, and shape. The particle composition can have a dramatic impact on the amount of incorporation obtained for a particular bath composition. For instance three times more TiO_2 than $A1_2O_3$ has reportedly been incorporated into a Ni matrix, under the same deposition conditions. The particle size also effects on the amount of co-deposited particles in the composite coatings. For example when the particle size in the electrolyte increases then amount of adsorbed ions on the surface increases, which leads to the increase in the migration velocity of the particles and also results in a higher columbic force of attraction, leads to increase in the amount of the particles. But the density of particles in the coating decreases as the particle size in the electrolyte increases [25].

The electrode/particle interaction will be affected by the particle properties, such as material type, shape, size, surface charge, concentration, and dispersion in the bath. The co-deposition of nano-sized particles produces a composite coating with a much higher hardness than that achieved with micron-sized particles. The smaller the particle, the more difficult the co-deposition into the metal matrix due to the high tendency of agglomeration. In addition, the smaller the particle size, the greater the effect of colloidal properties (van der Waals, electrostatic, and solvent interaction forces) [24].

Bath composition

The composition of the co-deposition bath is not only defined by the concentration and type of electrolyte used for depositing the matrix metal, but also by the particle loading in suspension, the pH, and the additives used. A variety of electrolytes have been used for the electro co-deposition process to form metal matrix of copper include acid copper sulphate bath, alkaline pyrophosphate bath. Electrolyte concentrations typically range between 100-600 g/l and the particle loading in suspension has ranged from 2-200 g/l [25].

According to Narasimman et al. [38], various additives can be used for preventing the agglomeration of particles, increasing the volume fraction of reinforcing particles in the deposit, and providing good dispersion and thus high hardness. The addition of surfactants plays a role in modifying the surface charge and reducing the particle agglomeration, thus improving their electrostatic adsorption on the cathode surface. As a result of decreasing the agglomeration of particles, the amount of effective particles would be significantly increased, resulting in higher amounts of the reinforcing particles. The addition of surfactants changes the zeta potential of the particles.

The effect of additives in the plating bath on the microstructure and physical properties of deposits was reported by many researchers. For example, the addition of saccharin to plating electrolyte was found to improve the ductility and brightness. The role of additives on a grain refining can be summarized as follows:

- As refiner, blocking the surface by complexation, decreasing the adsorbed ions diffusion on cathode, and thus inhibiting the growth of crystals.
- Controlling the evolution rate of hydrogen on cathode.
- Changing the cathodic overpotential [38].

5. TYPES OF NANOCOMPOSITES

5.1. Metal matrix nanocomposites

Various processes are used to manufacture MMCs which are described here. These processes are classified on the basis of temperature of the metallic matrix during processing. Accordingly, the processes can be classified into five categories: (1) liquid-phase processes, (2) solid–liquid processes, (3) two- phase (solid–liquid) processes and (4) in situ processes. (5) deposition techniques [39-40].

5.1.1. Electrodeposition of nickel nanocomposites

Nickel nanocomposite coatings are used in a wide variety of industrial and engineering applications, such as consumer electronics, chemical, computer and telecommunications industries in order to improve corrosion and wear resistance, modify magnetic and other properties. For nickel matrix electrodeposits, a great variety of particles have been used such as oxides i.e. TiO_2 [41, 42], Al_2O_3[43], CeO_2 [44, 45], ZrO_2 [46], graphene oxide GO [47], carbon nanotubes CNT; Al_2O_3/Y_2O_3/CNT [48], carbides like SiC [49-51], WC [52] and nitrides such as TiN [53] and Si_3N_4 [54].

The Ni-TiO_2 system was selected because nickel is an industrially important coating for corrosion protection [42]. Generally the volume fraction of co-deposited particles is limited for nanoparticles and usually it is inversely proportional to their size. For example, Shao et al. [43] studied the rate of incorporation of two different sizes of Al_2O_3 nanoparticles (50 nm and 300 nm) into a nickel deposit. Using similar operating parameters (1000 rpm, 20 mA cm^{-2}), it was found that the percentage volume fraction of the 300 nm Al_2O_3 in the nickel deposit was much higher compared to the 50 nm Al_2O_3. The presence of nanosized particles in a metal deposit may induce changes in the crystalline structure of the metallic coating.

Ni–CeO_2 nanocomposite coatings were prepared by co--deposition of Ni and CeO_2 nanoparticles with an average particle size of 7 nm onto pure Ni surfaces from a nickel sulphate. The as-codeposited Ni–CeO_2 nanocomposite coatings showed a superior oxidation resistance compared with the electrodeposited pure Ni coating at 800 °C. The co-deposited CeO_2 nanoparticles blocked the outward diffusion of nickel along the grain boundaries. However, the effects of CeO_2 particles on the oxidation resistance significantly decrease at 1050 °C and 1150 °C due to the outward-volume diffusion of nickel controlling the oxidation growth mechanism [45].

According to Zeng et al [46], increasing concentration of the CeO_2 nanoparticles in the bath increased the weight percent of CeO_2 particles in the nanocomposite coatings, and improved the micro- hardness, and the friction, corrosion, and wear behaviour of the coatings. However, excessive CeO_2 nano-particle loadings were detrimental to the coating properties.

Ni–Al_2O_3–SiC hybrid composite films with an acceptable homogeneity and granular structure having 9.2 and 7.7 % vol. Al_2O_3 and SiC nanoparticles, respectively were developed successfully by M. Masoudi et al [50]. Both micro hardness and wear resistance increased owing to dispersion and grain-refinement strengthening of nanoparticles. The oxidation resistance of the Ni–Al_2O_3–SiC hybrid composite coatings was measured to be approximately 41 % greater than the unreinforced Ni deposit and almost 30 % better than the Ni–Al_2O_3 composite coatings.

Ni-SiC nanocomposite coatings were applied on AZ91 magnesium alloy from Watts bath with different SiC content. Micro-hardness of specimens was measured and the results revealed a significant enhancement: from 74 Vickers for bare AZ91 magnesium alloy to 523 Vickers for coated specimen. The obtained data showed the superior corrosion resistance for the coated AZ91 magnesium alloy [51].

The effect of incorporation of Si_3N_4 particles in the Ni nanocomposite coating on the micro hardness, corrosion behaviour has been evaluated by Kasturibai et al. [54]. The micro hardness of the composite coatings (720 HV) was higher than that of pure nickel (310 HV) due to dispersion-strengthening and matrix grain refining and increased with the increase of incorporated Si_3N_4 particle content. The corrosion potential (E_{corr}) in the case of Ni–Si_3N_4 nano-composite had shown a negative shift, confirming the cathodic protective nature of the coating [54].

According to R. Abdel-Karim et al. [55], Ni–Mo nanocomposite coatings were prepared using a nickel salt bath containing suspended Mo nanoparticles using direct current. The crystallite size (18–32 nm) and the surface roughness increased by raising the current density. A remarkable deterioration in the corrosion resistance of Ni–Mo composites was observed with the increase of Mo content. This could be due to crystallite size-refining and surface roughness effect and correspondingly a large surface area. Resulted high surface roughness lead to improved electrocatalytic effect for hydrogen evolution.

5.1.2. Electrodeposition of copper nanocomposites

The conventionally used reinforcements in the Cu matrix such as oxides, carbide nanoparticles etc., have resulted in considerable improvement in the mechanical properties. Copper- TiO_2 nano composites were deposited from an acidic copper sulphate bath. Due to relation of optical properties and photo responsively of TiO_2 nanoparticles to nanoparticle size, surface area and morphology, optimization of these parameters in order to having efficient response have crucial importance. Good quality deposits (finer grain size and more homogeneous) were obtained at rather low pH [56].

According to Quayum et al. [57], Cu-NPs /ZnO / ITO composite film electrode has been prepared by electrodeposition of Cu nanoparticles (NPs), ZnO nanorods on indium tin oxide (ITO). Cu-NPs/ZnO composite electrode had high sensitivity and stability and showed higher catalytic current for glucose oxidation in the field of biosensors.

According to Li et al. [58], Cu/C composites were successfully fabricated by three step electro-deposition. The effects of hot pressure temperature and alloy element Fe on the interface characteristic of Cu/C composite were investigated. The addition of alloy element Fe not only improves the tensile strength and the lateral shear strength of the Cu/C composite, but also changes the interface bond type from the physical bond type to the chemical bond type.

The Cu/ZnO nanocomposite films have been synthesized by cathodic electrodeposition [59]. The SEM and TEM images reveal the formation of hexagonal two-dimensional ZnO sheets and Cu nanoparticles. The Cu/ZnO nanocomposite film showed good emission current stability.

Chrobak.et al. [60], studied both elastic and magnetic properties of the Cu+ Ni nanocomposite coatings with dispersed Ni nanopowder particles obtained by applying the electrolytic deposition method. The magnetization curves M (T) showed a superparamagnetic effect at $T<50$ K which depends on dispersion of magnetic particles in a nonmagnetic matrix. It was also shown that the observed decrease of the apparent Young's modulus due to an increase of coating roughness.

Pavithra et al. [61], demonstrated the synthesis of very hard Cu-Graphene composite foils by a simple, scalable and economical pulse reverse electrodeposition method with a well-designed pulse profile. Carbon as a reinforcement material, in the form of fibres, nanotubes etc., will result in superior mechanical, electrical properties and an extremely high thermal conductivity. The improved strength of metal matrix composites is due to its excellent mechanical properties of Graphene. In addition, graphene is also shown to block dislocation motion in a nanolayered metal-graphene composites resulting in ultrahigh strength [62].

5.1.3. Electrodeposition of cobalt nanocomposites

Nanocrystalline cobalt and cobalt-based alloys are good candidates for the replacement of the highly toxic electroplated hexavalent chromium. They have excellent mechanical and wear-resistant properties, high saturation magnetization, and good thermal stability. Nanocrystalline cobalt and its alloys have higher hardness over the polycrystalline counterparts. Electrochemically prepared Co nanodeposits have three to five times higher coercivity (Hc) than conventional polycrystalline Co [63].

Cobalt composites containing incorporated TiO_2 particles are interesting materials, due to the semiconducting properties of TiO_2, with applications as photocatalysts, particularly in the treatment of polluted water, but, in the same time, due to magnetic properties of Co matrix. More attention has been focused on ferromagnetism in Co-doped TiO_2 anatase films, nanocrystals, nanorods and nanotubes, with potential applications in spintronics. The inclusion of TiO_2 in a nanocomposite layer with increasing current density causes the decrease of saturation magnetization from 279.5 (a.u.) to 76.0 (a.u). Magnetic anisotropy of nanocomposite films depends on the concentration on morphology and magnetic properties of Co-TiO_2 electrodeposited nanocomposite films [63]. According to Sivaraman et al. [64], the electrodeposited the composites exhibited a partially amorphous/nanocrystalline character, with the crystalline fractions originating from the hexagonal-close packed structure of Co. A refinement of the Co crystallite size was observed in deposits containing higher weight percentage of yttrium compounds. The hardness increased with the yttrium content.

5.1.4. Electrodeposition of chromium nanocomposites

Deposition of thick Cr from Cr (III) bath is cumbersome and thin Cr does not have enough wear resistance. The electrodeposition of Cr–Al_2O_3 nanocomposite from Cr (III) bath appears a feasible way for improvement of wear resistance, hardness, lubricity and high temperature resistance of the deposited layer [65].

Cr-TiO_2 nano-composites were prepared by electrodeposition. The addition of TiO_2 in the coating led to improvement corrosion resistance of the composite coating as compared to the pure chromium coating. This improvement is due to the physical barriers produced by TiO_2 to the corrosion process by filling crevices, gaps and micron holes on the surface of the chromium coating. This excellent corrosion resistance of the composite coatings provides wide applications in modern industry [66].

Juneghani et al. [67] examined Cr–SiC nanocomposite coatings with various contents of SiC nanoparticles prepared by electrodeposition in optimized Cr plating bath containing different concentrations of SiC nanoparticles. The co-deposited SiC nanoparticles are uniformly distributed into the Cr matrix which improves the corrosion and wear performance of coating.

5.1.5. Electrodeposition of zinc nanocomposites

Zinc deposits provide good protection to iron and steel components due to its sacrificial nature, low cost and ease of application. Nanocrystalline zinc shows improved properties such as hardness, ductility, corrosion and wear resistance. The graphene based metal materials have been widely reported that the graphene as a support material for metal nanoparticles to obtain catalytic, optoelectronic and magnetic properties. The GO sheets were reduced to be rGO sheets after electrodeposition process while Zn^{2+} also reduced on the surface of rGO sheets. Due to the greatly increasing in the specific surface area, the Zn/rGO film may be used as efficient catalyst for the reaction of methanol generated by carbon dioxide and hydrogen (Figure 2) [68].

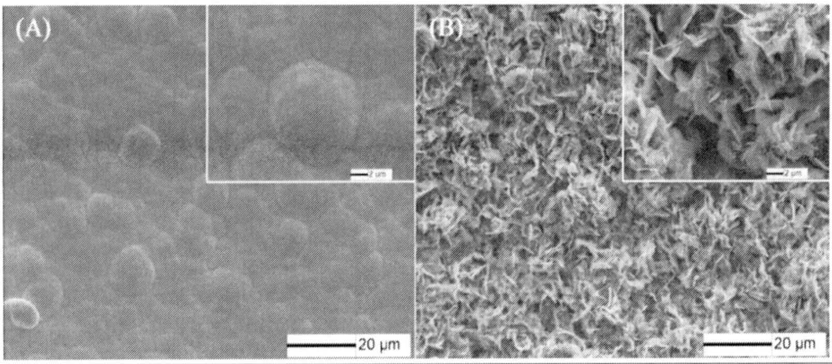

Figure 2. Representative SEM images of (A) Zn film and (B) Zn/rGO film. The insets are the corresponding high-resolution images [68].

Indeed, pure zinc coatings suffer from poor mechanical properties and the incorporation of a second hard phase during the electrodeposition process (e.g. ceramic nanoparticles such as ceria (Figure 3) [69], SiO_2 [70] or Al_2O_3 [71], is of great importance.

According to Gomes et al. [72], Zn-TiO_2 coatings were successfully prepared by co-deposition method. The film's surface is rough and the TiO_2 agglomerates are randomly distributed on the metallic matrix formed by staked hexagonal plates oriented in different direction. These unique characteristics are essential for the thermal conversion of Zn to ZnO. The photo-electrochemical degradation of organic molecules was achieved with this type of Ti/ZnO-TiO_2 photoanodes (Figure 4).

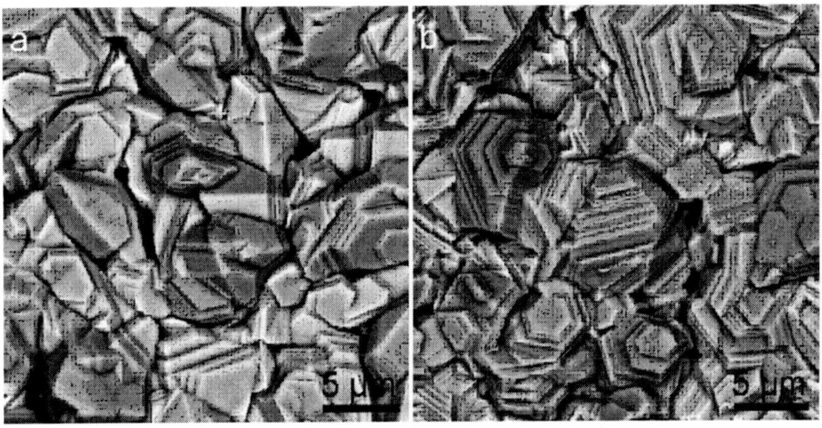

Figure 3. Surface SEM image of (a) pure zinc and (b) zinc-ceria nanocomposite coatings [69].

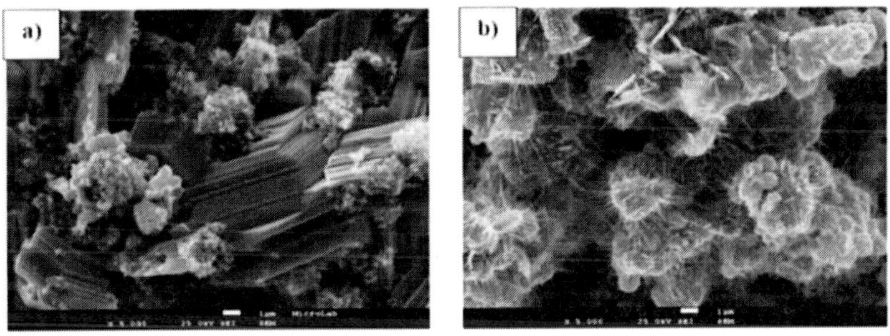

Figure 4. FEG-SEM images for (a) as-deposited and (b) annealed Zn-TiO2 nanocomposite films [72].

5.2. Ceramic matrix nanocomposites

Ceramic matrix composites (CMCs) can extensively overcome the intrinsic brittleness and lack or mechanical reliability of monolithic ceramics, which are otherwise attractive for their high stiffness and strength. In addition to mechanical effects, the reinforcing phase can improve electrical conductivity, thermal expansion coefficient, hardness and thermal shock resistance. The combination of these characteristics with intrinsic advantages of ceramic materials such as high-temperature stability, high corrosion resistance, light weight and electrical insulation, makes CMCs very attractive functional and structural materials for a variety of applications. A wide range of reinforcing fibers have been investigated, including those based on SiC, carbon, alumina, and mullite. However, carbon fibers are among the highest performance toughening elements investigated [73, 74].

Electrodeposition of ceramic films employing electrochemical methods is a fast evolving field. The methods for electrodeposition may be divided into two: electrophoretic and electrolytic deposition. Electrolytic deposition can be driven using cathodic currents by either reducing the metal ions, which causes their deposition, e.g., Cu_2O, or by driving a proton dependent reducing process, leading to an increase of the pH on the electrode surface and the subsequent metal hydroxide deposition. The deposition of ceramic nanocomposite materials is of utmost importance in different fields spanning from supercapacitors to anticorrosion films and from coatings of medical implants to advanced lubricant materials [75].

5.3. Polymeric matrix nanocomposites pncs

Cathodic electrodeposition of polymer-metal and polymer-ceramic materials is an area of intense interest. Such materials exhibit interesting mechanical, thermal, electrical, optical and magnetic properties whose applications have motivated the development of electrochemical techniques [76, 77]. PNCs are among the most promising class of new materials because they are light weight, flexible, low cost, anti-corrosive, and have good process ability and mechanical properties. Most polymers are electrical insulators and are used for a variety of insulating applications in industry. However, the inherent electrical insulation means polymers tend to hold the electrostatic charges, and allow the electromagnetic frequency interference (EMI) to travel through without loss [78].

Some polymer/nanoparticle (NP) composites have already been reported in the literatures, such as polyaniline/metal oxide NP composites, PPy/TiO_2, PPy/Ti, PPy/Au, PPy/Ag, PPy/Pt, and PPy/Pd, prepared by electrodeposition [79]. The incorporation of metal nanoparticles into the conducting polymer offers improved performance for both the host and the guest. They can be applied in application in electronics because incorporation of metal clusters is known to increase the conductivity of the polymer. The applications of these composites have also been extended to various fields such as, sensors, photovoltaic cells, memory devices, protective coatings against corrosion, and supercapacitors. The application of these composites in catalysis is of particular interest. The polymer allows the control of the environment around the metal center, thus influencing selectivity of the chemical reactions [80].

Conducting polymer/inorganic oxide nanocomposites have recently attracted great attention due to their unique microstructure, physiochemical and electro-optical properties, and a wide range of their potential usage in battery cathodes and in the construction of nanoscopic assemblies in sensors and microelectronics. Conductive polymer/ZnO nanocomposites coating were prepared on type-304 austenitic stainless steel (SS) using H_2SO_4 acid as electrolyte by potentiostatic methods. It was found that ZnO nanoparticles improve the barrier and electrochemical anticorrosive properties of polymer [81].

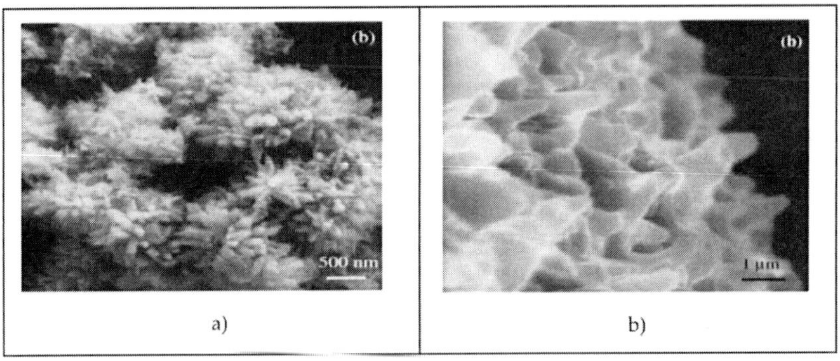

Figure 5. SEM micrographs of the deposits; (a) ZnO NPs, (b) ZnO NPs [81].

Electrically conducting polypyrrole (PPy)–iron group (Ni, Co, Fe and their alloys) composite films have been electrodeposited at cathodic process to give "tailored" soft and hard magnetism. The composites of the PPy doped with dodecylsulfate (DS) (PPy–DS) with NiFe alloy showed soft magnetism with coercivity HC, k <9 Oe, while the PPy–DS with CoMnP alloy showed hard magnetism with HC, k>1085 Oe. These results indicated the possibility conducting polymer–magnetic metals composites fabrication by electrodeposition [82].

The incorporation of metal nanoparticles into the conducting polymer offers enhanced performance for both the host and the guest. They have diverse application potentials in electronics because incorporation of metal clusters is known to increase the conductivity of the polymer. The applications of these composites have also been extended to various fields such as, sensors, photovoltaic cells, memory devices, protective coatings against corrosion, and supercapacitors. Of particular interest is the application of these composites in catalysis. The polymer allows the control of the environment around the metal center, thus influencing selectivity of the chemical reactions [83].

Y. Lattach et al. [84] stated that nanocomposite anode materials for water oxidation have been readily synthesized by electrodeposition of iridium oxide nanoparticles into poly (pyrrole-alkylammonium) films. Electroanalytical investigations have shown that the electrocatalytic activity of iridium oxide nanoparticles is fully maintained when they are incorporated in the polymer matrix.

Electrodeposition process was carried out by dispersing UHMWPE powders in an electrolytic solution of cobalt sulphate/cobalt chloride solution to obtain Co/UHMWPE composite coating on stainless steel substrate (304L). UHMWPE was selected as surface modifier element, due to its high biocompatibility and low coefficient of friction. This material can be used in many biomedical applications [85].

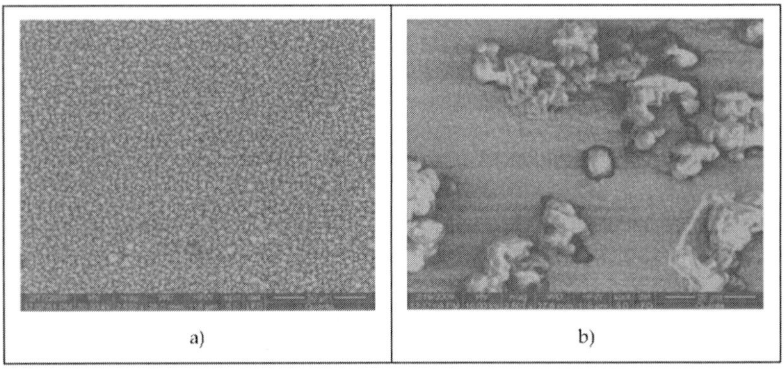

Figure 6. SEM surface morphology of coatings obtained at current density of 48 mA/cm^2 and deposition time of 30 min: (a) Pure Co; (b) Co/UHMWPE Composite Coatings with 30g/L [85].

6. CONCLUSIONS

In conclusion, materials science and engineering has experienced a tremendous growth in the field of nanocomposite development with enhanced chemical, mechanical, and physical properties. The progress in nanocomposites is varied and covers many industries. Nanocomposites with a variety of enhanced physical, thermal and other unique properties can be manufactured. They have properties that are superior to conventional microscale composites and can be synthesized using simple and inexpensive techniques. Theoretical aspects and mechanisms related to the electrodeposition process of nanocomposite films, from aqueous solutions, were discussed. The three types of matrix (metal, ceramic, polymers) nanocomposites were presented, underlining the needs for these materials and some recent results on structure, properties and potential applications.

REFERENCE

1. Gomes, I. Pereira, B. Fernandez and R. Pereiro, Advances in Nanocomposites - Synthesis, Characterization and Industrial Applications, (edit. B. S. R. Reddy) chapter 21, Electrodeposition of Metal Matrix Nanocomposites: Improvement of the Chemical Characterization Techniques, INTECH, 2011.
2. V. B. Singh and D. K. Singh, An Overview on the Preparation, Characterization and Properties of Electrodeposited-Metal Matrix Nanocomposites, Nanosci. Technol. 2014, 1(3), 1-20.
3. P. Henrique C. Camargo, K. G. Satyanarayan. F. Wypych, Synthesis, Structure, Properties and New Application Opportunities, Materials Research 2009, 12(1), 1-39.
4. V. Viswanathan, T. Laha, K. Balani, A. Agarwal, S. Seal, Challenges and Advances in Nanocomposite Processing Techniques, Materials Science and Engineering 2006, R 54 121–205.
5. R. Abdel-Karim and A.F. Waheed, Modern Surface Engineering Treatments, (edit. M. Aliofkhazraei), Chapter 6: Nanocoatings, INTECH, 2013.
6. X. Pang, Fabrication of Advanced Organic-Inorganic Nanocomposite Coatings for Biomedical Applications by Electrodeposition, Ph. D. thesis, McMaster University, Canada, 2008.
7. Riccardo Casati and Maurizio Vedani, Metal Matrix Composites Reinforced by Nano-Particles—A Review, Metals 2014, 4, 65-83.
8. .A. Avidko, Metal-Graphene Nanocomposite with Enhanced Mechanical Properties: A Review, Rev. Adv. Mater. Sci. 2014, 38, 190-200.
9. Cai and Y. Wang, Novel Nanocomposite Materials for Advanced Li-Ion Rechargeable Batteries, Materials 2009, 2, 1205-1238.
10. N. Silvestre, State-of-the-art Review on Carbon Nanotube Reinforced Metal Matrix Composites, International Journal of Composite Materials 2013, 3(6A), 28-44.
11. U. Erb, K. T. Aust, and G. Palumbo, Nanostructured Materials, (Edit: C. C. Koch), Chapter 6: Electrodeposited Nanocrystalline Materials, (Second Edition), Elsevier,2006.
12. T. Ohgai, Electrodeposited Nanowires and Their Applications, (edit: N. Lupu), Chapter 3: Fabrication of Functional Metallic Nanowires Using Electrodeposition Technique, INTECH, 2011.
13. M. S. Doolabi, S. K. Sadrnezhaad, D. S. Doolabi and M. Asadirad, Influence of Pulse Parameters on Electrocodeposition of Cr–Al2O3 Nanocomposite Coatings from Trivalent Chromium Bath, International Heat Treatment and Surface Engineering 2012, 6 (4), 178-184.
14. E. Beltowska-Lehman, A. Goral, P. Indyka, Electrodeposition and Characterization of Ni/Al2O3 Nanocomposite Coatings, Archives of Metallurgy and Materials 2011, 56(4), 919-931.
15. S. Mahdavi, S.R. Allahkaram, Composition, Characteristics and Tribological Behaviour of Cr, Co–Cr and Co–Cr/TiO2 Nano-composite Coatings Electrodeposited from Trivalent Chromium Based Baths, Journal of Alloys and Compounds 2015, 635(25), 150–157.
16. M. Praveen and T. V. Venkatesha, Electrodeposition and Corrosion Resistance Properties of Zn-Ni/TiO2 Nano composite Coatings, International Journal of Electrochemistry 2011, 2011, Article ID 261407, 4 pages.
17. Agarwal, S. R. Bakshi, D. Lahiri Carbon Nanotubes: Reinforced Metal Matrix Composites, Chapter 4: Metal-Carbon nanotube systems, CRC, 2011.

18. Sommers, Q. Wang, X. Han, C. T' Joen, Y. Parkd, A. Jacobi, Ceramics and Ceramic Matrix Composites for Heat Exchangers in Advanced Thermal Systems A review, Applied Thermal Engineering 2010, 30 (11), 1277-1291.
19. M. Abdullah, J. Ahmad, M. Mehmood, and M. Mujahid, Effect of CTAB Addition on Improvement of Properties of Al2O3 (w)–Al2O3 (n)–ZrO2 (n) (3mole% Yttria Stabilized Tetragonal Zirconia) Nanocomposite. International Journal of Information and Education Technology 2012, 2(5), 543-546.
20. M. Sathish Kumar, K. Manonmani, B. Parasuram, S. Karthikeyan, Composites Reinforced with Carbon Nanotubes - A Review, J. Environ. Nanotechnol. 2013, 2(3), 67-80.
21. L. Wei, N. Hu and Y. Zhang, Synthesis of Polymer—Mesoporous Silica Nanocomposites, Materials 2010, 3, 4066-4079.
22. G. Naveen Kumar, Y.V. Mohana Reddy, K. Hemachandra Reddy, Synthesis and Characterization of Iron Oxide Nanoparticles Reinforced Polyester/nanocomposites, International Journal of Scientific and Research Publications 2015, 5(8), 1-13.
23. K Nam, Y Tsutsumi, C Yoshikawa, Y Tanaka, R Fukaya, T Kimura, H Kobayashi, T Hanawa, and A Kishida, Preparation of Novel Polymer–Metal Oxide Nanocomposites with Nanophase Separated Hierarchical Structure, Bull. Mater. Sci. 2011, 34(7), 1289–1296.
24. Y. H. Ahmad and A. M. A. Mohamed, Electrodeposition of Nanostructured Nickel-Ceramic Composite Coatings: A review, Int. J. Electrochem. Sci. 2014, 9, 1942 – 1963.
25. Akarapu, Surface Property Modification of Copper by Nanocomposite Coating, Master thesis, National Institute of Technology Rourkela, India, 2011.
26. N. Saheb, N. Ul Qadir, M. U. Siddiqui, A. M. Arif, S. S. Akhtar and N. Al-Aqeeli Characterization of Nano-reinforcement Dispersion in Inorganic Nanocomposites: A Review, Materials 2014, 7, 4148- 4181.
27. N. Guglielmi, Kinetics of the Deposition of Inert Particles from Electrolytic Baths, J. Electrochem. Soc. 1972, 119, 1009-1012.
28. P. Celis, J.R. Roos, C. Buelens, A Mathematical Model For the Electrolytic Codeposition of Particle with a Metallic Matrix, J. Electrochem. Soc. 1987, 13, 1402- 1408.
29. T. R. Khan, Nanocomposite Coating: Codeposition of SiO2 Particles During Electrogalvanizing, Ph. D Thesis, Bochum University, Germany, 2011.
30. J. Fransaer, V. Bouet, J. P. Celis, C. Gabrielli, F. Huet and G. Maurin, Perturbation of the Flow of Current to a Disk Electrode by an Insulating Sphere, J. Electrochem. Soc. 1995, 142(12), 4181-4189.
31. J. Foster and A.M.J. Kariapper, Study of Mechanism of Formation of Electrodeposited Coatings, Trans. Inst. Met. Finish. 1973, 51, 27–31.
32. J. L. Valdes, Electrodeposition of Colloidal Particles, J. Electrochem. Soc. 1987, 134(4), 223C-225C.
33. J. Fransaer, J. P. Celis and J. R. Roos, Analysis of the Electrolytic Codeposition of Non-Brownian Particles with Metals, J. Electrochem. Soc. 1992, 139(2), 413-425.
34. P. M. Vereecken, I. Shao and P. C. Searson, Particle Codeposition in Nanocomposite Films, J. Electrochem. Soc. 2000, 147(7), 2572-2575.
35. Timoshkov, V. Kurmashev and V. Timoshkov, Advances in Nanocomposite Technology, (Edit. by Dr. A. Hashim), Chapter 3: Electroplated Nanocomposites of High Wear Resistance for Advanced Systems Application, INTECH, 2011.
36. Z. Abdel Hamid, Review Article: Composite and Nanocomposite Coatings, Journal of Metallurgical Engineering (ME) 2014, 3(1), 29-42.
37. T. Borkar, Electrodeposition of Nickel Composites Coatings, Mumbai University, India, 2007.
38. P. Narasimman, M. Pushpavanam and V.M. Periasamy, Effect of Surfactants on the Electrodeposition of Ni-SiC Composites, Portugaliae Electrochimica Acta 2012, 30(1), 1-14.
39. C. Kandpal, J. Kumar, H. Singh, Production Technologies of Metal Matrix Composite: A Review, IJRMET 2014, 4 (2), 27-32.
40. F. C. Walsh and C. Ponce de Leon, A Review of the Electrodeposition of Metal Matrix Composite Coatings by Inclusion of Particles in a Metal Layer: an Established and Diversifying Technology, Transactions of the IMF 2014, 92 (2), 83-98.

41. I. Palvlov, L. Benea, J. P. Celis, L. Vazouez, Influence of Nano-TiO2 Co-deposition on the Morphology, Microtopography and Cryttallinity of Ni/Nano-TiO2 Electrosynthesis Nanocomposite Coatings, Digest Journal of Nanomaterials and Biostructures 2013, 8(3), 1043 – 1050.
42. Y. Li, X. Sun, J. Qiao, Mechanical and Corrosion-Resistance Performance of Electrodeposited Titania–Nickel Nanocomposite Coatings, Surf. Coat. Technol. 2005, 192, 331– 335.
43. Shao, P.M. Vereecken, R.C. Cammarata, P.C. Searson, Kinetics of Particle Codeposition of Nanocomposites, J. Electrochem. Soc. 2002, 149, C610- C614.
44. H. Zhang, Y. Zhou, J. Sun, Preparation and Oxidation Behavior of Electrodeposited Ni–CeO2 Nanocomposite Coatings, Trans. Nonferrous Met. Soc. China 2013, 23, 2011–2020.
45. Y.B. Zeng, N.S. Qu and X.Y. Hu, Preparation and Characterization of Electrodeposited Ni-CeO2 Nano-composite Coatings with High Current Density, Int. J. Electrochem. Sci. 2014, 9, 8145 – 8154.
46. Zhitomirsky, M. Niewczas, and A. Petric, Synthesis and Magnetic Properties of Ni-Zirconia Composites, Materials and Manufacturing Processes 2003, 18(5), 719–730.
47. L. You, W. Guofeng, L. Qi, and Y. Mo, Ni/GO Nanocomposites and its Plasticity, Manufacturing Rev. 2015, 2, 8-13.
48. S. Mirzamohammadi, M. Kh. Aliov, A. R. Sabur, A. Hassanzahdeh-Tabrizi, Study of Wear resistance and Nanostructure of Tertiary Al2O3/Y2O3/CNT Pulsed Eectrodeposited Ni-Based Nanocomposite, Physicochemical Mechanics of Materials 2010, 1, 67- 75.
49. M. H. Fini, A. Amadeh, Nanomaterials: Applications and Properties (NAP-2011). Vol. 1, Part I, 2011.
50. M. Masoudi, M. Hashim, H. M. Kamari, Characterization of Novel Ni–Al2O3–SiC Nano-composite Coatings Synthesized by Co-electrodeposition, Appl Nanosci. 2014, 4, 649–656.
51. L. Shi, C. Sun, P. Gao, F. Zhou, W. Liu, Mechanical Properties and Wear and Corrosion Resistance of Electrodeposited Ni–Co/SiC Nanocomposite Coating, Applied Surface Science 2006, 252, 3591–3599.
52. k. Singh, M. K. Tripathi, V. B. Singh, Electrodeposition and Characterization of Ni-WC Composite Coating from Non – aqueous Bath, International Journal of Materials Science and Applications 2013, 2(2), 68-73.
53. J. Wang, F. Xia and M. Huang, Preparation and Mechanical Properties of Ni-TiN Composite Layers by Ultrasonic Electrode Position, Research Journal of Applied Sciences, Engineering and Technology 2013, 6(7), 1303-1308.
54. S. Kasturibai and G. P. Kaliagnan, Pulse Electrodeposition and Corrosion Properties of Ni–Si3N4 Nanocomposite Coatings, Bull. Mater. Sci. 2014, 37, 721–728.
55. R. Abdel-Karim, J. Halim, S. El-Raghy, M. Nabil, A. Waheed, Surface Morphology and Electrochemical Characterization of Electrodeposited Ni–Mo Nano-composites as Cathodes for Hydrogen Evolution, Journal of Alloys and Compounds 2012, 530, 85–90.
56. S. M. Delgosha, S. Sharifi, Influence of pH on the Electrochemical Deposition of Composite Coatings in Copper Matrix with TiO2 Nanoparticles, Optics 2014, 3(1), 1-4.
57. M. E. Quayum, B. Biswas and M. D. Khairul, H. Bhuiyan, Electro-synthesis of Cu / ZnO Nanocomposite Electrode on Ito Electrode and Its Application in Oxidation of Ascorbic Acid and Glicose, J. Asiat. Soc. Bangladesh, Sci. 2013, 39(2), 183-190.
58. Y. Li, P. Bai, Y. Li, Fabrication and Fibre Matrix Interface Characteristics of Cu/C (Fe) Composite, Science of Sintering 2009, 41, 193-198.
59. F. J. Sheini, J. Singh, O.N. Srivasatva, D. S. Joag, M. A. More, Electrochemical Synthesis of Cu/ZnO Nanocomposite Films and Their Efficient Field Emission Behavior, Applied Surface Science 2010, 256, 2110–2114.
60. Chrobak, M. Kubisztal, J. Kubisztal, E. Chrobak, G. Haneczok, Microstructure, Magnetic and Elastic Properties of Electrodeposited Cu+ Ni Nanocomposites Coatings, Journal of Achievements in Materials and Manufacturing Engineering 2011, 49(1), 17-26.
61. L. P. Pavithra, B. V. Sarada, K. V. Rajulapati, T. N. Rao & G. Sundararajan, A New Electrochemical Approach for the Synthesis of Copper-Graphene Nanocomposite Foils with High Hardness, Scientific Reports 2014, 4:4049, 1- 7.

62. M. Poiana, L. Vlad, P. Pascariu, A. V. Sandu, V. Nica, V. Georgescu, Effects of Current Density on Morphology and Magnetic Properties of Co-TiO2 Electrodeposited Nanocomposite Films, Optoelectronics and Advanced Materials– Rapid Communications 2012, 6 (3-4), 434 - 440.
63. R. Abdel-Karim, Handbook of Nanoelectrochemistry, (Edit.M.Aliofkhazraei, A.H. Makhlouf), Chapter 2: Electrochemical Fabrication of Nanostructures,Springer International Publishing, Switzerland, 2016.
64. K. M. Sivaraman, O. Ergeneman, S.Pané, E. Pellicer, J. Sort, K. Shou, S. Surinach, M.D. Baró, B.J. Nelson, Electrodeposition of Cobalt–yttrium Hydroxide/oxide Nanocomposite Films From Particle-free Aqueous Baths Containing Chloride Salts, Electrochimica Acta 2011, 56, 5142–5150.
65. M. S. Doolabi, S. K. Sadrnezhaad, D. Salehi Doolabi and M. Asadirad, Influence of Pulse Parameters on Electrocodeposition of Cr–Al2O3 Nanocomposite Coatings From Trivalent Chromium Bath, International Heat Treatment and Surface Engineering 2012, 6 (4), 178-184.
66. M. Noroozifar, M. Khorasani-Motlagh, Z. Yavari, Effect of Nano-TiO2 Particles on the Corrosion Behavior of Chromium-Based Coatings, Int. J. Nanosci. Nanotechnol. 2013, 9(2), 85-94.
67. M. A. Juneghani, M. Farzam, H. Zohdirad, Wear and Corrosion Resistance and Electroplating Characteristics of Electrodeposited Cr–SiC Nano-composite Coatings Trans. Nonferrous Met. Soc. China 2013, 23, 1993–2001.
68. H. Wenting, Z. Liqun, N. Haiyang, L. Huicong, and L. Weiping, Thin Film of Zn/rGO Nanocomposites Fabricated by Electrodeposition, International Journal of Materials Science and Engineering 2013, 1(1), 32-35.
69. L. Exbrayat, E. Calvi, T. Douillard, G. Marcos, C. Savall, C. Berziou, J. Creus, and P. Steyer, Role of Ceria Nanoparticles on the Electrodeposited Zinc Coating's Growth: Interest of a TEM-Scale Investigation, ECS Electrochemistry Letters 2014, 3 (9), D33-D35.
70. Y. Ullal, A. Chitharanjan Hegde, Corrosion Protection of Electrodeposited Multilayer Nanocomposite Zn-Ni-SiO2 Coatings, Электроннаяобработкаматериалов, 2013, 49(2), 73–79.
71. Constantinescu, F. Oprea, O. Mitoseriu, A Comparative Study of the Properties of Zinc-SiO2 and Zinc-Al2O3 Composite Layers, The 4th International Conference Advanced Composite Materials Engineering COMAT 2012, 18- 20 October 2012, Brasov, Romania.
72. Gomes, T. Frade and I. D. Nogueira, Morphological Characterization of Zn-Based Nanostructured Thin Films, Current Microscopy Contributions to Advances in Science and Technology, (Edit.A. Méndez-Vilas,), Formatex Research Center-Spain,2014.
73. J. Cho, A. R Boccaccini, M. SP Shaffer, Ceramic Matrix Composites Containing Carbon Nanotubes, Journal of Materials Science 2009, 44(8), 1934-1951.
74. M. Rosso, "Ceramic and Metal Matrix Composites: Routes and Properties", Journal of Materials Processing Technology 2006, 175, 364–375.
75. R. Toledano, R. Okner and D. Mandler, Electrochemical Codeposition of Ceramic Nanocomposite Films, ENS'07 Paris, France, 3-4 December 2007.
76. Zhitomirsky and X. Pang, Electrodeposition of Polymer-metal and Polymer-ceramic Nanocomposites, 207th ECS Meeting, 2006.
77. M. Oliveira and A.V. Machado, Nanocomposites: Synthesis, Characterization and Applications, (Edit. X. Wang), Chapter 4. Preparation of Polymer-Based Nanocomposites by Different Routes, Nova Science Publisher, 2013.
78. X. Luo, Electrically Conductive Metal Nanowire Polymer Nanocomposites, Ph. D. thesis, University of Alberta, USA, 2013.
79. B. Moghaddam, T. Nazari, J. Badraghi, M. Kazemzad, Synthesis of ZnO Nanoparticles and Electrodeposition of Polypyrrole/ZnO Nanocomposite Film, Int. J. Electrochem. Sci., 2009, 4, 247 – 257.
80. P. Satyananda, K. Balasubramanian, V. Thirukkallam, K. Varadarajan, Synthesis and Characterization of Metal Nanoparticle Embedded Conducting Polymer–Polyoxometalate Composites, Nanoscale Res Lett 2008, 3, 14–20.

81. Ganash, Anticorrosive Properties of Poly (o-phenylenediamine)/ZnO Nanocomposites Coated Stainless Steel, Journal of Nanomaterials 2014, (2014), Article ID 540276, 8 pages.
82. A. Ferreira, S.C. Domench and P.C. Lacaze, Synthesis and Characterization of Polypyrrole/TiO2 Composites on Mild Steel, Journal of Applied Electrochemistry 2001, 31, 49-56.
83. J. M. Ko, D.Y. Park, N.V. Myunga, J.S. Chungb, K. Nobe, Electrodeposited Conducting Polymer–magnetic Metal Composite Films, Synthetic Metals 2002, 128, 47–50.
84. Y. Lattach, J. F. Rivera, T. Bamine, A. Deronzier, and J. Moutet, Iridium Oxide–Polymer Nanocomposite Electrode Materials for Water Oxidation, ACS Appl. Mater. Interfaces 2014, 6, 12852–12859.
85. L. Benea, M. Mardare-Pralea, Electrodeposition of UHMWPE Particles with Cobalt for Biomedical Applications, Digest Journal of Nanomaterials and Biostructures 2011, 6(3), 1025-1034.

CHAPTER 12

Large-Scale Surfactant-Free Synthesis of p-Type SnTe Nanoparticles for Thermoelectric Applications

Guang Han [1], Ruizhi Zhang [2], Srinivas R. Popuri [3], Heather F. Greer [4], Michael J. Reece [2], Jan-Willem G. Bos [3], Wuzong Zhou [4], Andrew R. Knox [5] and Duncan H. Gregory [1,]*

[1]*WestCHEM, School of Chemistry, University of Glasgow, Glasgow G12 8QQ, UK*

[2]*School of Engineering & Materials Science, Queen Mary University of London, London E1 4NS, UK*

[3]*Institute of Chemical Sciences and Centre for Advanced Energy Storage and Recovery, School of Engineering and Physical Sciences, Heriot-Watt University, Edinburgh EH14 4AS, UK*

[4]*EaStCHEM, School of Chemistry, University of St. Andrews, St. Andrews, Fife KY16 9ST, UK*

[5]*School of Engineering, University of Glasgow, Glasgow G12 8QQ, UK*

ABSTRACT

A facile one-pot aqueous solution method has been developed for the fast and straightforward synthesis of SnTe nanoparticles in more than ten gram quantities per batch. The synthesis involves boiling an alkaline Na_2SnO_2solution and a NaHTe solution for short time scales, in which the NaOH concentration and reaction duration play vital roles in controlling the phase purity and particle size, respectively. Spark plasma sintering of the SnTe nanoparticles produces nanostructured compacts that have a comparable thermoelectric performance to bulk counterparts synthesised by more time- and energy-intensive methods. This approach, combining an energy-efficient, surfactant-free solution synthesis with spark plasma sintering, provides a simple, rapid, and inexpensive route to p-type SnTe nanostructured materials.

Keywords: tin telluride; synthesis; structure; thermoelectrics; nanomaterials

1. INTRODUCTION

Sustainable energy generation and conversion technologies have received extensive, world-wide attention, because of ever-growing global energy demands, decreasing fossil fuel reserves, and the negative influence arising from the combustion of these fuels on the environment [1,2]. Thermoelectrics are capable of directly converting thermal energy into electric energy and therefore, can be applied to the generation of useful electricity from both waste heat and solar heat, in various applications. The thermoelectric conversion efficiency of a material is governed by its dimensionless figure of merit, $ZT = S^2\sigma T/\kappa$, with S, σ, T, and κ representing the Seebeck coefficient, electrical conductivity, absolute temperature, and thermal conductivity, respectively [1,2,3,4,5,6,7]. Recently, tin chalcogenides have received extensive attention, due to their lead-free, less toxic nature [8,9,10,11,12]. Interestingly, SnTe resembles its Pb counterpart in terms of its crystal structure and band structure, suggesting that it has the potential to be an efficient thermoelectric material [13,14,15]. However, the high concentration of Sn vacancies in "pristine" SnTe creates a very high carrier concentration. This leads to p-type semiconductivity and, in fact, poor thermoelectric performance, with a low Seebeck coefficient and a high electronic component of the total thermal conductivity [16,17]. Recently, several approaches have been employed in order to significantly improve the thermoelectric performance of SnTe. For example, strategies such as band structure engineering, via the introduction of resonant impurity levels through In doping [18] and band convergence through Cd [19], Mg [20], Hg [21], Mn [22], or Ag alloying [23], have been applied to improve the power factor ($S^2\sigma$) of SnTe. Equally, solid solution alloying [24,25], nanostructuring [18,19,25,26,27,28,29,30], introducing interstitial atoms [31], and all-scale hierarchical architecturing [22], have been used to minimise the thermal conductivity by enhanced phonon scattering, pushing ZT to ~1.1–1.3 at ~823–873 K [18,19,20,29], and ~1.3–1.4 at ~900–923 K [21,22,26]. Nonetheless, the synthesis of SnTe requires time-consuming and energy-intensive processes, such as the heating, melting and annealing of precursor materials at high temperatures (~973–1423 K) [19,20,21,22,23,24,25,26,28,29,30,31,32,33]. Ahead of any potential utilisation of SnTe in thermoelectrics, it is critical to develop an inexpensive and scalable synthesis approach that would allow SnTe to be produced with control over the structure and performance.

Monoliths which are consolidated directly from metal chalcogenide nanomaterials have been demonstrated to be efficient thermoelectrics, mainly due to the enhanced phonon scattering by nanostructures [34,35,36,37]. In contrast, chemical vapour deposition (CVD) can produce one- and two-dimensional nanostructures of metal chalcogenides, such as SnTe [38,39,40,41], which can facilitate nanodevice fabrication. Conversely, however, it is problematic to then scale up this synthesis to produce thermoelectrics in large quantities. Bottom-up solution methods enable the energy-efficient synthesis of metal chalcogenide nanostructures to be realised, while simultaneously controlling their crystal structure, particle dimensions, and morphology [34,35,36,37,42,43]. Organic surfactants are commonly used in solution synthesis, to control the morphology of nanostructures through surface modification. Such surfactants are adsorbed on the surfaces of the synthesised nanostructures and influence their electrical conductivity [42,44]. In contrast, surfactant-free solution synthesis can produce chalcogenide nanomaterials with cleaner surfaces, ensuring a higher electrical conductivity [45,46]. To date, the reported solution approaches for SnTe normally use surfactants and only produce small amounts of samples (<500 mg) [47,48,49,50,51]. Further, there are limited reports of the thermoelectric performance of SnTe nanomaterials synthesised by solution methods [52,53]. Given the potential to improve the properties of the telluride, it is important to develop new solution approaches to SnTe nanostructures and to investigate the thermoelectric performance of the pellets consolidated from these nanomaterials.

Herein, we demonstrate a one-pot surfactant-free aqueous solution approach for the fast and straightforward preparation of >10 g of SnTe nanoparticles, with short reaction durations. The phase-pure, surfactant-free nanoparticles can be sintered into dense compacts with a stable p-type conducting behaviour and relatively low thermal conductivity.

2. RESULTS

2.1. Materials Synthesis

In a typical synthesis of SnTe nanoparticles, 100 mmol NaOH (Sigma-Aldrich Ltd., Dorset, UK, 99.99%) and 10 mmol $SnCl_2 \cdot 2H_2O$ (Sigma-Aldrich Ltd., Dorset, UK, 99.99%) were added to 50 mL deionised water (DIW) within a two-neck round-bottom flask, to yield a transparent Na_2SnO_2 solution. In parallel, 10 mmol Te (Sigma-Aldrich Ltd., Dorset, UK, 99.8%) and 20 mmol $NaBH_4$ (AlfaAesar, Haverhill, MA, USA, 98%) were added to 50 mL DIW within a single-neck round-bottom flask, to prepare a purple 0.2 mol·L^{-1} NaHTe solution:

$$2Te + 4NaBH_4 + 7H_2O \rightarrow 2NaHTe + Na_2B_4O_7 + 14H_2\uparrow. \quad (1)$$

After the Na_2SnO_2 solution was heated to its boiling temperature using an oil bath, the freshly prepared 50 mL NaHTe solution was promptly injected into the solution (1:1 Na_2SnO_2:NaHTe molar ratio for all experiments; Scheme 1a), leading to the immediate formation of a black precipitate (Scheme 1b). The mixture was heated to the boiling point, held for 2 h and allowed to cool to room temperature under Ar (BOC, 99.998%) on a Schlenk line. The products were collected by centrifuge, washed with DIW and ethanol several times and dried in an ambient atmosphere at 50 °C for 12 h. The amount of NaOH used (namely 0, 50, 100, and 150 mmol, corresponding to NaOH:$SnCl_2$ molar ratios of 0, 5, 10, and 15, respectively) and the duration of the boiling process (5 min, 2 h, 24 h), were independently changed to investigate their effects on the synthesis. Scaled-up syntheses were performed with five-fold precursor amounts, i.e., using 500 mmol NaOH, 50 mmol $SnCl_2 \cdot 2H_2O$, 50 mmol Te, and 100 mmol $NaBH_4$ (Scheme 1c); the products, with a typical yield of ~11.7 g (Scheme 1d), demonstrated phase purity and a morphology identical to the products synthesised on a lower scale. The synthesised samples used for the characterisation and performance evaluation were stored in an Ar-filled MBraun glove box (<0.5 ppm H_2O, <0.5 ppm O_2), to avoid possible reactions with ambient air.

2.2. Materials Characterisation

The phase composition and crystal structures of the as-prepared samples were investigated by powder X-ray diffraction (PXD), using a PANalytical X'pert Pro MPD diffractometer in Bragg-Brentano geometry (Cu Kα_1 radiation, λ = 1.5406 Å). Diffraction data were collected at room temperature with a step size of 0.017 ° over 10 ° ≤ 2θ ≤ 90 ° for 1 h. The crystal structures of the synthesised products were refined via the Rietveld method against PXD data, using the GSAS and EXPGUI software packages [54,55], with the previously published SnTe structure as a basis [56]. For the refinement, a shifted Chebyschev function (type 1 within GSAS) and a Pseudo-Voigt profile function (type 2 within GSAS) were applied to model the background and peak shape, respectively.

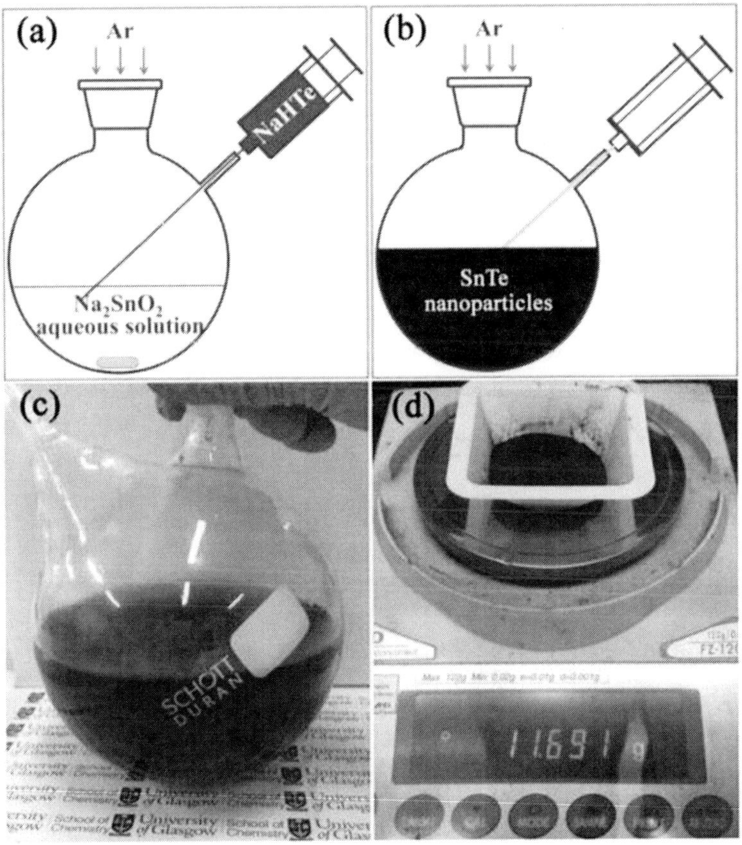

Scheme 1. Solution synthesis of SnTe nanoparticles: schematic drawings showing (**a**) injection of $NaHTe_{(aq)}$ into $Na_2SnO_{2(aq)}$ to trigger the reaction ($NaHTe + Na_2SnO_2 + H_2O \rightarrow SnTe + 3NaOH$) and (**b**) formation of SnTe nanoparticles; digital photographs showing (**c**) the nanoparticle solution after a scale-up synthesis and (**d**) a corresponding typical yield of SnTe nanoparticles (~11.7 g, ca. 95%) produced in such a one-pot synthesis.

The morphological and chemical characteristics of the synthesised products were investigated by scanning electron microscopy (SEM, Carl Zeiss Sigma, 5 and 20 kV for imaging and elemental analyses, respectively), equipped with energy-dispersive X-ray spectroscopy (EDS, Oxford Instruments X-Max 80, Oxford Instruments, Oxfordshire, UK). The synthesised powders were spread onto a conductive carbon tape that was mounted on a standard SEM sample stub. The microstructure and crystal structure were further characterised by transmission electron microscopy (TEM) using a JEOL JEM-2011 electron microscope fitted with a LaB_6 filament, operating at an accelerating voltage of 200 kV. TEM images, high resolution TEM (HRTEM) images and selected area electron diffraction (SAED) patterns were recorded using a Gatan 794 Multiscan CCD camera. The SnTe powders were dispersed in ethanol by sonication for 30 s, to obtain a uniform dispersion. Two to five drops of the suspension were dropped onto a 3 mm diameter holey C-coated Cu TEM grid. Thermogravimetric-differential thermal analysis (TG-DTA) of the samples was performed using a Netzsch STA 409 thermal analyser located in an Ar-filled MBraun glove box (<0.1 ppm H_2O, <0.1 ppm O_2). Approximately 25 mg of SnTe pellet was heated to 700 °C in an alumina pan under flowing Ar (60 $mL\cdot min^{-1}$), at a heating rate of 5 $°C\cdot min^{-1}$.

2.3. Materials Performance Evaluation

To measure the thermoelectric performance of SnTe nanoparticles, SnTe powders were loaded into a graphite die (diameter of 15 mm) within an Ar-filled glovebox, and then sintered using a spark plasma sintering (SPS) furnace (FCT HP D 25, FCT System GmbH) into dense pellets (relative density of >98%) at 450 °C for 5 min, in a vacuum (~4 Pa) with a uniaxial pressure of 50 MPa. The Seebeck coefficient (S) and electrical conductivity (σ) of the pellets were measured simultaneously, using a Linseis LSR-3 instrument under a helium atmosphere, within a temperature range of 300–530 K. The uncertainty in the measurement of S and σ are ~5%, leading to ~10% uncertainty for the thermoelectric power factor ($S^2\sigma$) measurement. The thermal conductivity (κ) of the pellets was calculated using $\kappa = DC_p\rho$, where D, C_p, and ρ are the thermal diffusivity coefficient, specific heat capacity, and density, respectively. D was measured using a Linseis LFA 1000 instrument in a vacuum, within a temperature range of 300–530 K. C_p values for pellets were calculated from the weighted average of the reported values of C_p for SnTe [18] and SnO_2 [57] and the Dulong-Petit limit of Te [58] (where the respective weights were determined from the refined phase fractions from the Rietveld refinements against PXD data). The density, ρ, of the pellets was measured by the Archimedes method. The uncertainty in the κ measurement is ~8%. The electronic thermal conductivity (κ_e) was estimated by the Wiedemann-Franz law ($\kappa_e = L\sigma T$, where L was estimated according to $L = exp\ (-S/116)$ [59]; S values are positive in this study) and the lattice thermal conductivity (κ_L) was calculated by subtracting κ_e from κ. ZT was calculated using $ZT = S^2\sigma T/\kappa$, with an uncertainty of ~15%.

3. RESULTS AND DISCUSSION

The injection of a NaHTe solution into aqueous Na_2SnO_2 leads to the immediate formation of a precipitate of SnTe (Scheme 1a,b). Boiling the suspension for 2 h generates >10 g crystalline, phase-pure SnTe nanoparticles (Scheme 1c,d). The PXD pattern (Figure 1a) shows that all of the reflections can be exclusively indexed to cubic SnTe (ICDD card No. 65-0319) [60]. Rietveld refinement against the PXD data (Figure 1a; Table 1 and Table 2) confirms that the SnTe product crystallises in the cubic space group *[Math Processing Error]*, with a = 6.3234(1) Å. SEM images (Figure 1b,c) reveal that the product exists as approximately isotropic nanoparticles, with individual diameters of 60–160 nm and an average diameter of ~95 nm. EDS spectra (Figure 1d) taken across the samples as point and area scans, consistently generate Sn:Te atomic ratios of 51.5(5):48.5(5).

TEM images (Figure 2a,b) confirm that the SnTe nanoparticles are 60–160 nm across. An exemplar SAED pattern (Figure 2c) collected from an isolated nanoparticle reveals the single crystalline nature of the particle and can be indexed along the [100] zone axis. A set of lattice spacings of 3.2 Å, intersecting at 90°, can be measured from the corresponding HRTEM image (Figure 2d), and correspond to the {002} lattice spacings, confirming the high crystallinity of the SnTe nanoparticles.

A controlled synthesis of phase-pure SnTe depends on certain, well-defined experimental parameters. For example, the NaOH concentration is a critical synthesis parameter for the synthesis of pure, crystalline SnTe nanoparticles. Without the addition of NaOH, the obtained product is poorly crystalline and contains a large fraction of Te impurity (Figure 3a). By contrast, when increasing the NaOH:$SnCl_2$ molar ratio to 5:1, the phase fraction of Te significantly decreases, accompanied by a sharp increase in the amount of SnTe (Figure 3a). At higher ratios (10:1, 15:1), phase-pure, highly crystalline SnTe can be obtained (Figure 3a). Clearly, increasing the NaOH concentration leads to SnTe with a higher purity. NaHTe can be easily oxidised into Te. Therefore, it is reasonable to deduce that NaOH can promote the disproportionation of Te into $Te^{2-}$and TeO_3^{2-}, to eliminate this Te impurity and to drive the formation of SnTe to completion [47,52], probably via:

$$3Te + 6NaOH \rightarrow 2Na_2Te + Na_2TeO_3 + 3H_2O. \quad (2)$$

SEM images (Figure 3b,c) reveal that the product synthesised with a NaOH:$SnCl_2$ ratio of 15:1, forms as nanoparticles with a mean size of ~115 nm across, which is slightly larger than the particle size obtained with a hydroxide:chloride ratio of 10:1 (Figure 1). As might be expected, the duration of the reaction provides another means to tune the SnTe particle size. Phase-pure SnTe persists over 5 min and 24 h reaction durations (Figure 3d), but the average particle sizes of the telluride can be tuned from ~65 to ~115 nm across, respectively (Figure 3e,f).

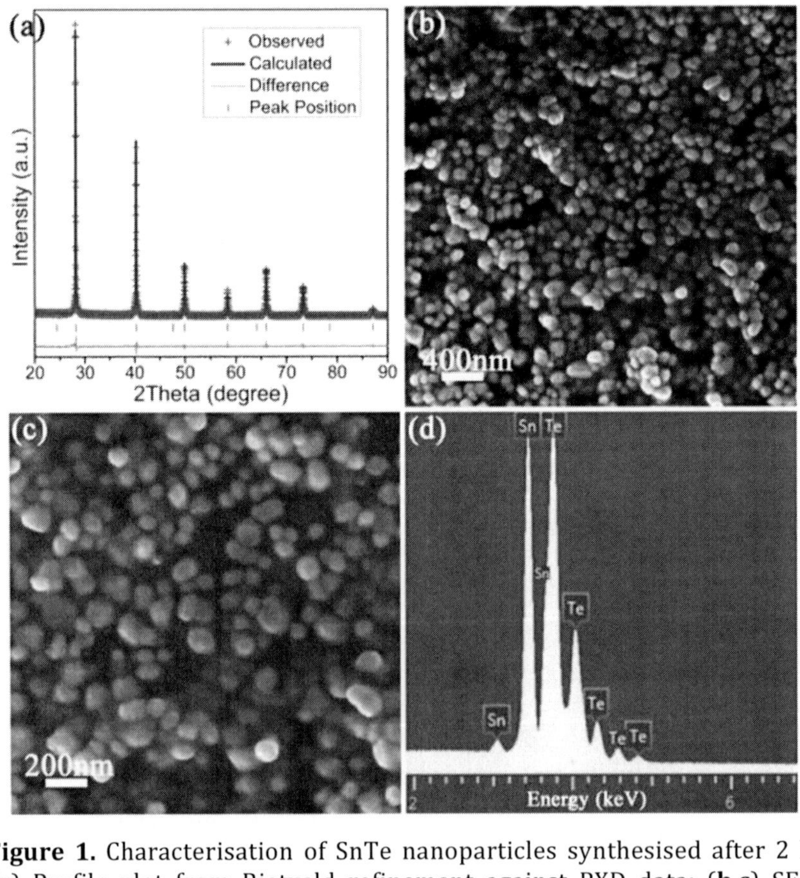

Figure 1. Characterisation of SnTe nanoparticles synthesised after 2 h: (**a**) Profile plot from Rietveld refinement against PXD data; (**b**,**c**) SEM images; (**d**) EDS spectrum.

Table 1. Crystallographic data for SnTe nanoparticles synthesised after 2 h boiling.

Chemical Formula	SnTe
Crystal System	Cubic
Space Group	$Fm\bar{3}m$
a (Å)	6.3234(1)
Volume (Å^3)	252.84(1)
Z	4
Formula Weight (g·mol^{-1})	246.29
Calculated density (g·cm^{-3})	6.470
R_{wp}	0.1181
R_p	0.0832
χ^2	1.449

Table 2. Atomic parameters for SnTe nanoparticles synthesised after 2 h boiling.

Atom	Wyckoff Symbol	x	y	z	100*U_{iso}/Å^2	Occupancy
Sn	4a	0	0	0	3.87(8)	1
Te	4b	0.5	0.5	0.5	2.49(7)	1

The ability to prepare SnTe nanostructures in >10 g quantities repeatedly allowed the facile fabrication of SnTe dense pellets, via SPS techniques. Pellets with ~98% of the theoretical SnTe density, sintered from 2 h solution-synthesised nanomaterials (10:1 NaOH:$SnCl_2$), were obtained. Rietveld refinement against the PXD data (Figure 4a, Table 3) shows that the pellets consist of a main phase of cubic SnTe and two minority phases of tetragonal SnO_2 and trigonal Te (72.3(1) wt % SnTe – 20.4(3) wt % SnO_2 – 7.3(2) wt % Te). This indicates that SnTe nanoparticles were partially oxidised and decomposed into SnO_2 and Te during the SPS process at a high sintering temperature, probably through the following reaction:

$$SnTe + O_2 \rightarrow SnO_2 + Te. \quad (3)$$

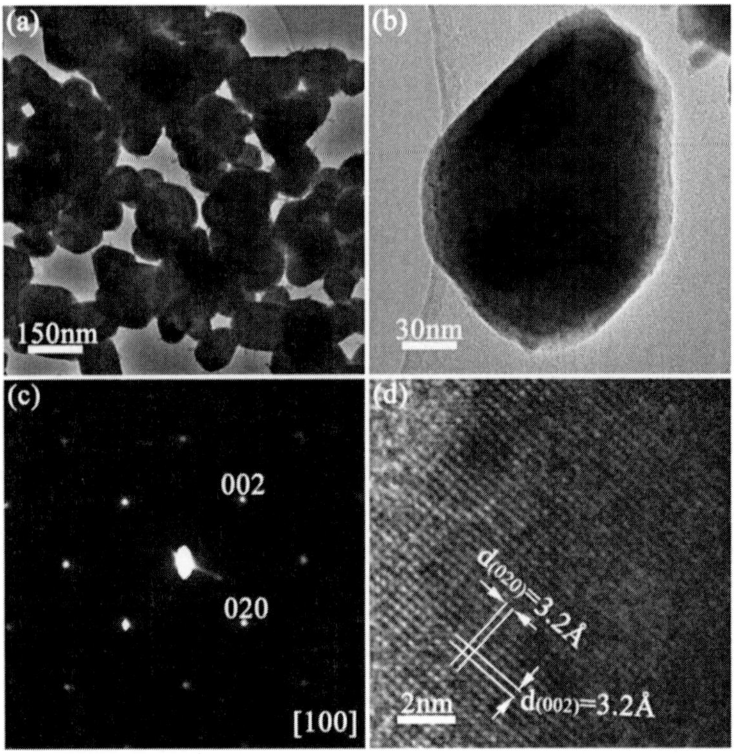

Figure 2. TEM characterisation of SnTe nanoparticles synthesised after 2 h: (**a**,**b**) TEM images of a group of nanoparticles and a single particle respectively; (**c**) SAED pattern collected from the particle in (**b**); (**d**) corresponding HRTEM image taken from the same particle.

Table 3. Crystallographic data for SnTe pellet.

Chemical Formula	SnTe	SnO_2	Te
Crystal System	Cubic	Tetragonal	Trigonal
Space Group	*Fm$\overline{3}$m (225)*	*$P4_2/mnm$ (136)*	*$P3_121$ (152)*
a (Å)		4.7610(10)	4.4836(16)
b (Å)	6.3441(7)		
c (Å)		3.2008(10)	5.9551(34)
Volume (Å^3)	255.33(8)	72.55(3)	103.67(5)
Z	4	2	3
Formula Weight (g·mol^{-1})	246.29	150.69	127.6
Calculated density (g·cm^{-3})	6.407	6.898	6.131
Phase fraction (wt %)	72.3(1)	20.4(3)	7.3(2)
R_{wp}		0.1497	
R_p		0.1120	
χ^2		2.359	

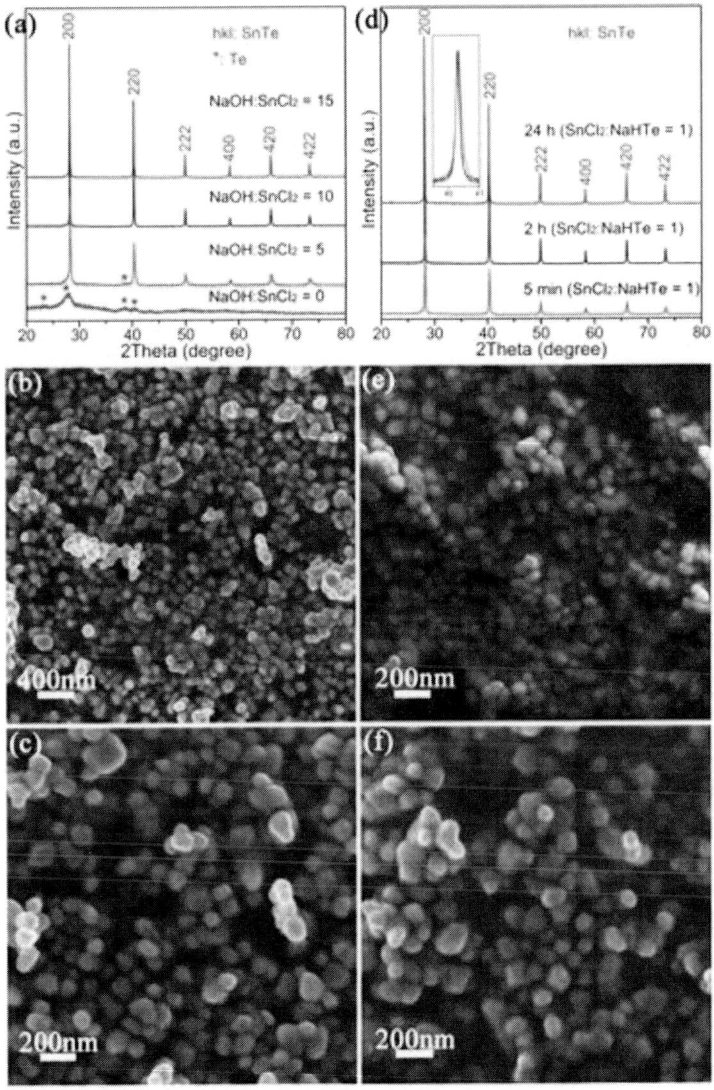

Figure 3. Characterisation of the products synthesised under varied conditions: (**a**) PXD patterns of products synthesised over 2 h with different NaOH:$SnCl_2$ precursor molar ratios (namely 0:1, 5:1, 10:1, 15:1); (**b**,**c**) SEM images of the SnTe nanoparticles synthesised over 2 h with a precursor molar ratio of 15:1; (**d**) PXD patterns of 10:1 NaOH:$SnCl_2$ products synthesised over different reaction durations (namely 5 min, 2 h, 24 h) (the inset shows the (220) peaks in a magnified section of the diffraction pattern); (**e**,**f**) SEM images of the 10:1 NaOH:$SnCl_2$ nanoparticles synthesised in 5 min and 24 h, respectively.

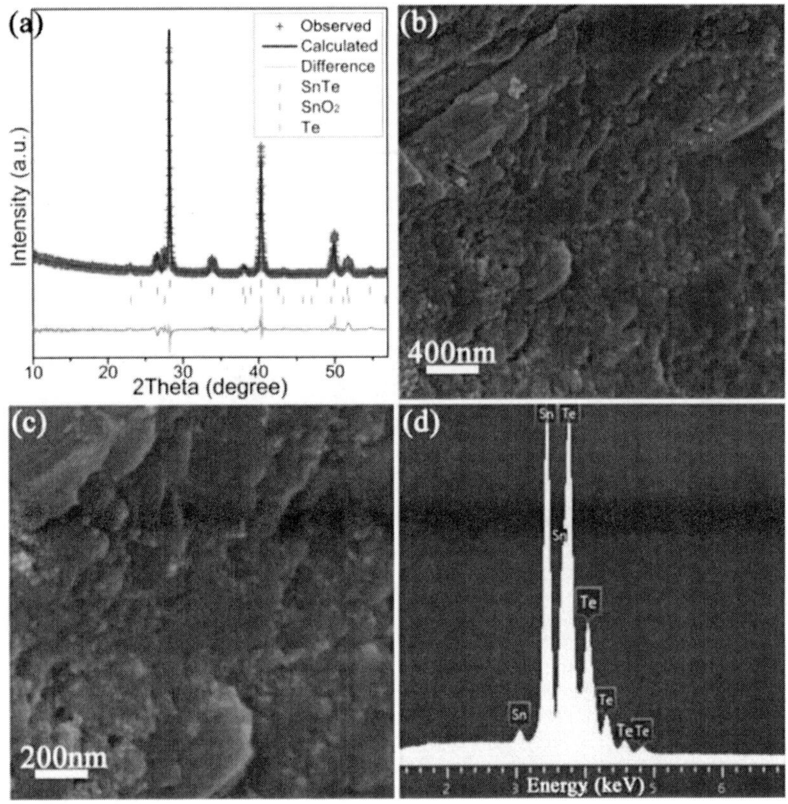

Figure 4. Characterisation of the SnTe pellet: (**a**) Profile plot from Rietveld refinement against PXD data; (**b**,**c**) SEM images; (**d**) EDS spectrum.

This exothermic oxidation process and an ensuing endothermic Te melting process have been detected at ~570 K and ~660 K, by thermogravimetric-differential thermal analysis (TG-DTA) of SnTe powder under an Ar flow (Figure 5a). The melting point of Te in this study is lower than that typically reported for bulk Te (~723 K), which is likely to be due to the small size of the Te product particles generated by the oxidation (Equation (3)) [61]. A similar oxidation/decomposition process has been observed in SnSe pellets which have been hot pressed from surfactant-free SnSe nanoparticles [43]. There is an expectation that the reactive surfaces of the small nanoparticles adsorb O_2, contributing to the oxidation to SnO_2 [43]. SEM images (Figure 4b,c) reveal that the pellets are constructed from a "matrix" of densely packed nanoparticles, with smaller (~20 nm) nanoparticles almost uniformly distributed within this "matrix". Similar nanoparticles with sizes of ~5–20 nm were formed when surfactant-free SnSe nanoparticles were hot pressed into dense pellets and these ~5–20 nm particles were identified as SnO_2 [43]. Considering that the as-synthesised SnTe nanoparticles are 60–160 nm across, we deduce that the smaller nanoparticles distributed across the internal structure of the sintered SnTe pellet most likely belong to $SnO_2$and/or Te. EDS otherwise confirms that the Sn:Te atomic ratio is ~55.5(5):45:5(5) (Figure 4d) in the pellet. Based on the phase fractions from the Rietveld refinement against the PXD data and the elemental analyses from EDS, it is estimated that about ~32% of the original SnTe nanoparticles decomposed/oxidised, with ~58% of the Te produced in this reaction (Equation (3)) vaporising during SPS. TG-DTA of an SnTe pellet under an Ar flow shows a negligible weight loss below 620 K, but suggests that thermal decomposition is initiated above this temperature (Figure 5b). As for the pellet sample, an endothermic peak was observed at ~660 K, which could be related to the melting of Te impurity in the pellet. In contrast to the TG-DTA of the powder sample, however, TG-DTA of the SnTe pellet presents no evidence of an exothermic peak at ~570 K, indicating no further oxidation and decomposition during heating of the SnTe pellet. PXD of the pellet after TG-DTA (Figure 5c) shows the presence of the same phases prior to thermal analysis, with very similar phase fractions. The maximum temperature for the thermoelectric measurements in this work was restricted to 530 K. By taking thermal and electrical data from 300 to 530 K, we could therefore guarantee that the measurement region represented one of stability for the material. Indeed, PXD data for the pellet demonstrated that the phase compositions and fractions were unchanged after thermoelectric evaluation (Figure 5c).

The electrical conductivity (σ) of the pellet decreases from ~2.53×10^5 $S{\cdot}m^{-1}$ at 300 K, to ~1.39×10^5 $S{\cdot}m^{-1}$ at 530 K (Figure 6a), demonstrating the behaviour typical of a degenerate semiconductor, that could be related to a low but significant concentration of intrinsic Sn vacancies in the SnTe structure [16,17]. S increases from ~34.5 $\mu V{\cdot}K^{-1}$ at 300 K, to ~55.8 $\mu V{\cdot}K^{-1}$at 530 K (Figure 6b). It should be noted that σ and S are relatively low and high, respectively, compared to those of SnTe bulk materials synthesised by high energy ball milling [18] and high-temperature methods [26,28,29], which could result from the influence of impurities (i.e., SnO_2 and Te) and/or the relatively low Sn vacancy concentration (no deviation in the Sn site occupancy from unity could be identified by Rietveld refinement, for example, which would support this premise). The power factor ($S^2\sigma$) of SnTe increases from ~0.30 $mW{\cdot}m^{-1}{\cdot}K^{-2}$ at 300 K, to ~0.43 $mW{\cdot}m^{-1}{\cdot}K^{-2}$ at 530 K (Figure 6c). These figures are comparable to those for SnTe synthesised by high energy ball milling [18] and via high-temperature methods [26,28,29] (Figure 6c).

The SnTe pellets display relatively low thermal conductivity (κ), and the values decrease from ~3.33 $W \cdot m^{-1} \cdot K^{-1}$ at 300 K, to ~2.84 $W \cdot m^{-1} \cdot K^{-1}$ at 530 K (Figure 6d). The corresponding lattice thermal conductivity (κ_L) decreases from ~1.63 $W \cdot m^{-1} \cdot K^{-1}$ at 300 K, to ~1.28 $W \cdot m^{-1} \cdot K^{-1}$ at 530 K (Figure 6e), which is lower when compared to bulk materials [18,26,28,29], and could be due to the nanostructuring of the SnTe pellets. *ZT* was derived from the electrical and thermal property results and increases from ~0.027 at 300 K, to ~0.081 at 530 K (Figure 6f). These *ZT* values are comparable to those obtained previously from SnTe bulk materials prepared via high temperature synthesis (e.g. heating, melting, and annealing precursors at ~973–1423 K for long durations) [26,28,29], or by high energy ball milling [18]. This demonstrates that our surfactant-free solution approach is an energy-efficient and large-scale method for synthesising SnTe, with a comparably high thermoelectric performance. Previous research has demonstrated that the power factor in SnTe can be significantly enhanced through band structure engineering, such as introducing resonant impurity levels through In doping [18] and band convergence through Cd [19], Mg [20], Hg [21], Mn [22] or Ag alloying [23]. Therefore, by introducing these dopants into SnTe nanoparticles as part of our solution synthesis process, improved values for $S^2\sigma$ and *ZT* can be expected. Judicious doping and substitution is also likely to provide a means to stabilise the cubic SnTe phase to decomposition at elevated temperature. These are aspects that we are currently investigating.

4. CONCLUSIONS

In summary, we have developed a simple, quick, and inexpensive one-pot solution route for synthesising SnTe nanoparticles in gram quantities (>10 g per run for a 2 h growth). Dense, nanostructured SnTe pellets have been fabricated by spark plasma sintering with a thermoelectric performance comparable to SnTe bulk materials synthesised by energy-intensive methods. As a similar solution method has already been applied to synthesise high-performance SnSe nanomaterials [45] and can be extended to SnS (to be reported soon), such a surfactant-free approach has been demonstrated to be an efficient, versatile method for synthesising efficient Sn chalcogenide thermoelectric nanomaterials.

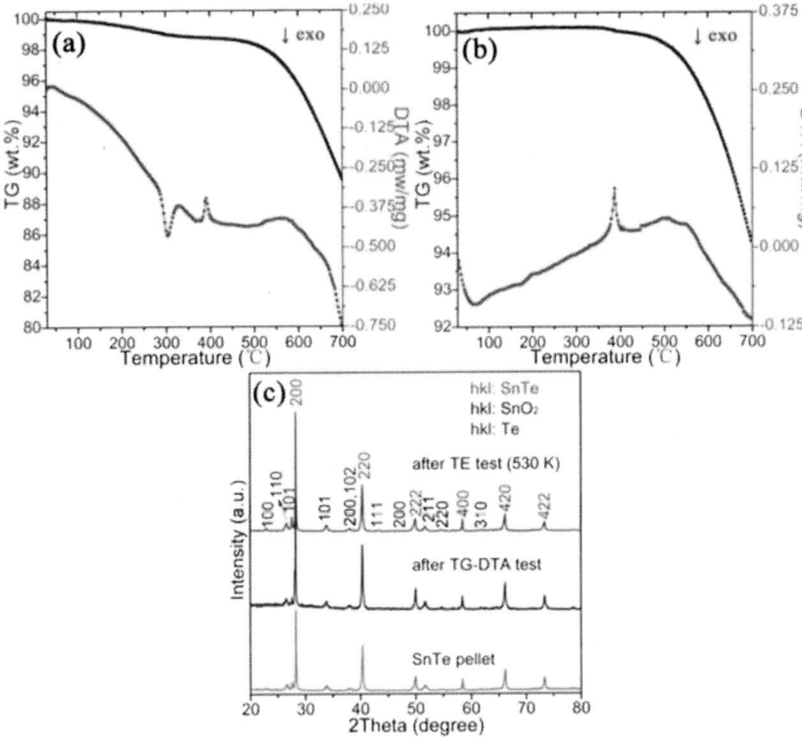

Figure 5. (**a**,**b**) TG-DTA profiles of the SnTe powder and pellet, respectively; (**c**) PXD patterns of an SnTe pellet before and after TG-DTA (to 973 K) and thermoelectric property testing (to 530 K).

ACKNOWLEDGMENTS

This work was financially supported by the EPSRC (EP/K022156/1 and EP/P510968/1). The authors thank Peter Chung for assistance with SEM. SRP and JWGB acknowledge the EPSRC for support (EP/N01717X/1). Michael J. Reece and Ruizhi Zhang would like to acknowledge the EPSRC for support under the DEFCOM grant (EP/N022726/1). Ruizhi Zhang acknowledges the support from a Marie Curie International Incoming Fellowship within the 7th European Community Framework Programme (Contract No. PIIF-GA-2013-624474).

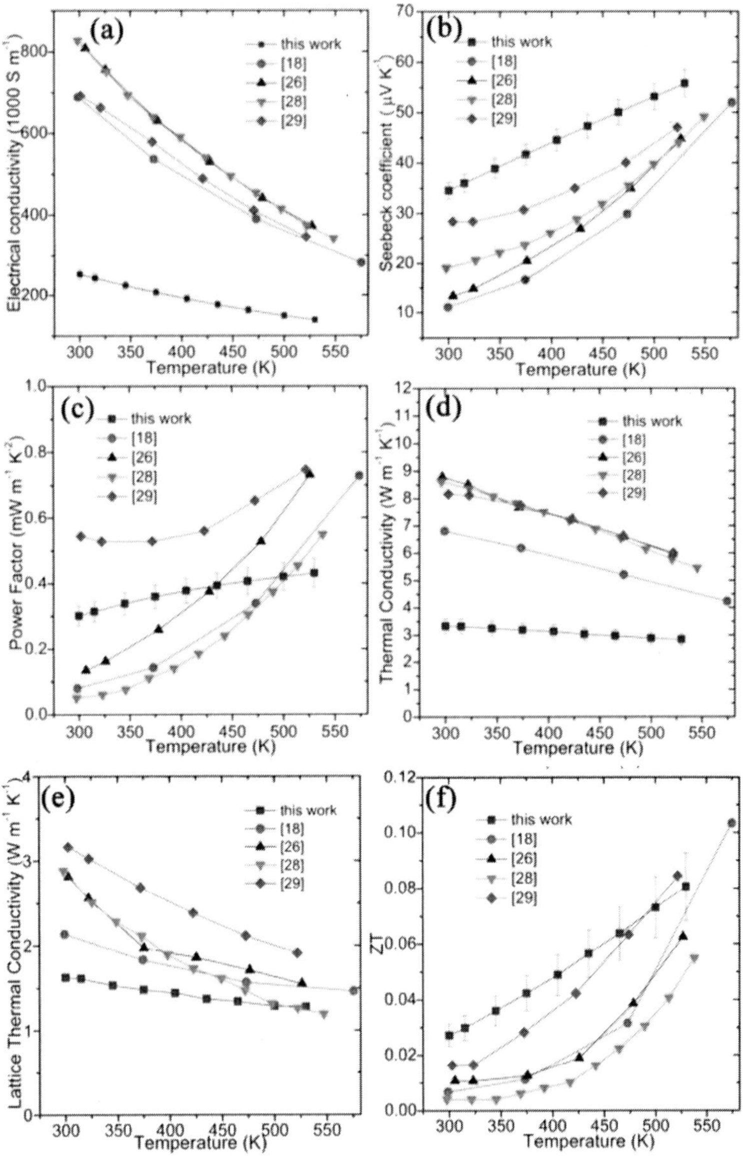

Figure 6. Thermoelectric properties of a SnTe pellet showing: (**a**) the electrical conductivity; (**b**) the Seebeck coefficient; (**c**) the power factor; (**d**) the thermal conductivity; (**e**) the lattice thermal conductivity and (**f**) *ZT* as a function of temperature, with comparisons with selected literature values for SnTe bulk materials synthesised by ball milling and high-temperature synthesis.

AUTHOR CONTRIBUTIONS

Guang Han and Duncan H. Gregory conceived and designed the experiments; Guang Han performed the materials synthesis, structural, and chemical characterisation; Ruizhi Zhang and Michael J. Reece fabricated the SnTe pellets using SPS and measured their electrical performance; Srinivas R. Popuri and Jan-Willem G. Bos measured the thermal diffusivity of the SnTe pellets; Heather F. Greer and Wuzong Zhou performed the TEM characterisation; Andrew R. Knox assisted in the performance measurement. Duncan H. Gregory led the project. Guang Han and Duncan H. Gregory analysed the data and wrote the paper with contributions from all of the authors.

CONFLICTS OF INTEREST

The authors declare no conflict of interest.

REFERENCES

1. Baxter, J.; Bian, Z.; Chen, G.; Danielson, D.; Dresselhaus, M.S.; Fedorov, A.G.; Fisher, T.S.; Jones, C.W.; Maginn, E.; Kortshagen, U.; et al. Nanoscale design to enable the revolution in renewable energy. Energy Environ. Sci. 2009, 2, 559–588.
2. Dresselhaus, M.S.; Chen, G.; Tang, M.Y.; Yang, R.G.; Lee, H.; Wang, D.Z.; Ren, Z.F.; Fleurial, J.P.; Gogna, P. New directions for low-dimensional thermoelectric materials. Adv. Mater. 2007, 19, 1043–1053.
3. Lan, Y.; Minnich, A.J.; Chen, G.; Ren, Z. Enhancement of thermoelectric figure-of-merit by a bulk nanostructuring approach. Adv. Funct. Mater. 2010, 20, 357–376.
4. Han, G.; Chen, Z.-G.; Drennan, J.; Zou, J. Indium selenides: Structural characteristics, synthesis and their thermoelectric performances. Small 2014, 10, 2747–2765.
5. Snyder, G.J.; Toberer, E.S. Complex thermoelectric materials. Nat. Mater. 2008, 7, 105–114.
6. Tan, G.J.; Zhao, L.D.; Kanatzidis, M.G. Rationally designing high-performance bulk thermoelectric materials. Chem. Rev. 2016, 116, 12123–12149.
7. Tritt, T.M.; Subramanian, M.A. Thermoelectric materials, phenomena, and applications: A bird's eye view. MRS Bull.2006, 31, 188–194.
8. Zhao, L.D.; Lo, S.H.; Zhang, Y.S.; Sun, H.; Tan, G.J.; Uher, C.; Wolverton, C.; Dravid, V.P.; Kanatzidis, M.G. Ultralow thermal conductivity and high thermoelectric figure of merit in SnSe crystals. Nature 2014, 508, 373–377.
9. Zhao, L.-D.; Tan, G.; Hao, S.; He, J.; Pei, Y.; Chi, H.; Wang, H.; Gong, S.; Xu, H.; Dravid, V.P.; et al. Ultrahigh power factor and thermoelectric performance in hole-doped single-crystal SnSe. Science 2015, 351, 141–144.
10. Peng, K.; Lu, X.; Zhan, H.; Hui, S.; Tang, X.; Wang, G.; Dai, J.; Uher, C.; Wang, G.; Zhou, X. Broad temperature plateau for high ZTs in heavily doped p-type SnSe single crystals. Energy Environ. Sci. 2016, 9, 454–460.
11. Han, M.K.; Androulakis, J.; Kim, S.J.; Kanatzidis, M.G. Lead-free thermoelectrics: High figure of merit in p-type AgSnmSbTem+2. Adv. Energy Mater. 2012, 2, 157–161.

12. Chen, Y.; Nielsen, M.D.; Gao, Y.B.; Zhu, T.J.; Zhao, X.B.; Heremans, J.P. SnTe-AgSbTe2 thermoelectric alloys. Adv. Energy Mater. 2012, 2, 58–62.
13. Brebrick, R.F.; Strauss, A.J. Anomalous thermoelectric power as evidence for two-valence bands in SnTe. Phys. Rev.1963, 131, 104–110.
14. Rogers, L.M. Valence band structure of SnTe. J. Phys. D Appl. Phys. 1968, 1, 845–852.
15. Santhanam, S.; Chaudhuri, A.K. Transport-properties of SnTe interpreted by means of a 2 valence band model. Mater. Res. Bull. 1981, 16, 911–917.
16. Brebrick, R.F. Deviations from stoichiometry and electrical properties in SnTe. J. Phys. Chem. Solids 1963, 24, 27–36.
17. Kafalas, J.A.; Brebrick, R.F.; Strauss, A.J. Evidence that SnTe is semiconductor. Appl. Phys. Lett. 1964, 4, 93–94.
18. Zhang, Q.; Liao, B.L.; Lan, Y.C.; Lukas, K.; Liu, W.S.; Esfarjani, K.; Opeil, C.; Broido, D.; Chen, G.; Ren, Z.F. High thermoelectric performance by resonant dopant indium in nanostructured SnTe. Proc. Natl. Acad. Sci. USA 2013, 110, 13261–13266.
19. Tan, G.J.; Zhao, L.D.; Shi, F.Y.; Doak, J.W.; Lo, S.H.; Sun, H.; Wolverton, C.; Dravid, V.P.; Uher, C.; Kanatzidis, M.G. High thermoelectric performance of p-type SnTe via a synergistic band engineering and nanostructuring approach. J. Am. Chem. Soc. 2014, 136, 7006–7017.
20. Banik, A.; Shenoy, U.S.; Anand, S.; Waghmare, U.V.; Biswas, K. Mg alloying in SnTe facilitates valence band convergence and optimizes thermoelectric properties. Chem. Mater. 2015, 27, 581–587.
21. Tan, G.J.; Shi, F.Y.; Doak, J.W.; Sun, H.; Zhao, L.D.; Wang, P.L.; Uher, C.; Wolverton, C.; Dravid, V.P.; Kanatzidis, M.G. Extraordinary role of Hg in enhancing the thermoelectric performance of p-type SnTe. Energy Environ. Sci. 2015, 8, 267–277.
22. Wu, H.J.; Chang, C.; Feng, D.; Xiao, Y.; Zhang, X.; Pei, Y.L.; Zheng, L.; Wu, D.; Gong, S.K.; Chen, Y.; et al. Synergistically optimized electrical and thermal transport properties of SnTe via alloying high-solubility MnTe. Energy Environ. Sci. 2015, 8, 3298–3312.
23. Banik, A.; Shenoy, U.S.; Saha, S.; Waghmare, U.V.; Biswas, K. High power factor and enhanced thermoelectric performance of SnTe-AgInTe2: Synergistic effect of resonance level and valence band convergence. J. Am. Chem. Soc.2016, 138, 13068–13075.
24. Banik, A.; Biswas, K. Lead-free thermoelectrics: Promising thermoelectric performance in p-type SnTe1-xSex system. J. Mater. Chem. A 2014, 2, 9620–9625.
25. Tan, G.J.; Shi, F.Y.; Sun, H.; Zhao, L.D.; Uher, C.; Dravid, V.P.; Kanatzidis, M.G. SnTe-AgBiTe2 as an efficient thermoelectric material with low thermal conductivity. J. Mater. Chem. A 2014, 2, 20849–20854.
26. Tan, G.J.; Shi, F.Y.; Hao, S.Q.; Chi, H.; Zhao, L.D.; Uher, C.; Wolverton, C.; Dravid, V.P.; Kanatzidis, M.G. Codoping in SnTe: Enhancement of thermoelectric performance through synergy of resonance levels and band convergence. J. Am. Chem. Soc. 2015, 137, 5100–5112.
27. Liang, T.; Su, X.L.; Tan, X.M.; Zheng, G.; She, X.Y.; Yan, Y.G.; Tang, X.F.; Uher, C. Ultra-fast non-equilibrium synthesis and phase segregation in InxSn1-xTe thermoelectrics by SHS-PAS processing. J. Mater. Chem. C 2015, 3, 8550–8558.
28. Banik, A.; Vishal, B.; Perumal, S.; Datta, R.; Biswas, K. The origin of low thermal conductivity in Sn1-xSbxTe: Phonon scattering via layered intergrowth nanostructures. Energy Environ. Sci. 2016, 9, 2011–2019.
29. Zhao, L.D.; Zhang, X.; Wu, H.J.; Tan, G.J.; Pei, Y.L.; Xiao, Y.; Chang, C.; Wu, D.; Chi, H.; Zheng, L.; et al. Enhanced thermoelectric properties in the counter-doped SnTe system with strained endotaxial SrTe. J. Am. Chem. Soc. 2016, 138, 2366–2373.
30. Zhang, L.J.; Wang, J.L.; Cheng, Z.X.; Sun, Q.; Li, Z.; Dou, S.X. Lead-free SnTe-based thermoelectrics: Enhancement of thermoelectric performance by doping with Gd/Ag. J. Mater. Chem. A 2016, 4, 7936–7942.
31. Pei, Y.Z.; Zheng, L.L.; Li, W.; Lin, S.Q.; Chen, Z.W.; Wang, Y.Y.; Xu, X.F.; Yu, H.L.; Chen, Y.; Ge, B.H. Interstitial point defect scattering contributing to high thermoelectric performance in SnTe. Adv. Electron. Mater. 2016, 2, 1600019.

32. Zhou, M.; Gibbs, Z.M.; Wang, H.; Han, Y.M.; Xin, C.N.; Li, L.F.; Snyder, G.J. Optimization of thermoelectric efficiency in SnTe: The case for the light band. Phys. Chem. Chem. Phys. 2014, 16, 20741–20748.
33. He, J.; Tan, X.J.; Xu, J.T.; Liu, G.Q.; Shao, H.Z.; Fu, Y.J.; Wang, X.; Liu, Z.; Xu, J.Q.; Jiang, H.C.; et al. Valence band engineering and thermoelectric performance optimization in SnTe by Mn-alloying via a zone-melting method. J. Mater. Chem. A 2015, 3, 19974–19979.
34. Mehta, R.J.; Zhang, Y.L.; Karthik, C.; Singh, B.; Siegel, R.W.; Borca-Tasciuc, T.; Ramanath, G. A new class of doped nanobulk high-figure-of-merit thermoelectrics by scalable bottom-up assembly. Nature Mater. 2012, 11, 233–240.
35. Zhang, G.; Fang, H.; Yang, H.; Jauregui, L.A.; Chen, Y.P.; Wu, Y. Design principle of telluride-based nanowire heterostructures for potential thermoelectric applications. Nano Lett. 2012, 12, 3627–3633.
36. Zhang, G.; Kirk, B.; Jauregui, L.A.; Yang, H.; Xu, X.; Chen, Y.P.; Wu, Y. Rational synthesis of ultrathin n-type Bi2Te3nanowires with enhanced thermoelectric properties. Nano Lett. 2011, 12, 56–60.
37. Han, G.; Chen, Z.-G.; Yang, L.; Hong, M.; Drennan, J.; Zou, J. Rational design of Bi2Te3 polycrystalline whiskers for thermoelectric applications. ACS Appl. Mater. Interfaces 2015, 7, 989–995.
38. Li, Z.; Shao, S.; Li, N.; McCall, K.; Wang, J.; Zhang, S.X. Single crystalline nanostructures of topological crystalline insulator SnTe with distinct facets and morphologies. Nano Lett. 2013, 13, 5443–5448.
39. Shen, J.; Jung, Y.; Disa, A.S.; Walker, F.J.; Ahn, C.H.; Cha, J.J. Synthesis of SnTe nanoplates with {100} and {111} surfaces. Nano Lett. 2014, 14, 4183–4188.
40. Safdar, M.; Wang, Q.S.; Wang, Z.M.; Zhan, X.Y.; Xu, K.; Wang, F.M.; Mirza, M.; He, J. Weak antilocalization effect of topological crystalline Insulator Pb1-xSnxTe nanowires with tunable composition and distinct {100} facets. Nano Lett 2015, 15, 2485–2490.
41. Wang, Q.S.; Cai, K.M.; Li, J.; Huang, Y.; Wang, Z.X.; Xu, K.; Wang, F.; Zhan, X.Y.; Wang, F.M.; Wang, K.Y.; et al. Rational design of ultralarge Pb1-xSnxTe nanoplates for exploring crystalline symmetry-protected topological transport. Adv. Mater. 2016, 28, 617–623.
42. Ibáñez, M.; Korkosz, R.J.; Luo, Z.S.; Riba, P.; Cadavid, D.; Ortega, S.; Cabot, A.; Kanatzidis, M.G. Electron doping in bottom-up engineered thermoelectric nanomaterials through HCl-mediated ligand displacement. J. Am. Chem. Soc.2015, 137, 4046–4049.
43. Han, G.; Popuri, S.R.; Greer, H.F.; Llin, L.F.; Bos, J.W.G.; Zhou, W.Z.; Paul, D.J.; Ménard, H.; Knox, A.R.; Montecucco, A.; Siviter, J.; Man, E.A.; Li, W.-G.; Paul, M.C.; Gao, M.; Sweet, T.; Freer, R.; Azough, F.; Baig, H.; Mallick, T.K.; Gregory, D.H. Chlorine-enabled electron doping in solution-synthesised SnSe thermoelectric nanomaterials. Adv. Energy Mater. 2017. (In press)
44. Cadavid, D.; Ibanez, M.; Shavel, A.; Dura, O.J.; Lopez de la Torre, M.A.; Cabot, A. Organic ligand displacement by metal salts to enhance nanoparticle functionality: Thermoelectric properties of Ag2Te. J. Mater. Chem. A 2013, 1, 4864–4870.
45. Han, G.; Popuri, S.R.; Greer, H.F.; Bos, J.W.G.; Zhou, W.Z.; Knox, A.R.; Montecucco, A.; Siviter, J.; Man, E.A.; Macauley, M.; et al. Facile surfactant-free synthesis of p-type SnSe nanoplates with exceptional thermoelectric power factors. Angew. Chem. Int. Ed. 2016, 55, 6433–6437.
46. Han, C.; Li, Z.; Lu, G.Q.; Dou, S.X. Robust scalable synthesis of surfactant-free thermoelectric metal chalcogenide nanostructures. Nano Energy 2015, 15, 193–204.
47. Zhang, W.X.; Cheng, Y.W.; Hu, J.Q.; Zhan, J.H.; Yu, W.C.; Yang, L.; Qian, Y.T. Low temperature synthesis of ultrafine SnTe powder in ethanol solvent. Chem. Lett. 2000, 29, 446–447.
48. An, C.H.; Tang, K.B.; Hai, B.; Shen, G.Z.; Wang, C.R.; Qian, Y.T. Solution-phase synthesis of monodispersed SnTe nanocrystallites at room temperature. Inorg. Chem. Commun. 2003, 6, 181–184.
49. Kovalenko, M.V.; Heiss, W.; Shevchenko, E.V.; Lee, J.S.; Schwinghammer, H.; Alivisatos, A.P.; Talapin, D.V. SnTe nanocrystals: A new example of narrow-gap semiconductor quantum dots. J. Am. Chem. Soc. 2007, 129, 11354–11355.

50. Ning, J.J.; Men, K.K.; Xiao, G.J.; Zou, B.; Wang, L.; Dai, Q.Q.; Liu, B.B.; Zou, G.T. Synthesis of narrow band gap SnTe nanocrystals: Nanoparticles and single crystal nanowires via oriented attachment. Cryst. Eng. Comm. 2010, 12, 4275–4279.
51. Xu, Y.; Al-Salim, N.; Hodgkiss, J.M.; Tilley, R.D. Solution synthesis and optical properties of SnTe nanocrystals. Cryst. Growth Des. 2011, 11, 2721–2723.
52. Li, Z.L.; Chen, Y.D.; Li, J.F.; Chen, H.; Wang, L.J.; Zheng, S.Q.; Lu, G.W. Systhesizing SnTe nanocrystals leading to thermoelectric performance enhancement via an ultra-fast microwave hydrothermal method. Nano Energy 2016, 28, 78–86.
53. Wang, L.; Zheng, S.; Chen, H. Enhanced electronic transport properties of Se-doped SnTe1−xSex nanoparticles by microwave-assisted solvothermal method. J. Electron. Mater. 2016.
54. Larson, A.C.; Von Dreele, R.B. General Structure Analysis System (GSAS); Los Alamos National Laboratory Report LAUR 86-748; Los Alamos National Laboratory: Los Alamos, NM, USA, 1994.
55. Toby, B.H. EXPGUI, a graphical user interface for GSAS. J. Appl. Crystallogr. 2001, 34, 210–213.
56. Bierly, J.N.; Muldawer, L.; Beckman, O. The continuous rhombohedral-gubic transformation in GeTe-SnTe alloys. Acta Metall. 1963, 11, 447–454.
57. Gamsjäger, H.; Gajda, T.; Sangster, J.; Saxena, S.; Voigt, W. Chemical thermodynamics of tin. In Nuclear Energy Agency Data Bank; Peronne, J., Ed.; Organisation for Economic Co-Operation and Development: Paris, France, 2012.
58. Lin, S.Q.; Li, W.; Chen, Z.W.; Shen, J.W.; Ge, B.H.; Pei, Y.Z. Tellurium as a high-performance elemental thermoelectric. Nature Commun. 2016, 7, 10287.
59. Kim, H.S.; Gibbs, Z.M.; Tang, Y.L.; Wang, H.; Snyder, G.J. Characterization of Lorenz number with Seebeck coefficient measurement. APL Mater. 2015, 3, 041506.
60. PDF-2 Release 2008. Joint Committee on Powder Diffraction Standards (JCPDS)-International Centre for Diffraction Data (ICDD). Avaliable online: http://www.icdd.com (accessed on 24 February 2017).
61. Zhang, B. Lead Chalcogenide Nano-Composites: Synthesis, Thermal and Electrical Transport Properties; Clemson University: Clemson, SC, USA, 2008.

CHAPTER 13

Radiation Influence on Edible Materials

Nelida Lucia del Mastro

ABSTRACT

Innovations in the food industry are shaped both by new technologies available and by society's requirements. A good knowledge on the chemistry and biological role of the macro-nutrients (proteins, carbohydrates, lipids and also energy and water) and micro-nutrients (minerals and vitamins) is required. Food production foundations include not only the design of the food products but also the materials, mechanics, ingredients, conversion and transformation all must be taken into consideration. Edible polymers arc polymeric materials that can be easily consumed by human beings or lower animals in whole or part via the oral cavity and given harmless effect to the health. An edible polymer is originated from natural products such as polysaccharides, proteins and lipids, with the addition of plasticizers and surfactants. Radiation-processing technologies are used currently for numerous applications of commercial and economic importance, but it is an emerging application with the use of ionizing radiation to enhance properties of edible polymers such as carbohydrates or proteins. This chapter aims at supplying the state of the art about the effects of ionizing radiation on edible polymers: starch and vegetal proteins and also on gelatin that comes from animal origin.

Keywords: ionizing radiation, polysaccharides, proteins, edible films, nanotechnology

1. INTRODUCTION

There is an increasing interest in green and bioactive materials that could be used in direct contact with aliments. Edible films can be produced from natural materials with film-forming ability that can be applied on food surfaces as a thin-layer edible film and can potentially extend the shelf life and improve the quality of food. Edible film as a solid sheet can be applied between food components or on the surface of the food system in order to inhibit migration moisture, oxygen, CO_2, aromas and lipids.

Edible polymer film is a thin layer of edible material formed on a food as a coating or placed (pre-formed) on or between food components; in this case, edible films can improve the quality of multicomponent foods. The sanitary condition of the edible packaging would need to be maintained during storage, transportation and marketing. The end result would be source reduction and or improved recyclability of the remaining elements of the packaging system.

One major advantage of using edible films and coatings is that several active ingredients (antimicrobials, antibrownings, texture enhancers and nutraceuticals) can be incorporated into the polymer matrix and consumed with the food, thus enhancing safety or even nutritional and sensory attributes [1].

Many potential uses of edible films can be described such as inhibit migration of moisture, oxygen, carbon dioxide, aromas; carry food ingredients (e.g., antioxidants, antimicrobials, flavor) and or improve mechanical integrity or handling characteristics of the food [2]. Practical uses of edible films include wrapping various products; individual protection of dried fruits, meat and fish; control of internal moisture transfer in pizzas, pies, which are based on the film's properties (e.g., sensory, mechanical, gas and solute barrier). Also, ingredients can be delivered to processors on water-soluble and edible packaging films in premeasured amounts of ingredients to food processors and foodservice operations. Future applications of the concept have been envisioned in consumer-sized pouches of dried products ready for reconstitution on water.

The functionality and performance of edible polymer mainly depend on their barrier, mechanical and color properties, which in turn depend on film composition and its formation process. Polysaccharide (cellulose, starch, dextrin, vegetable and other gums) and protein (gelatin, gluten, casein) based films can have suitable mechanical and sensory properties, while wax (beeswax, carnauba wax) and lipid or lipid derivative films have enhanced water vapor barrier properties. It was shown that the incorporation of oil between plasticized starch layers could reduce oxygen and water vapor permeability of the films [3]. The film-forming technology, solvent characteristics, plasticizing agents, temperature effects, solvent evaporation rate, coating operation and usage conditions of the film (relative humidity, temperature) can also substantially modify the ultimate properties of the film [4].

Important properties to be evaluated in an edible coating are its microbiological stability, adhesion, cohesion, wettability, solubility, transparency, mechanical properties, sensory and permeability to water vapor and gases. In order to choose the best edible film to be employed in each case, it is necessary to take in account the characteristics of the food intended to be protected. For coating a fresh fruit, for instance, it will be desired to induce low water vapor permeability to preserve texture and moderated O_2/CO_2 permeability to permit respiration. On the contrary, when it is intended to protect dried fruits and nuts, low water vapor permeability is required to maintain crispiness and low O_2permeability to avoid oxidation.

Proteins as components of edible films (gelatin, zein, casein) has usually better performance than polysaccharides (chitosan, starch, pectin); lipids have excellent water vapor barrier but do not form stand-alone films, and can be used as coatings, such as waxes on fruit surfaces. Besides the barrier efficiency, edible films and coatings have to be also sensory acceptable.

In the food industry, there are commonly used starches, dextrins, alginates, protein and lipid materials edible films as encapsulating materials [5]. A wide variety of foods are encapsulated: flavoring agents, acids, bases, artificial sweeteners, colorants, preservatives, leavening agents, antioxidants, agents with undesirable flavors, odors and nutrients, among others. In order to protect the viability of the probiotic bacteria, for instance, several types of biopolymers such as alginate, chitosan, gelatin, whey protein isolate, cellulose derivatives are used for encapsulation and several methods of encapsulation such as spray drying, extrusion, emulsion have been reported [6].

Active edible films represent one of the current and future trends in the development of new polymers for selected applications, particularly food packaging [7]. The concept of active packaging is described as a type of packaging that alters conditions surrounding the food to maintain product quality and freshness, improve sensory properties or enhance product safety and shelf life.

Edible and biodegradable polymer-based films offer alternative packaging without the environmental cost. In that case, they must fulfil the food packaging requirements as help to contain, protect, preserve, distribute and describe food from fresh to highly processed ones. The potential of biopolymers for packaging application is often limited due to their poor processability, so edible polymers originated from natural products are added to plasticizers and surfactants. A careful selection of the plasticizer allows biopolymer films with improved processability and high or low permeabilities to be manufactured [8].

Shit and Shah in 2014 distinguished the edible polymers into any of the four categories: (1) hydrocolloids, (2) polypeptides, (3) lipids, (4) synthetic and composite edible polymers [9].

In the following sections, the interaction of ionizing radiation with polysaccharides and proteins, which are considered the most important edible materials, as well as the role that radiation plays on the development of new edible materials will be addressed.

2. HIGH-ENERGY RADIATION AND POLYMERS

Gamma rays and electron beam are two commonly used ionizing radiation sources in industrial process. Gamma rays, 1.17 and 1.33 MeV, are emitted continuously from radioactive source such as cobalt-60, whereas electrons are generated from an accelerator to produce a stream of electrons called electron beam. The energy of electrons depends on the type of machines and can vary from 200 keV to 10 MeV. High-energy radiation using conventional gamma or e-beam sources is an adequate tool for the modification of polymer materials through degradation, grafting and cross-linking [10].

The absorbed dose of radiation is expressed in units of gray (Gy). The Gy was adopted by the International Commission of Radiation Units and measurements (ICRU) in 1975 as the special name for the standard international (SI) unit of absorbed dose, being 1Gy = 1 Joule/kg of matter. When processing food products very seldom the employed dose exceeded 20 kGy, and most of the cases remained below 10 kGy [11]. The adsorbed radiation dose is a consequence of the radiation exposure time and the distance between the product and the radiation source in the irradiation camera.

The long molecular chains of polymers can be broken by the absorption of a quantum of energy above the energy of the covalent bond of the main carbon chain, which typically is in the range of 5–10 eV. The energy of beta and gamma photons of 1 to 10 MeV surpasses by many orders of magnitude this minimum value, representing a high probability of affecting all kinds of polymers, naturals and synthetics alike [12]. Usually, there is a competition between cross-linking and chain scission reaction. If chain scission reaction predominates the material degrades.

Radiation, as a non-selective, highly efficient tool of ionization, may form excited sites, ions and free radicals in almost all kinds of materials. Radiation treatment of polymer mixtures, even if they are (partially) incompatible, gives a chance for bridge-forming bonds [13].

The key effect of radiation is the production of reactive oxygen species (ROS), which affects biomolecules (e.g., lipid, protein, polysaccharides). One of the important radiation-induced free-radical species is the hydroxyl radical, which indiscriminately attacks neighboring molecules often at near diffusion-controlled rates. Hydroxyl radicals are generated by ionizing radiation either directly by oxidation of water, or indirectly by the formation of secondary partially ROS. These may be subsequently converted to hydroxyl radicals by further reduction ("activation"). Secondary, radiation injury is therefore influenced by the surrounding antioxidant status and the amount and availability of activating mechanisms. The biological response to radiation may be modulated by alterations in factors affecting these secondary mechanisms [14].

3. RADIATION AND POLYSACCHARIDES

Carbohydrates or saccharides are polyhydroxy aldehydes or ketones that have the empirical formula $(CH_2O)_n$. Most of the carbohydrates found in nature occur as polysaccharides of high molecular weight. Polysaccharides are polymers of monosaccharides linked with glycosidic bonds containing usually no more than two kinds of residues. The most abundant monosaccharide is the six-carbon sugar D-glucose; it is the primordial monosaccharide from which all others are derived. D-Glucose is the most important fuel molecule for most organisms, and also the basic building block of the most abundant polysaccharides. The monomeric units contain many hydroxyl groups, which can engage in intra- and inter-molecular formation of hydrogen bonds. This hydrogen bonding keeps the chains together and contributes to the high tensile strength of the polymeric material. Other forms of functionalization can also occur.

Polysaccharide films are made from starch, alginate, cellulose ethers, chitosan, carageenan, or pectins and impart hardness, crispness, compactness, thickening quality, viscosity, adhesiveness and gel-forming ability to a variety of films. These films usually exhibit good gas permeability properties, resulting in desirable modified atmospheres that enhance the shelf life of the product without creating anaerobic conditions. Additionally, polysaccharide films and coatings can be used to extend the shelf life of muscle foods by preventing dehydration, oxidative rancidity and surface browning, but their hydrophilic nature makes them poor barriers for water vapor.

Starch contains only D-glucose as monomeric units and occurs in two forms, α-amylose (a long unbranched chain) and amylopectin (highly branched). Films produced from pure starch are generally brittle and difficult to handle, although there are some exceptional reports that starches such as cassava can make transparent and colorless flexible films without any previous chemical treatment [15].

Other polysaccharides like pectin, pullulan and chitosan, coming from different origins, present different capability of acting as edible active coatings [16].

The poor mechanical properties and water stability of starch have restricted its industrial applications. However, they can often be functionalized with carboxyl groups, phosphate groups and/or sulfuric ester groups, and the combination of plasticizers and surfactants can bring the enhancement of properties of the starch films [17, 18].

The development of biodegradable materials based on starch has become an attractive option and the production of starch-based plastics is gradually obtained considerable importance in the world. Chemical modifications (e.g., cross-linking) or using a second biopolymer in the starch-based composite have been studied as strategies to produce low water-sensitive and relatively high-strength starch-based materials [19–22].

Particularly, citric acid showed to cross-link starch and improve the tensile strength, thermal stability and decrease the dissolution of starch films in water and formic acid [23].

Starch is considered easily depolymerizable by radiation treatment. Gamma and electron beam irradiation are used to induce radiation degradation of chitosan, alginate, carrageenan, cellulose, pectin, for recycling these bio-resources and reducing the environmental pollution. These carbohydrates, when degraded by radiation, present various kinds of biological activities such as promotion of plant growth, anti-microbial activity, phytoalexins induction, biocontrol elicitors and also can be added to aquaculture and animal feed to enhance immunity of animals [24, 25].

The radiation starch or dextrin degradation is a clean and safety tool. The process is very efficient and can be easily controlled by choosing a proper irradiation dose. A dose of 4.4 kGy, for instance, was enough to decrease the molar mass of wheat starches by one order of magnitude [26]. Other works describe the action of gamma radiation on potato, bean and maize starches [27–29].

Physical and structural characteristics of rice flour and starch obtained from gamma-irradiated white rice were determined. Pasting viscosities of the rice flour and starch decreased continuously with the increase in irradiation dosage. Gamma irradiation had no significant effect on the amylopectin branch chains, but produced more branch chains when the irradiation dosage was less than 9 kGy. It might be deduced that gamma irradiation caused the breakage of the amylopectin chains at the amorphous regions, but had little effects on the crystalline regions of starch granules, especially at low-dosage irradiation [30].

Several other reports were published about the beneficial use of gamma radiation on the improvement of properties of different biopolymeric materials. Ionizing radiation was employed for the synthesis of polysaccharide derivatives to be used as oral delivery system for a colon-specific drug carrier [31]; other works mention the increase of mechanical and barrier properties of polysaccharide materials and the induction of grafting of methylcellulose-based films [32].

Some authors reported the development of edible films from potato peel. High-pressure, gamma rays and ultrasound were applied to potato peel solutions to break down biopolymer particles in the solution small enough to allow for biopolymer film formation. Film properties, including moisture barrier and tensile properties, color and microstructures, were investigated from the films formed with different concentrations of plasticizer (glycerol) and emulsifier (soy lecithin). The authors concluded that the concentrations of both plasticizer (glycerol) and emulsifier (soy lecithin) were important variables in producing biopolymer films from potato peel [33].

4. RADIATION AND PROTEINS

All proteins are polymers and their monomeric units are α-amino acids. Twenty chemically different amino acids are incorporated in proteins. Proteins in foods serve dual roles as nutrients and structural building blocks. The concept of protein functionality has historically been restricted to non-nutritive functions—such as creating emulsions, foams and gels—but this places sole emphasis on food quality considerations and potentially overlooks modifications that may also alter nutritional quality or allergenicity. Foegeding proposed recently a new model that addresses the function of proteins in foods based on the length scale(s) responsible for the function. Properties such as flavor binding, color, allergenicity and digestibility are explained based on the structure of individual molecules, placing this functionality at the nano/molecular scale. At the next higher scale, applications in foods involving gelation, emulsification and foam formation are based on how proteins form secondary structures that are seen at the nano- and microlength scales, collectively called the mesoscale. The macroscale structure represents the arrangements of molecules and mesoscale structures in a food. Macroscale properties determine the overall product appearance, stability and texture. For applications in food products, protein functionality should start with the identification of functional needs a0 scales. Those needs are then evaluated relative to how processing and other ingredients could alter desired molecular scale properties, or proper formation of mesoscale structures. This allows for a comprehensive approach to achieving the desired function of proteins in foods [34].

Proteins have long been empirically used to make biodegradable, renewable and edible packaging materials. Numerous cereal and vegetable proteins, such as corn zein, wheat gluten and soy proteins, and animal proteins (such as milk proteins, collagen, gelatin, keratin and myofibrillar proteins) are commonly used to form agricultural packaging materials.

Lacroix and Vu, among others, had described the methods to produce protein-based films and coatings, the properties of them and their applications in food systems. They focus on selected proteins originated from animal and plant sources consisting of caseins, whey proteins, collagen and gelatin, plasma proteins, myofibrillar proteins, egg white proteins, soy protein, wheat gluten and zein [35–37].

Soy protein isolate (SPI), a protein with good biocompatibility, biodegradability and processability, has a significant potential in the food industry, agriculture, bioscience and biotechnology. Up to now, several technologies have been applied to prepare SPI-based materials with equivalent or superior physical and mechanical properties with petroleum-based materials [38].

Among protein-based films, those made of gelatin are specially important and can be produced from a variety of origins, with or without the addition of other components as plasticizers, surfactants or mixture with others substances [39–42].

Fish gelatin is a potential alternative to current mammalian (beef and pork) gelatin. However, its physical and thermal properties limit its use in many applications. The treatment of microbial transglutaminase as a cross-linking agent could be a practical way to increase the use of fish gelatin films in various applications [43].

Edible protein film characteristics can be enhanced with chemical and enzymatic methods, combining with hydrophobic material or some polymers or using a physical method, and the resulting film properties will depend on modification methods and conditions [44].

The enzyme and chemical modifications are efficient in lowering water vapor permeability. Composite edible protein films in combination with lipids can result in better functionality than the films produced with only proteins, especially with respect to their barrier properties. Of the lipids, waxes produce the best water vapor barrier properties, but produce fragile and/or brittle films.

The preparation of gelatin and glycerol mixture with microbial transglutaminase as the cross-linking agent was described [45]. A composite casein-gelatin was prepared using also transglutaminase [46]. In some cases, however, the chemical/enzymatical cross-linking was not satisfactory [47].

Application of radiation has been extended to the modification of proteins. Depending on the adsorbed radiation dose or radiation exposure time, various effects can be achieved resulting in the polymerization (cross-linking) or depolymerization of protein molecules. Most food proteins, however, undergo irradiation-induced cross-linking and subsequent improvement on the film properties [48].

Radiation-induced lipid oxidation can be readily detected by our body's olfactory cells—rancidity, off-flavor and, sometimes on the positive aspect, aroma; similarly, it is easy to notice discoloration that results from the oxidation of pigments. In contrast, protein oxidation occurs undetected by sensory organs; instrumental analysis is therefore required. Yet, proteins are very susceptible to reactive oxygen species (ROS) and impart both desirable and undesirable consequences when oxidatively modified. For example, oxidant-initiated disulfide bond formation among gluten molecules is responsible for the desirable rheology and sponginess of bread. And functional myosin (or actomyosin) aggregates produced by the reaction of cysteine, lysine and tyrosine residues with low concentrations of free radicals promote protein gel networks in processed muscle foods, hence the products' firmness and mouthfeel.

Chemically, protein oxidation that can result from radiation treatment in the presence of O_2 involves the initial modification of amino acid side chain groups by ROS. The radiation sensitivity of the amino acids is the highest for cysteine and decrease following the sequence: cysteine, methionine, tyrosine, tryptophan, phenylalanine, valine, leucine, histidine, glutamyl, proline, threonine, arginine and lysine. The initial modification of amino acid side chain groups by ROS leads to the conversion to carbonyl and other derivatives. Electron-deficient carbonyl groups are highly reactive with amines and thiol moieties to produce cross-links between polypeptides or segments within the same protein molecule. Disulfide and dityrosine are other covalent linkages in oxidatively stressed proteins. Protein radicals, which usually have a long half-life, are precursors of polymers as well [49].

The field of health science has pioneered protein oxidation research. Similarly, research in food science over the past two decades has established ubiquitous occurrences of protein oxidation in both fresh and processed foods, especially meat products. Most of the studies point to the negative aspect because uncontrolled oxidation leads to deleterious consequences: tissue hardening (as in frozen fish), loss of water-binding potential and off-flavor due to thiol oxidation. For that reason, developments about novel antioxidants, such as phenolic derivatives, peptides/protein hydrolysates, phospholipids and polysaccharides, and their role in food quality preservation are going on [50].

Many peptides and protein hydrolysates, being preferred targets of ROS and excellent functional compounds (water-binding, foaming, rheology and so on), are qualified as "multi-functional" natural food additives. Aside from the negative impacts of oxidation, mildly oxidized lipids give us a complex, highly desirable aroma of foods as that coming from frying. Today, the beneficial effects of limited protein oxidation are also no longer ignored. That must be taking in account whenever any protein food is submitted to oxidative processes such as irradiation.

Gamma-irradiation affects proteins by causing conformational changes, oxidation of amino acids, rupture of covalent bonds and formation of protein free radicals that can be beneficial for specific further applications. Chemical changes in the proteins that are caused by gamma irradiation are fragmentation, cross-linking, aggregation and oxidation by oxygen radicals that are generated in the radiolysis of water. For example, the hydroxyl and super oxide anion radicals that are generated by radiation of film-forming solution could modify the molecular properties of the proteins, which results in the alteration of protein films by covalent cross-linkages formed in protein solution after irradiation.

Using gamma irradiation to induce cross-linking was found to be an effective method for the improvement of both barrier and mechanical properties of the edible films and coatings based on proteins. There are plenty of examples in the literature that corroborate the efficiency of radiation application on natural films, like those made of zein, the protein found in maize [51]. Different types of gelatin-based films with enhanced properties can be produced by means of application of ionizing radiation [52, 53].

The effect of electron beam accelerator doses on properties of plasticized fish gelatin film was studied. The electron spin resonance spectra indicated free radical formation during irradiation, which might induce intermolecular cross-linking. Tensile strength for gelatin film significantly increased after irradiation (improved by 30% for 60 kGy). The vapor permeability was weakly affected by irradiation. Surface tension and its polar component increased significantly in accordance with the increase of wettability. So, the authors suggest that irradiation may change the orientation of polar groups of gelatin at the film surface and cross-link the hydrophobic amino acids. They did not observed modification of the crystallinity of the film. So, they conclude that structure changes only occurs in the amorphous phase of the gelatin matrix. It is also observed that irradiation enhances the thermal stability of the gelatin film, by increasing the glass transition temperature and the degradation temperature [54].

Cross-linked copolymers of gelatin and poly(vinyl alcohol) (PVA) with excellent water absorption and water retention abilities were successfully synthesized using Co-60 gamma radiation [55]. Also, gamma irradiation was applied on milk protein to improve characteristics of milk protein films [56–59].

5. RADIATION AND NANO TECHNOLOGY ON SYNTHETIC AND COMPOSITE EDIBLE POLYMERS

Two classes of biodegradable polymers can be distinguished: synthetic or natural polymers. Recent developments in biopolymer-based food packaging materials include natural biopolymers (such as starches and proteins), synthetic biopolymers (such as poly lactic acid), biopolymer blends and nanocomposites based on natural and synthetic biopolymers [60]. The combination of synthetic and natural polymers to form films are of remarkable importance in producing composite polymers with permeability or mechanical properties according to the need of a specific application that can be applied either in the form of an emulsion, suspension or dispersion of the non-miscible constituents, or in successive layers, or in the form of a solution in a common solvent [61].

Biodegradable synthetic aliphatic polyester, like polylactide (PLA), has been studied extensively for a number of applications like drug delivery system. Frequently renewable resource-generated monomers possess better mechanical properties and easy processability by conventional methods like thermoforming, injection and blow molding with non-toxic degradation products, which have made it superior than the other conventional thermoplastics [62].

In addition to traditional plant materials for biodegradable polymer production, it is worthy to mention the advances in synthesizing novel polymers within transgenic plants, especially those in the polyhydroxyalkanoate class [63].

Application of nanoscience and nanotechnology to the agriculture and food sector is relatively recent compared with their use in drug delivery and pharmaceuticals [64, 65].

In the food industry, nanotechnology can be utilized in order to enhance the delivery of food ingredients to target sites, increase flavor, inhibit bacterial growth, extend product shelf life and improve food safety.

Applications of nanomaterials that do not involve direct addition of nanoparticles to consumed foods, and thus more likely to be marketed to the public in the short term, are related to food packaging and food safety.

These applications include polymer/clay nanocomposites as high barrier packaging materials, silver nanoparticles as potent antimicrobial agents and nanosensors and nanomaterial-based assays for the detection of food-relevant analytes (gases, small organic molecules and food-borne pathogens) [66].

Other emerging topics of nanotechnology for food and agriculture are smart delivery of nutrients, bioseparation of proteins, rapid sampling of biological and chemical contaminants and nanoencapsulation of nutraceuticals, as well as advances in technologies, such as DNA microarrays, microelectromechanical systems and microfluidics [67, 68].

Nano-scale cellulose fiber materials (e.g., microfibrillated cellulose and bacterial cellulose) are promising candidates for bio-nanocomposite production due to their abundance, high strength and stiffness, low weight and biodegradability [69].

Several reinforcing nanoparticles such as clays, silica or silver have been used for industrial applications, but cellulose nanocrystals (CNCs) are a better choice for food industry due to their biodegradable and biocompatible nature as well as their outstanding potential in improving mechanical and barrier properties of nanocomposites [70–72]. Also, cellulose nanofibers (CNF) reinforcement improved mechanical properties, except elongation of mango puree edible films [73].

The development of nanoscale systems for the encapsulation, protection and delivery of lipophilic nutrients, vitamins and nutraceuticals was recently reported [74]. A promising route to the synthesis of protein-mimetic materials that are capable of complex functions, such as molecular recognition and catalysis, is provided by peptoid nanosheets polymers structurally related to biologically occurring polypeptides [75].

The conjunction of radiation and nanotechnology for the improvement of biomaterials are drawing special interest [76]. Gamma radiation and nanocrystalline cellulose was used for the reinforcement of poly(caprolactone) composites [77]. Also, the radiation synthesis of gelatin hydrogels containing Ag nanoparticles was reported [78]. Grasielli and his group synthesized nanoparticles of seroalbumin via intramolecular cross-linking using gamma rays technology [79]. The radiation synthesis of a composite prepared from the algae polysaccharide alginate and clay nanocomposite was also described [80].

6. CONCLUDING REMARKS

To meet the growing demand of recyclable or natural packaging materials and consumer demands for safer and better quality foods, new and novel food-grade packaging materials or technologies have been, and continue to be, developed. Examples of these packaging materials include bio-based polymers, bioplastic or biopolymer packaging products made from raw materials originating from agricultural or marine sources. Under that view, the use of ionizing radiation to synthesize and or enhance properties of edible polymers based on polysaccharides or proteins appears as an emerging promising application. The use of edible coating previously cross-linked by irradiation has permitted a better control of the release of active compounds. So far, the industrial effectiveness of the processing accumulates enough evidences within the huge spectrum of possibilities that radiation technology can offer. In the meanwhile, in order to increase further applications of edible films prepared using radiation technology, an update of the sensory implications, regulatory status and future perspectives on innovative food preservation must still be examined carefully.

REFERENCE

1. Dhall RK. Advances in edible coatings for fresh fruits and vegetables: a review. Critical Reviews in Food Science and Nutrition. 2013;53(5):435-450.
2. Zhu D, Jin L, Wang Y, Ren H. Water barrier property of gelatin-based films cross-linked with microbial transglutaminase. Advanced Materials Research. 2011;335-336:1391-1395. 10.4028/www.scientific.net/AMR.335-336.1391.
3. Basiak E, Debeaufort F, Lenart A. Effect of oil lamination between plasticized starch layers on film properties. Food Chemistry. 2016;195:56-63.
4. Guilbert S, Gontard N, Cuq B. Technology and applications of edible protective films. Packaging Technology and Science. 1995;8(6):339-346. Article first published online: 28 APR 2006.
5. Gibbs BF, Kermasha S, Alli I, Mulligan CN. Encapsulation in the food industry: a review. International Journal of Food Sciences and Nutrition. 1999;50(3):213-224.
6. Huq T, Khan A, Khan RA, Riedl B, Lacroix M. Encapsulation of probiotic bacteria in biopolymeric system. Critical Reviews of Food Science Nutrition. 2013;53(9):909-916.
7. Mellinas C, Valdés A, Ramos M, Burgos N, Garrigós MC, Jiménez A. Active edible films: current state and future trends. Journal of Applied Polymer Science. 2016; app.42631,
8. Jost V, Stramm C. Influence of plasticizers on the mechanical and barrier properties of cast biopolymer films. Journal of Applied Polymer Science. 2015;133(2).
9. Shit SC, Shah PM. Edible polymers: challenges and opportunities. Journal of Polymers. 2014;2014(2014), Article ID 427259, 13 p.
10. Clough RL. High-energy radiation and polymers: a review of commercial processes and emerging applications. Nuclear Instruments & Methods in Physics Research. Section B-Beam Interactions with Materials and Atoms. 2001;185:8-33.

11. Mastro NL. Role of irradiation treatment on food industry. International Journal of Nuclear Governance, Economy and Ecology. 2011;3(3):266-273. DOI:http:// dx.doi.org/10.1504/IJNGEE.2011.042213.
12. Czvikovszky T. Degradation effects in polymers. In: IAEA. Advances in Radiation Chemistry of Polymers. TECDOC-1420, Vienna, IAEA; 2004. pp. 91-102. @ISBN 92–0–112504–6 @ISSN 1011–4289.
13. Czvikovszky T. Radiation assisted compatibilization of polymers. In: IAEA. Advances in Radiation Chemistry of Polymers. TECDOC-1420, IAEA, Vienna, IAEA; 2004. pp. 61-74. @ISBN 92–0–112504–6 @ISSN 1011–4289.
14. Riley PA. Free radicals in biology: oxidative stress and the effects of ionizing radiation. International Journal of Radiation Biology. 1994;65 (1):27-33.
15. Vicentini NM, Dupuy N, Leitzelman M, Cereda MP, Sobral PJA. Prediction of cassava starch edible film properties by chemometric analysis of infrared spectra. Spectroscopy Letters. 2005:38(6):749-767.
16. Treviño-Garza MZ, García S, del Socorro Flores-González M, Arévalo-Niño K. Edible active coatings based on pectin, pullulan, and chitosan increase quality and shelf life of strawberries (Fragaria ananassa). Journal of Food Science. 2015;80(8):M1823-M1830.
17. Gutierrez MC, Nunez-Santiago MD, Romero-Bastida CA, Martinez-Bustos F. Effects of coconut oil concentration as a plasticizer and Yucca schidigera extract as a surfactant in the preparation of extruded corn starch films. Starch-Starke. 2014;66(11-12):1079–1088.
18. Nobrega MM, Olivato JB, Grossmann MEV, Bona E, Yamashita F. Effects of the incorporation of saturated fatty acids on the mechanical and barrier properties of biodegradable films. Journal of Applied Polymer Science. 2012;124(5):3695-3703. DOI:10.1002/app.35250.
19. Ghanbarzadeh B, Almasi H, Entezami AA. Improving the barrier and mechanical properties of corn starch-based edible films: effect of citric acid and carboxymethyl cellulose. Industrial Crops and Products. 2011;33:229-235.
20. Ghanbarzadeh B, Almasi H, Entezami AA. Physical properties of edible modified starch/carboxymethyl cellulose films. Innovative Food Science & Emerging Technologies. 2010;11(4):697-702.
21. Dick M, Pagno CH, Costa TMH, Gomaa A, Subirade M, Rios AO, Flores SH. Edible films based on chia flour: development and characterization. Journal of Applied Polymer Science. 2016;133(2) Article Number 42455.
22. Fakhouri FM, Fontes LCB, Gonçalves PVM, Milanez CR, Steel CJ, Collares-Queiroz FP. Films and edible coatings based on native starches and gelatin in the conservation and sensory acceptance of Crimson grapes. Ciênc. Tecnol. Aliment. Campinas (Food Sc Technol. Campinas. 2007;27(2).
23. Reddy N, Yang Y. Citric acid cross-linking of starch films. Food Chemistry. 2010;118(3):702-711.
24. Makuuchi K, Cheng S. Radiation Processing of Polymer Materials and Its Industrial Applications. Hoboken, New Jersey, John Wiley and Sons, Inc. 2012. 325 p. ISBN978-0470-58769-0.
25. Kume T, Nagasawa N, Yoshii F. Utilization of carbohydrates by radiation processing. Radiation Physics and Chemistry. 2002;63(3-6):625-627.
26. Hu B, Huang M, Yin S, Zi M, Wen Q. Effects of electron-beam irradiation on physicochemical properties of starches separated from stored wheat. Starch/Staerke. 2011;63(3):121-127.
27. Chung HJ, Liu Q. Molecular structure and physicochemical properties of potato and bean starches as affected by gamma-irradiation. International Journal of Biological Macromolecules. 2010;47(2):214-222.
28. Gani A, Bashir M, Wani SM, Masoodi FA. Modification of bean starch by γ-irradiation: effects on functional and morphological properties. LWT - Food Science and Technology. 2012;49(1):162-169.
29. Liu T, Ma Y, Xue S, Shi J. Modifications of structure and physicochemical properties of maize starch by γ-irradiation treatments. LWT - Food Science and Technology. 2012;46(1):156-163.

30. Bao JS, Ao Z, Jane JL. Characterization of physical properties of flour and starch obtained from gamma-irradiated white rice. Starch/Stärke. 2005 57:480–487.
31. Ali AEH, AlArifi A. Characterization and in vitro evaluation of starch-based hydrogels as carriers. Carbohydrate Polymers. 2009;78:725-730.
32. Salmieri S, Khan RA, Safrany A, Lacroix M. Gamma rays-induced 2-hydroxyethyl methacrylate graft copolymerization on methylcellulose-based films: structure analysis and physicochemical properties. Industrial Crops and Products. 2015;70:64-71.
33. Kang HJ, Min SC. Potato peel-based biopolymer film development using high-pressure homogenization, irradiation, and ultrasound. LWT—Food Science and Technology. 2010;43(6):903-909.
34. Foegeding EA. Food protein functionality—a new model. Journal of Food Science. 2015;80(12):C2670-C2677.
35. Lacroix M, Vu KD. Edible Coating and Film Materials: Proteins—Chapter 11. In: Han JH (Ed). Innovations in Food Packaging (2nd Edition) 2014, Elsevier, pp. 277-304.
36. Su JF, Huang Z, Yuan XY, Wuang XY, Li, M. Structure and properties of carboxymethyl cellulose/soy protein isolate blend edible films crosslinked by Maillard reactions. Carbohydrate Polymers. 2010;79(1):145-153.
37. Rakotonirainy AM, Padua GW. Effects of lamination and coating with drying oils on tensile and barrier properties of zein films. Journal of Agriculture and Food Chemistry. 2001;49(6):2860-2863.
38. Song F, Tang DL, Wang XL, Wang YZ. Biodegradable soy protein isolate-based materials: a review. Biomacromolecules. 2011;12:3369-3380.
39. Kim DY, Min SC. Trout skin gelatin-based edible film development. Journal of Food Science. 2012;77(9):E240-E246.
40. Ma H, Shen JJ, Yang Q, Zhou J, Xia SH, Cao J. Effect of the introduction of fish collagen on the thermal and mechanical properties of poly(lactic acid). Industrial Engineering and Chemistry Research. 2015;54(43):10945-10951.
41. Guerrero P, Beatty E, Kerry JP, de la Caba K. Extrusion of soy protein with gelatin and sugars at low moisture content. Journal of Food Engineering. 2012;110:53-59.
42. Soradech S, Nunthanid J, Limmatvapirat S, Luangtanaanan M. An approach for the enhancement of the mechanical properties and film coating efficiency of shellac by the formation of composite films based on shellac and gelatin. Journal of Food Engineering. 2012;108(1):94-102.
43. Yi JB, Kim YT, Bae HJ, Whiteside WS, Park HJ. Influence of transglutaminase-induced cross-linking on properties of fish gelatin films. Journal of Food Science. 2006;71(9):E376-E383.
44. Arabestani A, Kadivar M, Shahedi M, Goli, SAH, Porta R. The effect of oxidized ferulic acid on physicochemical properties of bitter vetch (Vicia ervilia) protein-based films. Journal of Applied Polymer Science. 2015;33(2).
45. Zhu D, Jin L, Wang Y, Ren H. Water barrier property of gelatin-based films cross-linked with microbial transglutaminase. Advanced Materials Research. 2011;335-336:1391-1395. 10.4028/www.scientific.net/AMR.335-336.1391.
46. Zhang YN, Liu N, Zhao XH. A study on the preparation and some functional properties of a cross-linked casein-gelatin composite by a microbial transglutaminase. International Journal of Food Science & Technology. 2011;46(12):2641-2647.
47. Alves PMA, Carvalho RA, Moraes ICF, Luciano CG, Bittante AMQB, Sobral PJA. Development of films based on blends of gelatin and poly(vinyl alcohol) cross linked with glutaraldehyde. Food Hydrocolloids. 2011;25(7):1751-1757.
48. Kuan YH, Bhat R, Patras A, Karim AA. Radiation processing of food proteins-A review on the recent developments. Trends in Food Science & Technology. 2013;30:105-120.
49. Xiong Y, Foegeding EA. Protein oxidation—the less appreciated sibling of lipid oxidation. Editorial. Journal of Food Science. 2015;80(7):iii.
50. Shahidi F, Zhong Y. Novel antioxidants in food quality preservation and health promotion. European Journal of Lipid Technology. 2010;112(9):930-940.
51. Lacroix M, Khan R, Senna M, Sharmin N, Salmieri S, Safrany A. Radiation grafting on natural films. Radiation Physics and Chemistry. 2014;94:88-92.

52. Khan MA, Rahman N, Rahman M. Preparation and characterization of gamma radiation cured gelatin-pva bio-blend. Multi-functional materials and structures III, Parts 1 and 2. In: Lee JH. Advanced Materials Research. 2010;123-125:347-350. The 3rd International Conference on Multi-Functional Materials and Structures, Chonbuk Natl Univ, Jeollabuk do, Jeonju, South Korea, September 14–18, 2010.
53. Vu KD, Hollingsworth RG, Salmieri S, Takala PN, Lacroix M. Development of bioactive coatings based on γ-irradiated proteins to preserve strawberries. Radiation Physics and Chemistry. 2012;81(8):1211-1214.
54. Benbettaieb N, Karbowiak T, Brachais CH, Debeaufort F. Impact of electron beam irradiation on fish gelatin film properties. Food Chemistry. 2016;195:11-18.
55. Kaur I, Bhati P, Bala K, Kanwar SS. Effective immobilization of lipase onto a porous gelatin-co-poly (vinyl alcohol) copolymer and evaluation of its hydrolytic properties. Journal of Applied Polymer Science. 2014;131(10), article # 39622.
56. Vachon C, Yu HL, Yefsah R, Alain R, St-Gelais D, Lacroix M. Mechanical and structural properties of milk protein edible films cross-linked by heating and γ-irradiation. Journal of Agriculture and Food Chemistry. 2000;48(8):3202-3209.
57. Ciesla K, Salmieri S, Lacroix M. Gamma-irradiation influence on the structure and properties of calcium caseinate–whey protein isolate based films. Part 1. Radiation effect on proteins gels and films. Journal of Agriculture and Food Chemistry. 2006;54(17):6374-6384.
58. Ciesla K, Salmieri S, Lacroix M. Gamma-irradiation influence on the structure and properties of calcium caseinate–whey protein isolate based films. Part 2. Influence of polysaccharide addition and radiation treatment on the structure and functional properties of the films. Journal of Agriculture and Food Chemistry. 2006;54(23):8899-8908.
59. Ramos OL, Fernandes JC, Silva SI, Pintado ME, Malcata FX. Edible films and coatings from whey proteins: a review on formulation, and on mechanical and bioactive properties. Critical Reviews in Food Science and Nutrition. 2012;52(6):533-552.
60. Tang XZ, Kumar P, Alavi S, Sandeep KP. Recent advances in biopolymers and biopolymer-based nanocomposites for food packaging materials. Critical Reviews in Food Science and Nutrition. 2012;52(5):426-444.
61. Arvanitoyannis IS. Totally and partially biodegradable polymer blends based on natural and synthetic macromolecules: preparation, physical properties, and potential as food packaging materials. Journal of Macromolecular Science. 1999;39(2):205-271.
62. Gupta AP, Kumar V. New emerging trends in synthetic biodegradable polymers – polylactide: a critique. European Polymer Journal. 2007;43(10):4053-4074.
63. Mooney BP. The second green revolution? Production of plant-based biodegradable plastics. Biochemistry Journal. 2009;418(2):219-232.
64. Sanguansri P, Augustin MA. Nanoscale materials development: a food industry perspective. Trends in Food Science and Technology. 2006;17:547-556.
65. Karim S, Dufresne A, Tahir PM, Karimi A, Abdulkhani A. Biodegradable starch-based composites: effect of micro and nanoreinforcements on composite properties. Journal of Materials Science. 2014;49(13):4513-4521.
66. Duncan TV. Applications of nanotechnology in food packaging and food safety: barrier materials, antimicrobials and sensors. Journal of Colloid and Interface Science. 2011;363(1):1-24.
67. Sozer N, Kokini JL. Nanotechnology and its applications in the food sector. Trends in Biotechnology. 2009;27(2):82-89.
68. Hannon JC, Kerry J, Cruz-Romero M, Morris M, Cummins E. Advances and challenges for the use of engineered nanoparticles in food contact materials. Trends in Food Science & Technology. 2015;43(1):43-62.
69. Siró I, Plackett D. Microfibrillated cellulose and new nanocomposite materials: a review. Cellulose. 2010;17(3):459-494.
70. Criado P, Fraschini C, Salmieri S, Lacroix M. Modification of nanocrystalline cellulose for bioactive loaded films. Journal of Research Updates in Polymer Science. 2014;3:122-135.

71. Khan A, Huq T, Khan RA, Riedl B, Lacroix M. Nanocellulose-based composites and bioactive agents for food packaging. Critical Reviews of Food Science Nutrition. 2014;54(2):163-174.
72. Silva JBA, Pereira FV, Druzian JI. Cassava starch-based films plasticized with sucrose and inverted sugar and reinforced with cellulose nanocrystals. Journal of Food Science. 2012; 77(6):N14–N19.
73. Azeredo HMC, Mattoso LHC, Wood D, Williams TG, Avena-Bustillos RJ, Mchugh TH. Nanocomposite edible films from mango puree reinforced with cellulose nanofiber. Journal of Food Science. 2009;74(5):N31-N35.
74. McClements DJ. Nanoscale nutrient delivery systems for food applications: improving bioactive dispersibility, stability and bioavailability. Journal of Food Science. 2015:80(7):N1602-N1611.
75. Mannige RV, Haxton TK, Proulx C, Robertson EJ, Batigelli A, Butterfoss GL, Zuckerman RN, Whitelam S. Peptoid nanosheets exhibit a new secondary-structure motif. Nature. 2015;526:415-420..
76. Kharisov BI, Kharissova OV, Mendez UO, editors. Radiation Synthesis of Materials and Compounds. Boca Raton, 2013, CRC Press, Taylor & Francis Group; 560 p. ISBN 13.978-1-4665-0523-0.
77. Khan RA, Beck S, Dussault D, Salmieri S, Lacroix M. Mechanical and barrier properties of nanocrystalline cellulose reinforced poly(caprolactone) composites: effect of gamma radiation. Journal of Applied Polymer Science. 2013;129(5):3038-3046.
78. Leawhiran N, Pavasant P, Soontornvipart K, Supaphol P. Gamma irradiation synthesis and characterization of AgNP/gelatin/PVA hydrogels for antibacterial wound dressings. Journal of Applied Polymer Science. 2014;131(23), article #41138.
79. Espinoza SLS, Sanchez ML, Risso VA, Grasielli M. Radiation synthesis of seroalbumin nanoparticles. Radiation Physics and Chemistry. 2012;81(9):1417-1421.
80. Taleb MFA, Hegazy DE, Ismail SA. Radiation synthesis, characterization and dye adsorption of alginate–organophilic montmorillonite nanocomposite. Carbohydrate Polymers. 2012;87(3):2263-2269.

CHAPTER 14

Nanostructuring of Material Surfaces by Laser Ablation

Cinthya Toro Salazar, María Laura Azcárate and Carlos Alberto Rinaldi

ABSTRACT

Irradiation of materials such as iron and silicon with single nanosecond laser pulses produces nanostructures on its surfaces. Nevertheless, the deposition before irradiation of thin films on the surface of the silicon wafers can modify the shapes of these structures. Upon laser irradiation, different effects are produced on the surfaces of monocrystalline silicon wafers coated with a thin film of Si_3N_4 than on that of bare ones. After irradiation with a Nd:YAG laser pulse of 532 nm, the coated silicon surface presents a nanostructure that, due to its hydrophobic behavior, can be used for biological applications such as cell growth. On the other hand, the nanostructures formed on the surface of metals, such as iron, make them more resistant to oxidation processes by changing their oxidation potentials.

Keywords: nanosecond laser ablation, nanoparticles, surface roughness, nanostructures, laser micromachining

1. INTRODUCTION

The surface nanostructuring and the generation of nanoparticles by laser ablation with nanosecond lasers are subjects that have gained importance in the last 15 years [1]. These issues were present since the early experiments with pulsed lasers because they are inherent to pulsed laser interaction with matter, and researchers have made great efforts to get rid of them. In laser ablation micromachining, these effects are known as HAZ (heat affected zone) and are undesirable [1, 3]. It was not until the explosive spring forth of nanotechnology in technological applications in several areas such as medicine and microelectronics among many others that the appearance of different nanostructures in a wide variety of materials started to be reported in the literature.

In this chapter, we will present a review of the nanostructures found by our research group during the quality control of the surfaces of micromachined devices. On the other hand, a method for generating metallic nanoparticles which was developed on the basis of the analysis of the material ejected during the micromachining ablation process will be reported. Nanoparticles were then generated in different media such as air, deionized water, isopropyl alcohol and sodium dodecyl sulfate (SDS) solution. Preliminary results of the generation of Fe@Au nanoparticles (core–shell nanoparticles) will be also discussed.

A brief summary of the fundamentals of laser ablation of solid substrates and of the theory of the resulting ablation plume will be given in the next section. A large variety of studies in this field including basic research and applications have been carried out worldwide by many researchers. However, we will focus on the area of micromachining by laser ablation [1–4].

1.1. Laser ablation

The laser ablation mechanism is one of the most complex phenomena observed when laser radiation interacts with a solid material [1]. Ablation may be produced either with pulsed or with intense continuous wave (cw) lasers [1, 3]. For a given type of material, the onset of ablation takes place around a threshold fluence ($\emptyset\ th$), which depends on absorption mechanisms and particular properties of the material, such as its microstructure, its morphology, and the presence of defects as well as on laser parameters (intensity, wavelength, and pulse duration) [1, 2, 5, 6]. When irradiation is performed with several consecutive pulses impinging on the same sample's area, $\emptyset\ th$ can vary due to the accumulation of defects in the previously irradiated zone. At a given wavelength, the amount of material removed per pulse usually shows a logarithmic increase with fluence in accordance with the Lambert–Beer law. On the other hand, Villagran-Muniz et al. [7] reported heuristic equations to describe the fluence dependence of the amount of removed material per pulse of a given substrate with parameters that can be related to physical properties of the substrate.

In particular, for laser pulse durations larger than 10 ps, the laser beam will also interact with the ablation plume as in the case of 10 ns Nd:YAG lasers. In these conditions, the plume will also absorb and scatter radiation. In consequence, the amount of energy reaching the substrate will be less since part of it will be absorbed by the plasma, generating a hot plasma. This excited plasma will then expand and create shock waves in the molten substrate. Frozen shock waves can be observed in the material when the irradiated zone cools [8]. Another effect produced by the excited plume is the explosion of the remnant molten material, which produces splashes. The expelled liquid will inevitably solidify around the irradiated area together with plume's material condensation. In addition, the rapid generation of large thermal gradients may induce excessive thermal stress and thermoelastic excitation of acoustic waves. These stresses may add hardening work, buckling, or cracking [1] to the response of the material, effects which must be taken into account in each particular application.

However, this is not a general feature of all materials. The dynamics of the laser–matter interaction starts with the electronic excitation and relaxation. For weak electric fields, ionization of an atom occurs when the energy of the incident photon exceeds the binding energy of the valence electron. Thus, unlike metals, in a large variety of materials with valence bands, such as dielectrics, electrons are not directly excited into the conduction band. In Chapter 8 of reference [1], an extensive explanation of the differences in the absorption processes of between metals and dielectrics is given.

Only the final results of the ablation process and perhaps some insights of the different phenomena that can possibly occur are experimentally observed. When the material's processing involves a laser, the light–matter interaction mechanisms that simultaneously occur will depend on the pulse duration. Thus, the most relevant mechanisms that take place on a substrate according to the laser pulse duration are as follows: material melting with cw and millisecond pulsed lasers, material vaporizing with nanosecond pulses, and finally, sublimation of material with femtosecond pulses [1]. It is worthy to note that short pulses (1 ns–1 μs) emitted by lasers with Q-switch devices also reduce the thermal impact on the material.

1.1.1. Applications: nanostructuring and manufacturing of nanoparticles

Nanomanufacturing was developed in the last decade driven by the nanotechnology progress. Sub-micrometric structures in the irradiation zones or resulting from the ablation plume were observed since the beginning of pulsed lasers applications to materials' processing. However, at that time, these were a problem and many attempts have been made to mitigate them. It was only in the last decade that they were considered useful [9–12].

The large scope of applications of the sub-micrometric structures' with unique properties that have been lately found [13–15] contributed to this change of perspective. Although the manufacture of nanometric structures and particles with pulsed lasers is quite simple and can be applied to a large variety of materials, the phenomena involved are difficult to analyze and simulate, resulting in a theoretical and experimental research field in continuous growth. In particular, in this chapter, we will present a brief overview of the heuristic functions which describe the laser fluence dependence of the removed material in laser ablation processes of different substrates as well as the nanostructures and nanoparticles found in the analysis with an electronic microscope of the quality of micromachined devices and of the material ejected in the ablation plume, respectively.

1.1.1.1. Surface modification – nanostructuring

Nanostructuring by interaction with pulsed lasers is known since the beginning of lasers' applications to machining. However, in recent years, it is an issue that is gaining large importance due to its potential applications in various fields. Different types of nanostructures (NS), named laser-induced periodic surface structures (LIPSS), have been reported. The low spatial frequency LIPSS (LSFL) are characterized by a spatial period Λ of about the laser wavelength (λ), that is, $\Lambda \sim \lambda$. Structures called high spatial frequency LIPSS (HSFL) have a spatial period Λ shorter than the laser wavelength, $\Lambda \ll \lambda$. So far, it is accepted that Λ depends on the laser wavelength and on the radiation incidence angle. These LIPPS are generally described by their period Λ and their orientation with respect to the polarization of the laser [9]. However, the origin of the HSFL is still under debate and some theories have been proposed.

Zhang et al. [10] report NS formed with ultrashort lasers pulses (Ti: Sapphire, 1 kHz, 800 nm, 120 fs). They describe the formation of both types of ripples: LSFL and HSFL. They explain that LSFL arise from optical interference effects due to the coherent interaction between the incident radiation and the electromagnetic wave scattered by the surface. Two main mechanisms are proposed for HSFL formation. One is related to the interaction of the laser pulse with the surface plasma produced by the incident laser. The other mechanism is associated to a combination of interference effects and second harmonic generation. For example, they have experimentally found that the period of the ripples decreases with the increase of the laser fluence on the surface. They report ripples generated perpendicular to the laser polarization in the single-crystal superalloy

CMSX-4. According to the experiments of these authors, HSFL ($\Lambda \ll \lambda$) are generated from LSFL ($\Lambda \sim \lambda$) with a period about half of the LSFL spatial period. That is, while the period of the crests of LSFL is about the laser wavelength (λL = 800 nm, $\Lambda LSFl \sim$ 760 nm), that of the HSFL is $\Lambda HSFl \sim$ 360 nm, which is about half the laser wavelength.

NS in silicon were observed in areas up to 100 mm^2 using a versatile laser machining station. Experiences show that the morphology and frequency of the ripples of these structures do not depend either on the focusing geometry, energy, scan rate, polarization, or incident angle of the radiation [11]. It can be thus inferred that laser-based synthesis of nanomaterials is an area that is in its beginnings.

1.1.1.2. Manufacturing of nanoparticles

The manufacture of nanoparticles by pulsed laser ablation is framed within what is known as laser synthesis of nanomaterials (NM). The nanomaterials have extensive applications in electronics, medicine, and energy generation and storage. The physical and chemical properties of these materials strongly depend on their dimensions and shape. The most investigated nanostructures are spheres (nanoparticles (NPs) and "quantum-dots") and cylinders (nanotubes, nanowires, nanorods). In particular, the nanostructures are made by pulsed laser ablation (PLA). In the PLA of a solid, the NPs produced have the same composition of the solid from which they come. If it takes place in air or vacuum, then thin coatings of NPs formed from the ejected clusters can be generated on the surface. When PLA takes place in a liquid, then a colloidal solution is obtained. On the other hand, there is also evidence that the interaction of the laser radiation with nano- and micro-materials in solutions and/or in suspensions in liquids/gases can produce new structures: nanostructures or alloys, as reported in Chapter 7 of reference [1].

Pulsed laser ablation of materials was introduced with the advent of the ruby laser around 1960 [16]. Since then laser for machining research has been directed to obtain new emission wavelengths, increase the repetition rate and output energy and, of course, decrease the pulse duration, nowadays up to attoseconds (10^{-18} s). This research was mainly encouraged by the fact that the laser processing is a powerful tool that does not wear during machining and, in addition, does not produce pollution. On the other hand, in PLA, materials are subjected to high temperatures and pressures, giving rise to a very particular chemistry with the production of various compounds of oxides, carbides, and nitrides. Besides, since cooling rate is very high due to the rapid expansion of the plasma plume, metastable nanomaterials which are difficult to obtain by other techniques can be produced.

PLA is conceptually simple. However, the material removal mechanisms extend over very long intervals, from fs to ms, at least nine orders of magnitude. When the laser is focused on the surface of a dielectric solid, wavelength dependent non-thermal effects take place. When the solids are metals and semiconductors, thermal processes are also present. The mechanisms and the characteristics of the ejected species strongly depend on the parameters of the laser radiation (wavelength, pulse duration, and fluence).

With a nanosecond laser, PLA is basically a photothermal phenomenon. The incident energy excites the electronic and vibrational levels of the substrate and, in consequence, the material is heated, melted, and evaporated in the first 10–100 ps following the pulse arrival. For relatively low fluence radiation (≤0.3–1 J/cm^2), the ejected material, by desorption and evaporation, is mostly atomic size vapor. Then, the resultant vapor plume expands vertically and is ionized by the photons that keep coming. For fluences near the ablation threshold, the amount of ionized material can be calculated by the Saha equation [17]. If the fluence is larger than the ionization threshold, then optical breakdown occurs and the degree of ionization can be estimated from the Saha-Boltzmann equation.

The plasma role during the laser–matter interaction is not yet fully understood and has been discussed over the past two decades. The plasma is composed of a set of atoms and electrons that remain confined in a high energy field. The laser can produce plasma because during a very short time lapse the photons that interact with the atoms of the solid target can strip off about 15 electrons from each atom. Therefore, the photons that arrive after the solid has been vaporized, contribute to increase the degree of ionization [1]. Ionization strongly influences the dynamics of the plume's condensation.

As the laser fluence increases ($\Phi \geq 10$ J/cm^2), there is a remarkable increase in the amount of ablated material suggesting that a different material ejection mechanism is taking place. Experimentally, it is observed that the removed material consists of a mixture of vapor and micron size droplets (≥10 μm). Different mechanisms have been proposed to explain this behavior. Some suggest that temperature gets close to the critical temperature of the material and then a phase explosion follows [1].

In general, it can be said that in the nanomaterials' synthesis, clusters' ejection and gas to particles condensation, take place. The literature reveals that their final size varies within four orders of magnitude [1]. However, Hubental et al. propose a *tailor*ing method which can be used to homogenize NPs' size [18].

Regardless of the confinement in which the NPs are generated, certain steps taking place in their formation may be outlined:

1) Vaporization (Nd:YAG lasers) and atomization (fs lasers).
2) Phase explosion.
3) Fragmentation.
4) Mechanical exfoliation or chipping.
5) Hydrodynamic spraying.

The high temperature and density of the ejected material near the target's surface develops a pressure exceeding in several orders of magnitude the atmospheric pressure, leading to an expansion of the vapor. During the adiabatic expansion (occurring at low pressure or vacuum) that follows, the thermal energy is converted into kinetic energy, causing a very fast cooling of the plasma (from 10^4 to 10^5 K in 1 μs for a ns laser). The extreme cooling rate leads the plasma to a supersaturation condition in which nucleation becomes energetically favorable [19].

From nucleation theory [20], it is known that the nucleation barrier to form a spherical cluster depends on the cohesive forces between the atoms in the liquid phase, the energetic barrier due to the surface tension, and the plasma ionization (in dielectric materials the electric field leads to polarization). If the radius of the particles with packed atoms is very small, the particle will continue growing, while if it is large, the particle will stop growing and may even break up. The laser beam polarizes the atoms and this effect tends to pack them, so that the critical radius will be even smaller. So, the nucleation energy decreases and a higher amount of nuclei to generate the nanoparticles are produced [21].

Nucleation is said to be homogeneous if clusters are produced from the vaporized material, and the NPs are composed of a few tens of atoms. On the other hand, if clusters are already present during the vapor plume condensation, it is considered a heterogeneous condensation. The already existing clusters are considered nucleation centers and play a predominant role in the condensation stage.

The number of particles is decreased both by collisions and coalescence producing an increase of the NPs' average size. Coalescence in a vapor–liquid medium spontaneously occurs as a reduction of the total surface area during this process, corresponding to a reduction of Gibbs free energy. It takes place up to a few ms after the laser pulse. Then, the NPs cluster due to Van der Waals and electrostatic forces. Clustering is a characteristic feature of NPs' synthesis by laser ablation in gases or liquids that do not contain added stabilizing agents [21–24].

NPs suspended in transparent liquid media can be irradiated, too. Experiments show that when NPs' suspensions are irradiated with pulsed lasers, new structures [25] such as disks, segments, cubes, and pyramids of nanoscale dimensions [26–29] are generated.

2. RESEARCH METHODS

The nanostructures described in this chapter were found when machining micro-devices with a micromachining station based on laser ablation [30]. In particular, the monocrystalline silicon nanostructure was obtained in wafers coated with a sacrificial layer. The nanoparticles were found while analyzing the material ejected during the micro-devices' machining. Later, adequate containers for their generation were developed.

The micromachining station was composed of a laser and a substrates' positioning system. The laser was a frequency doubled Nd:YAG laser (532 nm) emitting pulses of 8 ns FWHM at a repetition rate of 10 Hz. The maximum achievable fluence, for this wavelength, was 250 mJ/cm^2. The substrates' positioning system consisted of eight motors: six stepper motors with micrometric resolution and two piezoelectric motors with nanometric resolution. A diagram of the micromachining station is shown in **Figure 1**.

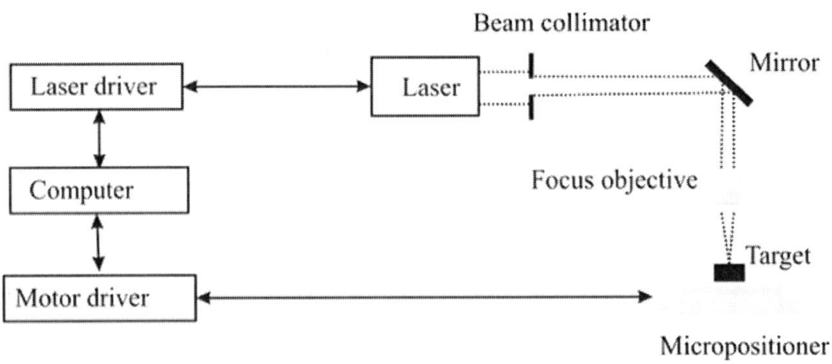

Figure 1. Micromachining system diagram.

Different samples of 2 cm × 2 cm of monocrystalline silicon were cut from c-Si wafers of 4″ diameter and 525 μm thickness. These samples were later coated with different films: Si_2O, Si_3N_4, positive photoresins, resins with dyes, or in some cases, a combination of these films.

The plasma-enhanced chemical vapor deposition (PECVD) technique [30–32] was used to perform Si_2O and Si_3N_4 films coatings. The effect of laser ablation on films of different thicknesses such as 110, 300, 430, 510, 1220, and 2400 nm was studied. The c-Si samples coated with Si_2O were named c-Si + $Si_2O(x)$, where x indicates the film thickness in nm. The thicknesses of the Si_3N_4 films were 180, 420, 500, 570, and 960 nm, and the naming: Si + $Si_3N_4(x)$, x indicating the film thickness in nm. Commercial grade c-Si wafers were also used. These wafers have a 50 nm thick Si_3N_4 film coating on either one or both sides.

Metals such as Fe (99.9%), Ag (99.9%), Au (99.9%), Ta (99.9%), and Cu were other materials processed. Results obtained on Tantalum coated with a 2.5 μm thick AZ1518 resin with the spin-coating technique [33] will be also presented.

2.1. Heuristic equation and its correlation with the physical properties of the irradiated substrate

Since the description of the different regimes observed in the ablation process was found to be a logistic function of the laser fluence, an ad hoc plasma detector was designed based on the works of Bredicce et al. [34] and M. Villagrán et al. [7]. This device, the working principles of which are detailed in reference [30], was used to detect the plasma during the machining. We found that for c-Si, the maximum value of the recorded signals was proportional to the amount of extracted material in each laser shot during the ablation regime according to references [7, 34]. **Figure 2** shows a plot of the induced electric signal maximum amplitude as a function of the incident fluence for c-Si and c-Si + $SiO_2(2400)$.

The results were fitted with the heuristic function described by the following Eq.(1) that relates its parameters to the physical properties of irradiated substrates [34]:

$$MJ(\emptyset)=\left[A1(\emptyset)-A2(\emptyset)1+(\emptyset\emptyset C)SC \right]+A2(\emptyset)MJ\emptyset=A1\emptyset\text{-}A2\emptyset1+\emptyset\emptyset CSC+A2\emptyset$$

where: $M\,J\,(\emptyset)$ is the amount of material removed per pulse which is proportional to the electric signal. Its units are [mass/(area)(time)], and it is a function of the laser fluence.

$\emptyset$ is in the laser fluence in mJ/cm^2.

A_1 and A_2 represent the lower and higher plateau values related to the onset of the absorption and ablation regimes, respectively. Their units are [mass/(area)(time)].

SC is a dimensionless fitting parameter related to the surface roughness.

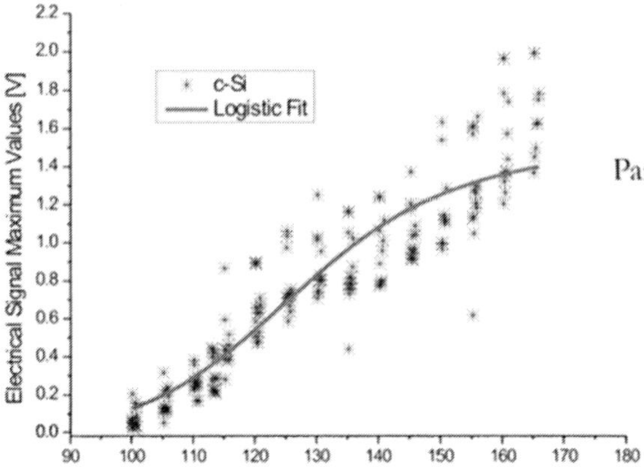

Parameter	Value	Error
A_1	0	0
A_2	1.5	0
Φ_C	127.2	0.8
S_C	9.9	0.6

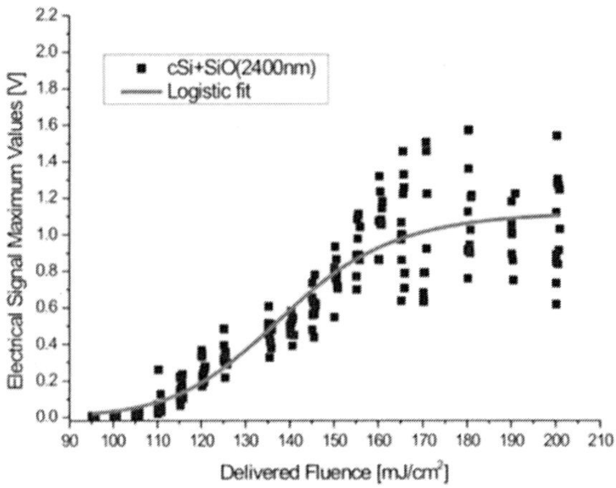

Parameter	Value	Error
A_1	0	0
A_2	1.12	0.04
Φ_C	139	2
S_C	11	1.5

Figure 2. mplitude of the electric signal versus laser fluence fitted by the logistic function of Eq. (1). (a) For c-Si and (b) for c-Si + SiO_2(2400) along with the values of the fitting functions' parameters.

ØC is the critical fluence. Its value indicates the onset of the true ablation process, that is, the fluence value from which on the material removal occurs mainly through ablation. For example, values of $ØC$ = 127 mJ/cm^2 and $ØC$ = 139 mJ/cm^2 were found for c-Si and c-Si + Si_2O(2400), respectively (**Figure 2**).

Two plateaus were found for c-Si coated with thin films <1 μm. We postulated that these values indicated whether the laser radiation had reached the substrate or not. **Figures 3** and **4** show the results obtained for two different thicknesses of two kinds of films: c-Si + Si_3N_4(70) and c-Si + Si_3N_4(350), and, c-Si + SiO_2(110) and c-Si + SiO_2(220), respectively, along with the parameters' values.

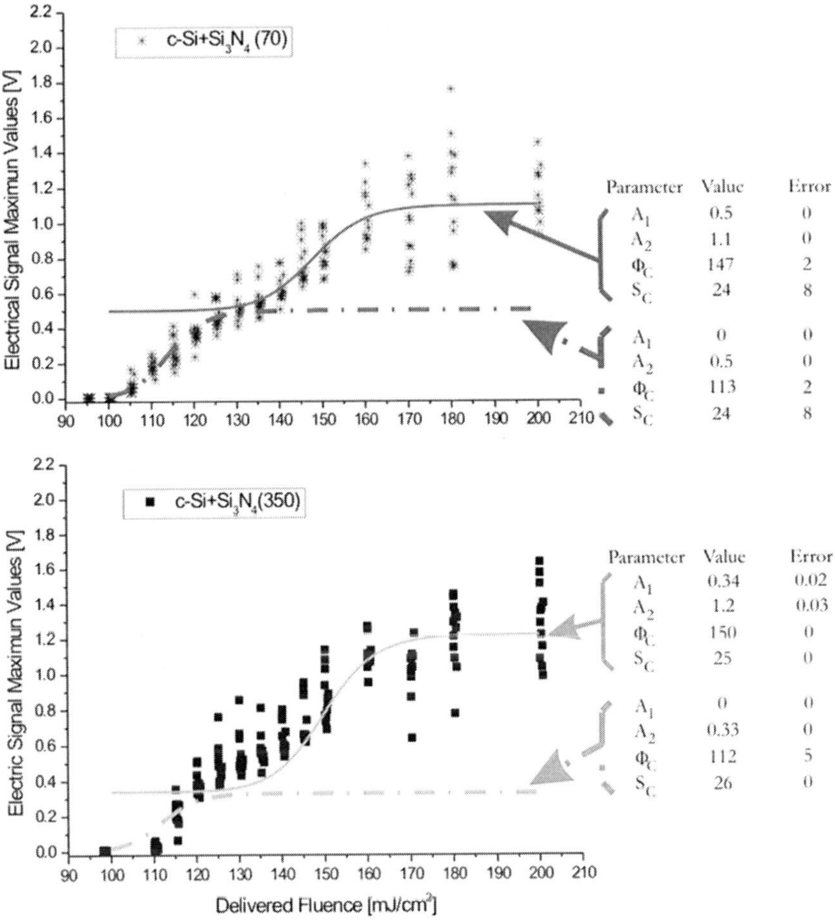

Figure 3. Amplitude of the electric signal versus laser fluence fitted by the logistic function, Eq. (1) for Si_3N_4thin films thicknesses deposited with the EPCVD technique on monocrystalline silicon: 70 and 350 nm. The parameters of the logistic function are shown too.

The fits show that the effect of the film is to change the surface roughness. This is evidenced by the value of the Sc parameter. The larger the value of this parameter, the lower is the roughness of the surface [7]. Thus, the film acts as a filter until the critical fluence, $\boldsymbol{\emptyset\ C}$, is reached. At that moment, the material is ejected from the surface through ablation and generates a nanostructure on the substrate.

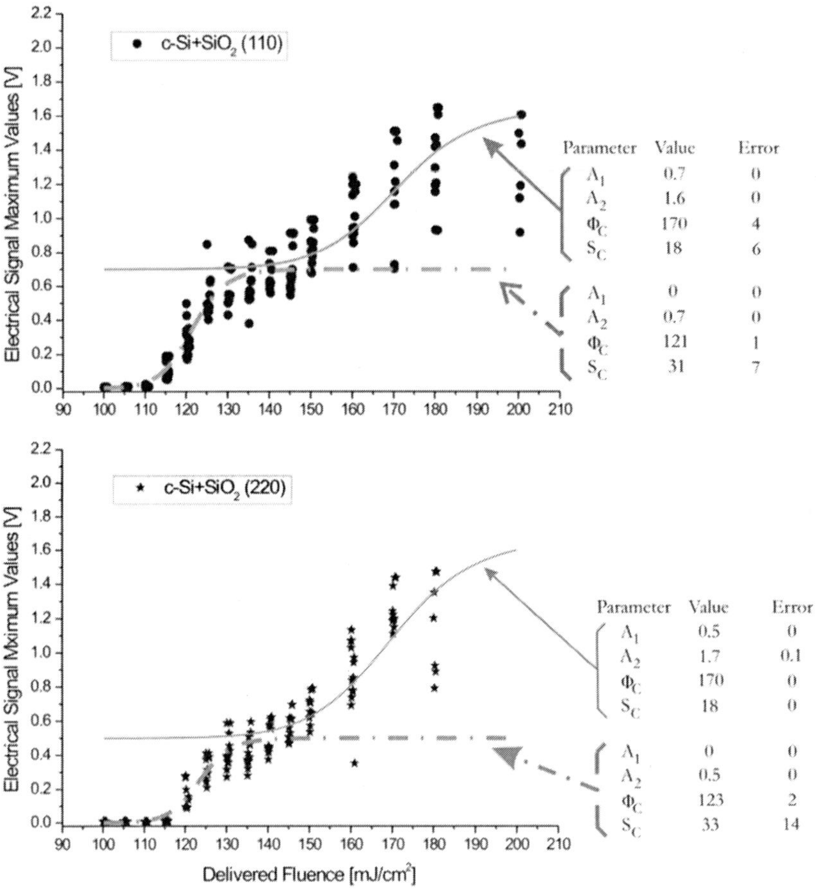

Figure 4. Amplitude of the electric signal versus laser fluence fitted by the logistic function, of Eq. (1) for two silicon dioxide thin films thicknesses deposited with the EPCVD technique on monocrystalline silicon: 110 and 220 nm. The parameters of logistic function are shown too.

2.2. Nanostructuring and nanoparticles manufacturing or generation

2.2.1. Nanostructuring on monocrystalline silicon (c-si)

Nanostructuring on c-Si was observed during the manufacture of prototypes of nanopores and micro-cavities used as molecules' detectors, drug-delivery control, 2D microstructures, replication matrices, etc. in biological applications. These devices were manufactured using the following techniques in the specified order: (1) laser ablation micromachining (one laser shot cavities), (2) selective chemical attack, and (3) ultrasound cleaning (sonication) [35, 36]. The c-Si nanostructuring was observed after the ns laser pulse impacts on the c-Si + Si_3N_4(370) sample. The cavity formed and its nanostructured base can be observed in the SEM (scanning electron microscope) micrographies shown in **Figure 5**. c-Si nanostructuring has been reported in the literature only with sub-picosecond laser pulses [36–38]. The nanostructures produced in the irradiated zones reduce the reflection of the silicon surface. This effect would be of great interest in the manufacture of silicon photovoltaic cells.

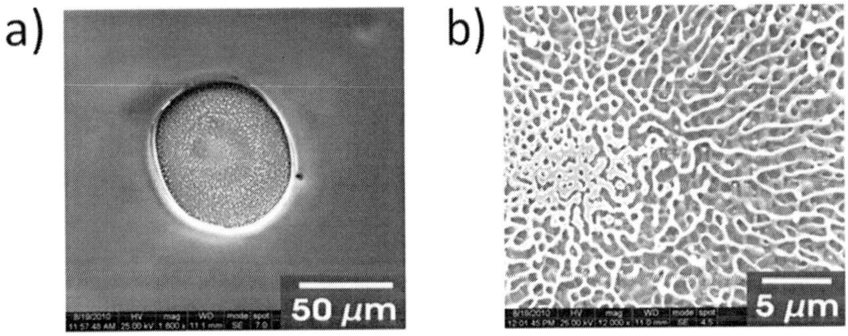

Figure 5. c-Si + Si_3N_4(370) nanostructuring obtained with a ns laser pulse (Nd:YAG, 8 ns, 532 nm). (a) Complete cavity. (b) Magnification of the cavity base.

Nanostructuring was also observed in the machining of a micro-chromatographic column in a c-Si + Si_3N_4(50) commercial wafer [30]. The roughness of the grooves, inherent to the c-Si—laser interaction—was found to be very useful to increase the sample concentration. The micro-column was manufactured as a sequence of straight lines. Part of this micro-chromatographic column and the roughness of the walls of the grooves is shown in **Figure 6**. The larger channels' depth was 50 μm. Devices used for separating substances require roughnesses similar to those of the sample's channels [39].

Finally, microstructures and nanostructures were obtained on the walls of cylindrical and conical cavities. For this purpose, the machining station was used with either both percussion and trepanning drilling modes.

In the percussion drilling mode, the laser pulse always impacts in the same sample's area removing material until the desired depth is reached. Grooves of several microns width are obtained as can be seen in **Figure 7a**.

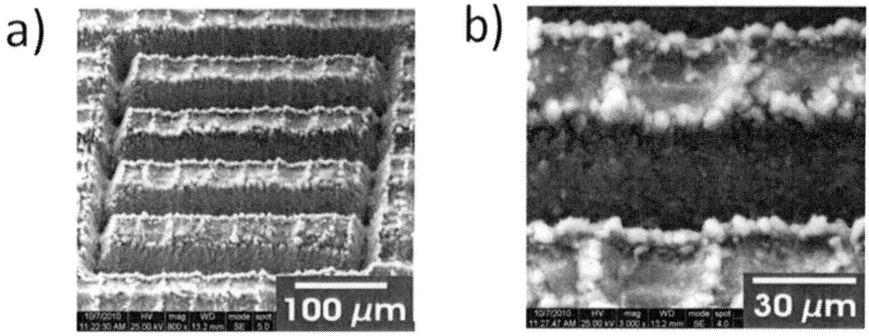

Figure 6. SEM micrographies of (a) the c-Si sample surface after machining. (b) Magnification of part of the machined channels to show the deeper channels. In order to observe the roughness of the channel's wall, the sample's surface was placed forming an angle of 20° with electron beam.

In the trepanning technique, a rotating movement is generated by applying sine wave voltages to the piezoelectric motors with coordinate phases associated to the XY axis of the positioning system. This circular nanometric movement produces a sand-down effect which decreases the drilled cavity walls' roughness. The misalignments of the laser beam focusing system as well as the beam's inhomogeneities are averaged and holes with controlled walls' shapes and roughness can be drilled. As a result, well-controlled circular holes with sub-micrometric structures on the wall can be drilled.

The SEM micrographies of **Figure 7** show cavities drilled on a sample of c-Si + Si_2O(300) with different number of laser pulses of 110 mJ/cm^2. A drilling performed with 2400 pulses with the percussion method can be seen in **Figure 7a**. The grooves on the walls and the debris can be also observed. The debris tower several microns over the silicon surface. A cavity drilled by trepanning with 200 pulses is presented in **Figure 7b**. Marks left by single pulses can be also seen on the sides. In this case, the field of view of electronic microscope allows observing the cavity bottom. Micrography (**Figure 7c**) corresponds to a cavity drilled by trepanning with 2000 laser pulses, the bottom of which can no longer be seen due to its depth. The walls of the cavities of **Figure 7b,c**, show the typical smoothness of the trepanning mode.

2.2.2. Nanostructuring in metals

The micromachining experiments in metals were performed with the machining station of **Figure 1**. Nanostructuring on the surface of the devices machined on metallic substrates was observed by electronic microscopy [30]. In this section, the most relevant results obtained are presented.

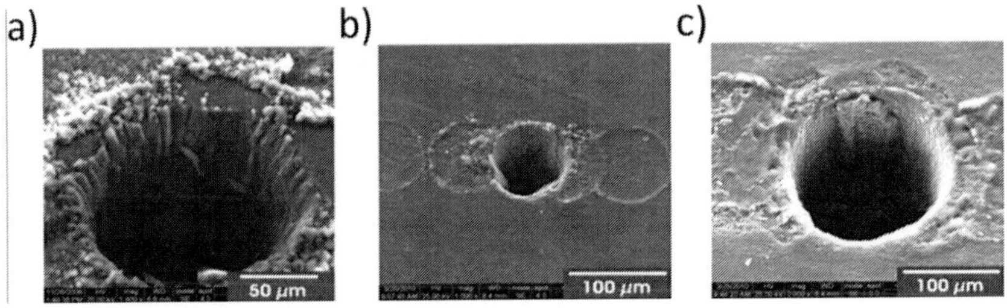

Figure 7. SEM micrographies of (a) c-Si wafer drilled by percussion with 2400 pulses. The sample was inclined 45° to observe the cavity's walls were grooves of ~10 μm are seen. (b, c) Micrographies of cavities drilled by trepanning with 200 and 2000 pulses, respectively. The effect of increasing the laser pulse number for a fixed sample-lens distance and a laser fluence of 110 mJ/cm^2 can be compared. A sine wave voltage of 10 V and 3 Hz was applied to the piezoelectric motors.

2.2.2.1. Cavities in ta(99.9%)

Porous tantalum is an alternative metal for the manufacture of total joint arthroplasty components that offer several unique properties since it has excellent biocompatibility and is safe to use in vivo [40]. Upon direct irradiation with several pulses impacting on the same area of a Ta(99.9%) sample, formation of air bubbles of sub-micrometric dimensions gives rise to a porous coating of tantalum and tantalum oxide of the surface. **Figure 8** shows cavities drilled on a Ta sample with 1, 500, and 1000 laser pulses of 200 mJ/cm^2.

2.2.2.2. Nanostructuring in cu

Micro-ionizer prototypes to be used in an ion mobility spectrometer [41] were machined on a printed circuit board (PCB). The thickness of the copper coating of these boards is about 300 μm. Direct irradiation of the Cu coating of the PCB was performed by the trepanning method with 200 mJ/cm^2. The SEM micrographies are shown in **Figure 9a**. It shows a sequence of contiguous cavities each drilled with 25 laser pulses each. **Figure 9b**,c shows the details of the cavities at different magnifications.

The appearance of a porous coating of copper and copper oxide is observed as the number of laser pulses impinging on the same sample area is increased. The lower density of porous copper makes it suitable to be used as a lighter material for electrodes and catalysis.

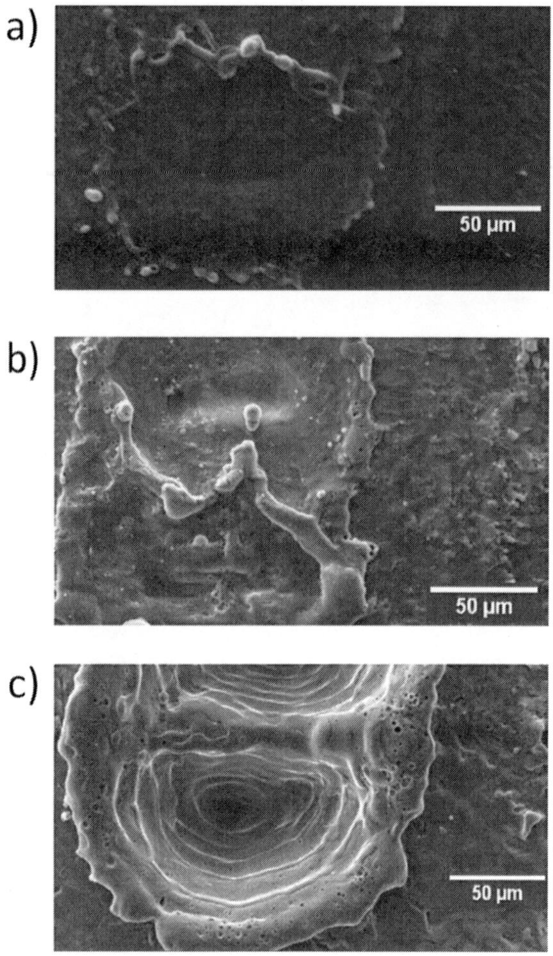

Figure 8. SEM micrographies of cavities drilled on a Ta(99.9%) sample with (a) 1, (b) 500, and (c) 1000 laser pulses of 200 mJ/cm^2.

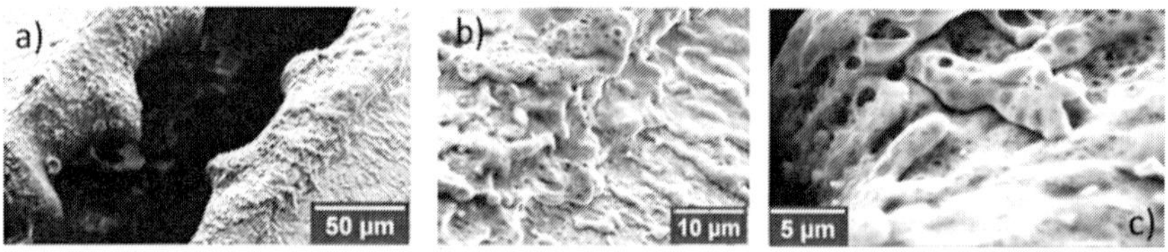

Figure 9. SEM micrographies (a) contiguous cavities machined on the copper coating of a PCB with 25 pulses each. (b, c) Micrographies at larger magnifications to show the porosity.

Figure 10 shows the superficial modifications originated by laser irradiation of sub-micrometric structures of copper oxide. **Figures 10a,b** are SEM micrographies of the surface before irradiation; **Figure 10c,d** after irradiation with one square laser shot.

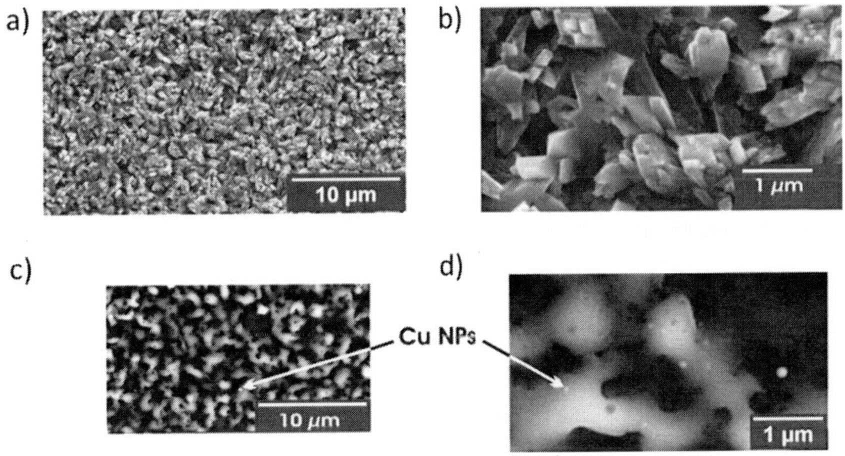

Figure 10. SEM micrographies of a copper oxide surface: (a, b) before irradiation; (c, d) after irradiation with one square laser shot. Cu nanoparticles of lighter color than the rest of the surface due to their larger conductivity can be appreciated.

2.2.2.3. Nanostructuring in fe(99.9%)

The nanostructuring found in Ta and Cu encouraged the irradiation of Fe(99.9%) samples. Only sub-micrometric structures in Fe obtained with sub-picosecond laser pulses have been reported in the literature. The HAZ effect produced on the surface of steel plates is used as anticorrosive. In reference [41], Yang et al. report that the zones in the surroundings of a folding axis are more liable to corrosion due to the larger mechanical strains to which they are subjected. On the other hand, the authors found that steel plates which had been irradiated with laser pulses before being folded became more resistant to oxidation. **Figure 11** shows the surface of a Fe(99.99%) plate after irradiation with laser pulses of 200 mJ/cm^2 in which details of the induced surface nanostructuring can be observed.

Figure 11. Fe (99.99%) plate surface after irradiation laser pulses of 200 mJ/cm^2. The magnification of the electronic microscope (SEM) increases from left to right disclosing details of the surface nanostructuring.

2.2.3. Fe(99.9%), au(99.9%), ag(99.9%) nanoparticles in different media

The manufacture method of NPs of different metals and the diverse analysis performed will be described in the following. The first NPs were generated in air in an ad hoc chamber suitable for obtaining metallic NPs in gaseous media. The substrate was fixed to the container which was then fixed to the machining system. The laser was focused on the surface of the substrate, and the movement of the positioning system of the substrate was controlled by a PC. To ease the subsequent analysis with an electronic microscope, the NPs formed were stuck on carbon tapes.

Figure 12 shows the Fe NPs generated in air at ambient temperature and pressure with two different irradiation processes. Those shown in **Figure 12a** were generated by drawing lines with a speed of 10 μm/s with pulses of 70 μm spot size. The irradiated zones and the number of pulses are detailed in **Figure 12b**. The NPs shown in **Figure 12c**,d were generated by drilling holes by percussion with 1000 pulses. The difference between both methods is the generation of clusters of NPs. The size of the NPs generated by overlapping pulses on the substrate's surface (**Figure 12a**) is ~200 nm, and, as can be seen, they are grouped in clusters. On the other hand, the size of the NPs generated by drilling is ~50 nm and they are also grouped in clusters. In addition to NPs clusters, the formation of dense spherical microparticles is observed (**Figure 12c**). This effect could be produced by the change of the superficial crystalline structure introduced by the ns 532 nm laser pulse on the metal surface, Fe(99.9%) in this case [43, 44].

The generation of NPs in liquid media was performed at ambient temperature and pressure. The substrate was immersed in the liquid contained in a beaker. Laser pulses were focused on the substrate's surface, while the position of the substrate was continuously being changed so that each pulse would impact in a non-irradiated surface area. Metallic NPs dispersions in different liquid media such as ultrapure H_2O, isopropanol, and a solution of sodium dodecyl sulfate (SDS) were obtained following this method.

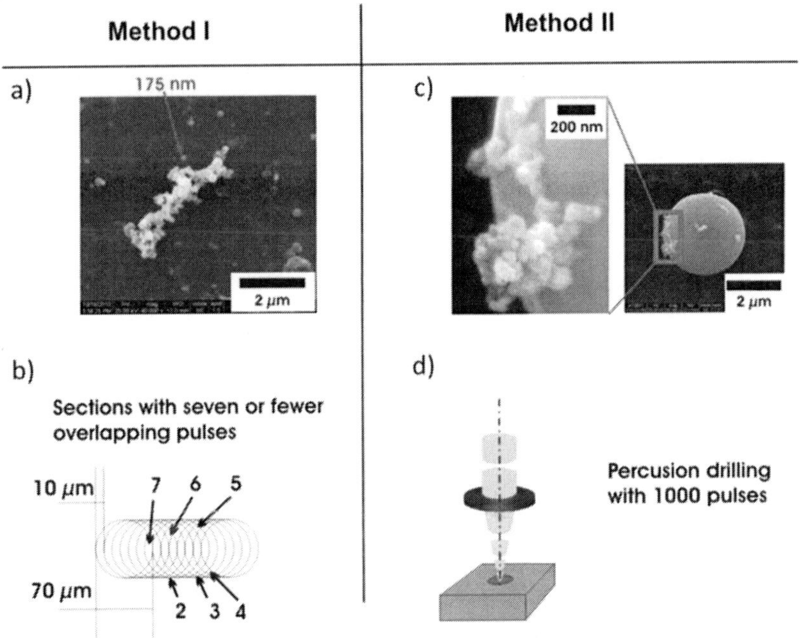

Figure 12. (a) Fe NPs manufactured by drawing lines with up to seven contiguous pulses, (b) Fe NPs generated by percussion (1000 pulses in the same sample area). In both processes, the laser fluence was 220 mJ/cm^2. All micrographies were obtained by SEM.

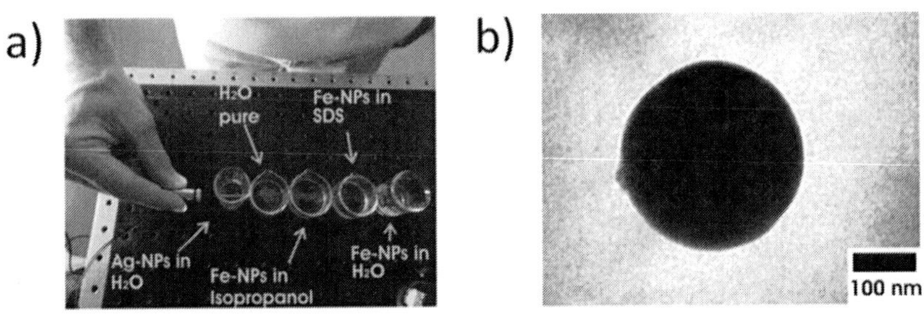

Figure 13. NPs generated in air and in different liquid media. (a) Tyndall effect in dispersions and solutions: (from left to right) Ag NPs in ultrapure H_2O, ultrapure H_2O, Fe NPs in isopropanol, Fe NPs in SDS, and Fe NPs in ultrapure H_2O. (b) Characterization of the Fe NPs in a SDS solution by microscopy (HR-TEM).

Different methods were used for the NPs characterization:

1) Tyndall effect for a rapid verification of the NPs' generation in liquid media. In order to observe this effect, the dispersion is illuminated with a cw laser. For example, in **Figure 13a**, a 98 mW cw laser @405 nm was used. The presence of NPs disperses the laser beam and makes it visible at simple sight. This effect is not observed in solutions or as in the case of ultrapure H_2O in the second beaker from left to right of **Figure 13a**.
2) High-resolution transmission electron microscopy (HR-TEM), which allowed studying the NPs' appearance, form and size (**Figure 13b**).
3) UV–visible light absorption spectrophotometry for the analysis of suspensions.

The advantage of this last method is that it allows preserving the NPs' properties *in situ*, avoiding alterations due to drying effects. Clustering or oxidation effects usually appear when the NPs are separated from the medium in which they were originated (**Figure 14**).

Finally, the size distribution of the NPs in their original medium was estimated from the absorbance spectra of the dispersions and their comparison with those calculated with the MiePlot algorithm (Appendix E, reference [30]).

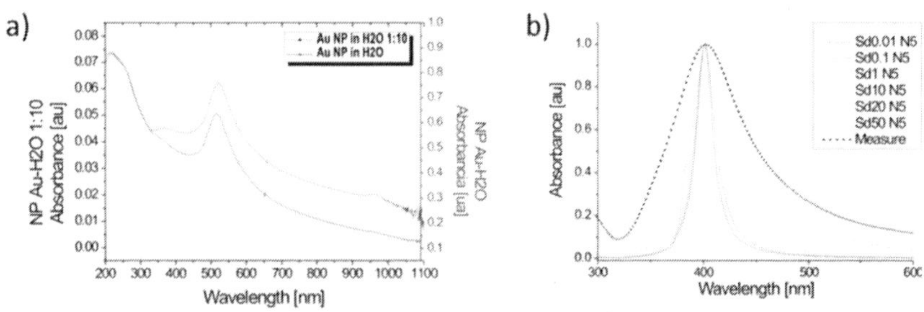

Figure 14. (a) Absorbance spectrum of a dispersion of Au NPs in ultrapure H_2O compared to that of the same dispersion diluted 10 times, (b) comparison of the absorbance spectrum of Ag NPs in H_2O experimentally obtained with that calculated for Ag NPs of 20 nm of radius in H_2O with a log-normal distribution with different standard deviations.

2.2.3.1. Core–shell nanoparticles (preliminary results)

Nanoparticles of different nucleus' materials and *core–shell* were produced as prototypes for *drug delivery*. The particles were required to have a ferromagnetic nucleus to be able to be delivered with magnets to the desired zones and a *core–shell* of a material assimilable by the organism so as not to be rejected by it.

The following method was used for the synthesis of *core–shell* NPs. First, a suspension of Fe NPs in ultrapure H_2O was obtained by laser ablation of a Fe(99.9%) plate. An Au(99.9%) plate was subsequently irradiated in the Fe NPs suspension. The Fe NPs acted as nucleation centers for Au crystallization according to the nucleation principles described in Section 1.1.1.2.

The absorbance spectrum of the final suspension obtained, shown in **Figure 15**, resulted in very good agreement with those reported by other authors in references [45, 46].

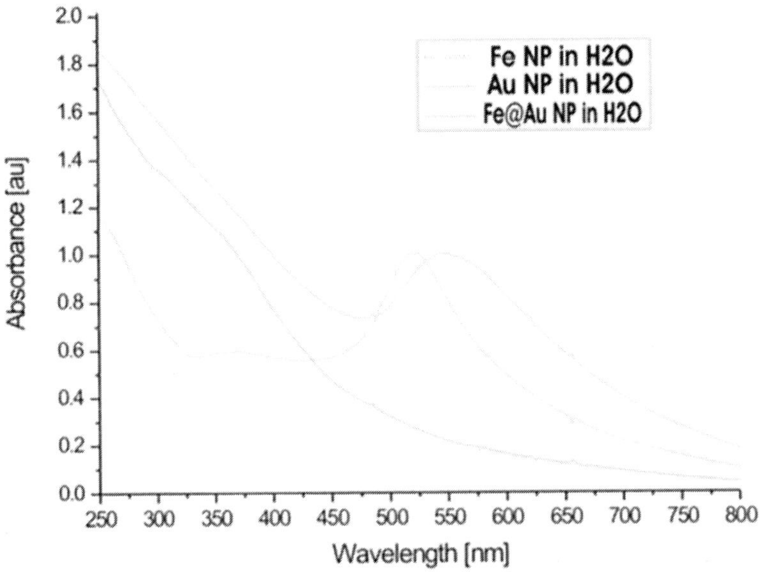

Figure 15. Absorbance spectrum of dispersions of Fe NPs, Au NPs, and Fe@Au NPs in ultrapure H_2O.

3. CONCLUSIONS

Laser ablation allows numerous technological developments ranging from laser-induced breakdown spectroscopy (LIBS), pulsed laser deposition (PLD), laser propulsion, to surface modification and generation of nanoparticles, NPs. In this chapter, we have presented a review of the results obtained with these two latter processes. However, it should be noted that these effects or phenomena were found to occur during the implementation of the micromachining processes and have gained relevance since their observation by means of electron microscopy in its different forms SEM and TEM. The possibility of producing nanostructures on surfaces opens up new possibilities for studying their effects on physicochemical surface phenomena. These effects, mentioned in the Section 1.1.1, as laser-induced periodic surface structures and the increase of roughness in the generation of channels for chromatographic columns will allow the production of cheaper systems based on these principles. On the other hand, the heuristic model presented in Section 2.1 to explain the phenomenon of laser ablation allows us to understand the effect of surface modification and use it as a tool to transform surfaces by coating them with thin films. Another very important issue is the possibility of producing core–shell nanoparticles of the type shown in the Section 2.2, using different metals. In our case, the production of Fe@Au NPs resulted a very convenient way to unify in a same material the magnetic properties of Fe with the biological properties of Au. As a final remark, as it was our intention to put forward throughout this chapter, we would like to emphasize that nanosecond lasers are still a very important source for the modification of surfaces for technological uses.

REFERENCE

1. K. Sugioka, M. Meunier and Alberto Piqué. Laser Precision Microfabrication. Springer Series in Materials Science. 2010.
2. N.B. Dahotre and S.P. Harimkar. Laser Fabrication and Machining of Materials. Springer. 2008.
3. W. M. Steen. Laser Material Processing. Springer-Verlag. 1991.
4. E. Gentili, L. Tabaglio and F. Aggogeri. Review on Micromachining Techniques. Italy: Department of Mechanical Engineering, University of Brescia. 2005.
5. M.B. Mason (BP Solar) and J. Fieret (Exited Ltd). Advanced Laser Processing for Industrial Solar Manufacturing. Contractor BP Solar Limited. 2006.
6. M.M.S. Amera, M.A. El-Ashrya, L.R. Dosserb, K.E. Hixb, J.F. Maguirec and B. Irwin. Femtosecond versus nanosecond laser machining: comparison of induced stresses and structural changes in silicon wafers. Applied Surface Science. 2005;242:162–167.
7. M. Villagran-Muniz, H. Sobral, C. Rinaldi, I. Cabanillas-Vidosa and J. Ferrero. Optical emission and energy disposal characterization of the laser ablation process of CaF2, BaF2, and NaCl at 1064 nm. Journal of Applied Physics. 2008;104:103112–103117.
8. C. Toro, C. Lasorsa, C. Sánchez Aké, M. Villagran-Muñiz and C. Rinaldi. New method for nanosecond laser machining. Journal of Laser Micro Nanoengineering. 2012;7:269–273.
9. J. Eichstädt, G.R.B.E. Römer and A.J. Huisin 't Veld. Determination of irradiation parameters for laser-induced periodic surface structures. Applied Surface Science. 2013;264:79–87.
10. W. Zhang, G. Cheng, Q. Feng, L. Cao, F. Wang and R. Hui. Abrupt transition from wavelength structure to subwavelength structure in a single-crystal superalloy induced by femtosecond laser. Applied Surface Science. 2011;257:4321–4324.

11. R. Le Harzic, D. Dörr, D. Sauer, M. Neumeier, M. Epple, H. Zimmermann and F. Stracke. Formation of periodic nanoripples on silicon and germanium induced by femtosecond laser pulses. Physics Procedia. 2011;12:29–36.
12. T.Q. Jia, H.X. Chen, M. Huang, and F.L. Zhao, J.R. Qiu, R.X. Li and Z.Z. Xu, X. K. He, J. Zhang and H. Kuroda. Formation of nanogratings on the surface of a ZnSe crystal irradiated by femtosecond laser pulses. Physical Review B. 2005;72:125429.
13. Available from: http://www.nasa.gov/centers/ames/research/technologyonepagers/nanotechnology-landing.html.
14. Available from: http://www.cnea.gov.ar/institutos/nanociencia_nanotecnologia.php.
15. Available from: http://www.fan.org.ar.
16. T.H. Maiman. Nature. 1960;187:493.
17. G. Bekefi. Principles of Laser Plasmas. 1st ed. New York: Wiley. 1976.
18. F. Hubenthal. Vol. 1, Sección 1.13. In: Comprehensive Nanoscience and Technology. 2011;375–435.
19. D. Kashchiev. Nucleation: Basic Theory with Applications. Oxford: Butterworth-Heinemann. 2000.
20. B.S. Luk'yanchuk, W. Marine, S.I. Anisimov and G.A. Simakina, Condensation of vapor and nanoclusters formation within the vapor plume, produced by ns-laser ablation of Si, Ge and C, Proc. SPIE 3618, 434. 1999.
21. M.S. Tillack, D.W. Blair and S.S. Harilal. The effect of ionization on cluster formation in laser ablation plumes. Nanotechnology. 2004;15:390–403.
22. J. Perrière, É. Millon, É. Fogarassy, A.V. Kabashin and M. Meunier. Recent advances in laser processing of materials. Amsterdam: European Materials Research Society: Elsevier. 2006.
23. R. Hergenroder. Laser-generated aerosols in laser ablation for inductively coupled plasma spectrometry. Spectrochimica Acta Part B: Atomic Spectroscopy. 2006;61(3).
24. P.S. Liu, W.P. Cai and H.B. Zeng. Journal of Physical Chemistry C. 2008;112:3261.
25. K.Y. Niu, J. Yang, S.A. Kulinich, J. Sun, H. Li and X.W. Du. Journal of the American Chemical Society. 2010;132:9814.
26. Z. Yan and D.B. Chrisey. Pulsed laser ablation in liquid for micro-/nanostructure generation. Journal of Photochemistry and Photobiology C: Photochemistry Reviews. 2010;doi:10.1016/j.jphotochemrev.2012.04.004.
27. D. Werner, S. Hashimoto, T. Tomita, S. Matsuo and Y. Makita. Journal of Physical Chemistry C. 2008;112:1321.
28. Z.J. Yan, R.Q. Bao and D.B. Chrisey. Langmuir. 2011;27:851.
29. A.V. Simakin, V.V. Voronov, G.A. Shafeev, R. Brayner and F. Bozon-Verduraz. Chemical Physics Letters. 2001;348:182.
30. C.T. Salazar. Micromecanizado por ablación láser en silicio, en silicio cubierto por películas y en metales [thesis]. Buenos Aires, Argentina: Facultad de Ingeniería, Universidad de Buenos Aires. 2015;129 p.
31. C. Toro, C. Lasorsa and C. Rinaldi. Detecting plasma produced by laser in a micromachining system as in-process control. Journal of Laser Micro Nanoengineering. 2011;6:220–224.
A. Rodrigo, et al. Dominant plasma species for TiN film formation by plasma CVD. Journal of Physics D: Applied Physics. 1997;30:2397.
32. Spin coating theory. Louisville: Micro/Nano Technology Center, University of Louisville. 2013.
33. H. Sobral, M. Villagrán-Muniz and F. Bredice. Energy balance in laser ablation of metal targets. Journal of Applied Physics. 2005;98:083305.
34. B. Lerner, M. Perez, C. Toro, C. Lasorsa, C.A. Rinaldi, A. Boselli and A. Lamagna. Generation of cavities in silicon wafers by laser ablation using silicon nitride as sacrificial layer. Applied Surface Science, Amsterdam: Elsevier Science BV. 2012;258:2914–2919.
35. M. Halbwaxa, T. Sarneta, Ph. Delaportea, M. Sentisa, H. Etienneb, F. Torregrosab, V. Vervischb, I. Perichaudc and S. Martinuzzic. Micro and nano-structuration of silicon by femtosecond laser: application to silicon photovoltaic cells fabrication. doi:10.1016/j.tsf.2007.12.117.

36. T. Baldacchini, J.E. Carey, M. Zhou and E. Mazur. Superhydrophobic surfaces prepared by microstructuring of silicon using a femtosecond laser. Langmuir. 2006;22:4917–4919.
37. B.R. Tull, J.E. Carey, E.Mazur, J.P. McDonald and S.M. Yalisove. Silicon surface morphologies after femtosecond laser irradiation. Mrs Bulletin. 2006;31:626–633.
38. E.H.M. Camara, P. Breuil, D. Briand, L. Guillot, C. Pijolat and N.F. de Rooij. Review: Micro gas preconcentrator in porous silicon filled with a carbon absorbent. Sensors and Actuators B. 2010;148:610–619.
39. B. Levine, C.J. Della Valle and J.J. Jacobs, Applications of porous tantalum in total hip arthroplasty. Journal of the American Academy of Orthopaedic Surgeons. 2006; 14(12):646–655.
40. N.G. Boggio, P. Alonso Castillo, J.J. Ortiz, A. Lamagna, A. Boselli and C.A. Rinaldi. Desarrollo de un espectrómetro por movilidad iónica para la detección de compuestos orgánicos (explosivos, drogas y contaminantes). Anales AFA.
41. L.J. Yang, J. Tang, M.L. Wang, Y. Wang and Y.B. Chen. Surface characteristic of stainless steel sheet after pulsed laser forming. Applied Surface Science. 2010;256:7018–7026.
42. C. Toro, E.D. Cabanillas and C.A. Rinaldi. New method for micro and nanoparticles fabrication. In: Japan Laser Processing Society, editor. The 12th International Symposium on Laser Precision Microfabrication; Japan, Takamatsu, Kagawa. 2011.
43. C. Toro, E.D. Cabanillas and C.A. Rinaldi. Generación de nanopartículas y modificación de superficies de Fe con ablación láser. In: Instituto de Química del Noroeste Argentino, editor. XVII Congreso Argentino de Fisicoquímica y Química Inorgánica; Argentina. Córdoba. 2011.
44. S. Kayal and R.V. Ramanujan. Anti-cancer drug loaded iron–gold core–shell nanoparticles (Fe@Au) for magnetic drug targeting. Journal of Nanoscience and Nanotechnology. 2010;10:1–13.
45. Journal of Magnetism and Magnetic Materials. 2007;310:2369–2371.

CHAPTER 15

Mechanical Behavior of Articular Cartilage

Nancy S. Landínez-Parra[1, 2], Diego A. Garzón-Alvarado and Juan Carlos Vanegas-Acosta[1]

[1] Group of Mathematical Modeling and Numerical Methods GNUM-UN, Faculty of Engineering, National University of Colombia,, Colombia

[2] Human Corporal Movement Department, Faculty of Medicine, National University of Colombia,, Colombia

1. INTRODUCTION

Articular Cartilage (AC) is a poro-elastic biological material that allows the distribution of mechanical loads and joint movements. As a biphasic material, in the presence of load, the articular cartilage deforms its solid matrix and modifies the fluid hydrostatic pressure within. The aim of this chapter is to present a mathematical model that predicts the mechanical behavior of articular cartilage, taking into account the duality between the solid matrix and articular liquid, and its poro-elastic characteristics. Using a finite element method approach, the response of a piece of articular cartilage in one and two dimensions has been simulated, with tensile, compressive and oscillatory mechanical loads. The analysis of the results allows a qualitative validation of the poro-elastic behavior of the model due to the solid matrix deformation and the fluid outflow that causes variations of pressures inside the articular cartilage in accordance with reported trials. The mathematical model allows for prediction of articular cartilage's biomechanical behavior. These results contribute to the research processes in fields of study such as biomechanics and tissue engineering.

2. BACKGROUND

One of the pathologic entities that most often affect quality of life of individuals is osteoarthrosis (OA), which is caused by the deterioration of cartilage in synovial joints. In the U.S. in the early nineties it was estimated that 37.9 million people (which constituted 15% of the population) suffered from one of the various existing musculoskeletal diseases. OA was present in 21 million individuals (Lawrence et al., 1998).

OA compromises skeletal muscle function, causing pain and difficulty in basic activities of daily living. Several studies have shown that the forces exerted on cartilage can modify its structure and composition, resulting in a change in the biomechanical behavior of the same (Wu & Kirk, 2001). The onset and progression of OA are commonly affected by mechanical factors associated with either joint loading or local contact stress (Andriacchi et al., 2004).

The mechanical environment of cartilage cells (chondrocytes) is an important factor influencing joint health and function. Chondrocytes in articular cartilage utilize mechanical signals in conjunction with other environmental, genetic, extrinsic and intrinsic hormonal and/or paracrine or autocrine factors to regulate their metabolic activity. This capability provides the means by which articular cartilage may alter its structure and composition to meet the physical demands of the body (Boschetti et al., 2004).

The permeability of cartilage contributes to many tissue functions such as transport of food to the chondrocytes, the ability to withstand high loads, and maintaining a fluid film to lubricate opposing joint surfaces (Guilak et al., 1999). Measuring the cartilage permeability considering its depth, by the behavior of a fluid flow induced by the application of a pressure gradient, can show a decrease in permeability in relation to depth for each level of the applied pressure difference.

Here is a review of the anatomy, morphology and physiology of articular cartilage for the purpose of more clearly understanding its responses to load and its relationship to the deformation processes and the changes of fluid pressure within.

2.1. Biological tissues

Biological materials are generally multiphase, coexisting in a solid and a fluid phase (Hubertus Frijns, 2000; Doblaré, 2005; Haider & Schugart, 2006). They also have a strong microstructure that gives them a clear heterogeneous and anisotropic condition and, in addition, its mechanical behavior is strongly nonlinear nonlinear (Wilson et al., 2004; Ateshian et al., 1997; Chan et al., 2004). All of this, without taking into account important aspects such as the strong dependence on age, sex, metabolism, and in particular history and diseases and, ultimately, cell activity and its interaction with the particular environment in which it develops.

In this sense, it can be seen as soft biological tissues in which the elastic modulus is approximately equal to the stresses to which they are subjected. Some typical cases correspond to the arteries and veins, cartilage, ligaments, tendons, muscles or skin. In general, they are composites, made up of organic matrix reinforced by fibers of collagen and elastin. Its behavior again depends on its structural composition, especially the percentage of fibers, their characteristics and the type of grouping. Thus, tissues specialized for tensile strength (e.g. ligaments) are rich in fiber and their orientation essentially coincides with the direction of stress to which they are subjected, while the elastic absorbing the compressive forces (e.g. cartilage) are rich in proteoglycans and fibers distributed in various directions. The cartilage then can be referred to as hydrated tissue, which has a highly compressible behavior corresponding to a saturated biphasic material (if one considers the four-phase ion diffusion) with the possibility of the escape of the fluid inside. The main purpose of this behavior is to provide optimum synovial joint lubrication conditions, friction, wear, shock absorption and load distribution. Interstitial fluid flow in these tissues is crucial both in their viscoelastic properties as well as in the lubrication mechanism; contemplating this is necessary to model the behavior of such tissues. Furthermore, they are again heterogeneous and anisotropic due to the preferred orientation of collagen fibers (Wilson et al., 2004).

Understanding the importance of in-depth knowledge of the composition and behavior of articular cartilage, in the next section this tissue will be described.

2.2. Articular cartilage

Cartilage is categorized as an avascular, aneural and alinfatic tissue. It's composed of cells called chondrocytes, surrounded by an extracellular matrix (ECM) that they secrete. It's formed by an abundant extracellular matrix in which the chondrocytes are located in spaces called gaps (See Fig 1). The chondrocytes synthesize and secrete the organic components of the extracellular matrix, which are essentially collagen, hyaluronic acid, proteoglycans and glycoproteins. Hyaline and fibrous cartilage are distinguished by the characteristics of the matrix. There is also elastic cartilage, in which the elastin is a part of the extracellular matrix (Koenig, 2011; Sopena-Juncosa et al., 2000; Nordin & Frankel, 2004).

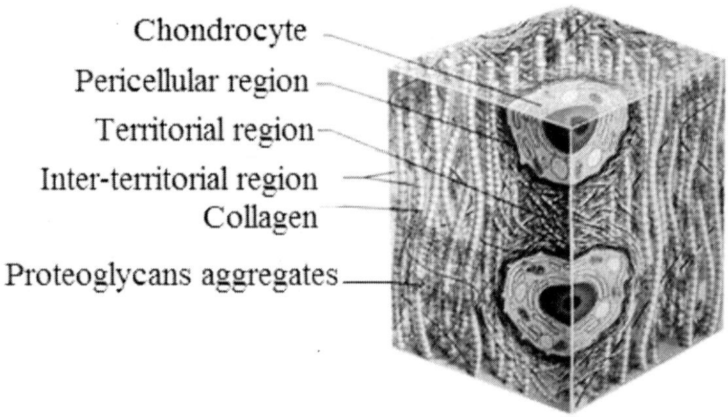

Figure 1. Representation of a fragment of Articular Cartilage. (Sopena-Juncosa et al., 2000)

For a better understanding, each of the structures mentioned above will be defined. Initially chondrocytes will be described, then subsequently the ECM and each of its components.

2.2.1. Chondrocytes

They are cells sparsely distributed in the tissue. They constitute approximately 10 percent of cartilage volume. They are located in some gaps within the ECM, to which they adapt. They produce the adjacent ECM, but they are also capable of depolymerizing and removing it to broaden its gaps. (This fact is very evident in the process of endochondral ossification) The key feature of the intercellular substance is its hyper-hydration state (80 percent), and combined with proteoglycans, it forms a gel. Greater or lesser elasticity of the cartilage depends on its water content (Koenig, 2011; Sopena-Juncosa et al., 2000; Nordin & Frankel, 2004).

2.2.2. Extracellular matrix - ecm

This consists of an organized and dense network of thin collagen fibers embedded in a concentrated solution of proteoglycans. It is responsible for the mechanical properties of cartilage (Koenig, 2011; Sopena-Juncosa et al., 2000; Nordin & Frankel, 2004). This matrix is composed of:

Water: (60-80 percent) Water is the main component of cartilage, which contributes to its damping properties, cartilage nutrition and the articular lubrication processes. It allows for the deformation of the cartilage in response to mechanical loads, flowing inside and outside them. Collagen:

(10-20percent). Predominantly type II (90-95 percent), giving the cartilage great tensile strength.Proteoglycans: (PGs) (10-15 percent). These are complex macromolecules, responsible for the resistance to compression of the cartilage. They are secreted by chondrocytes and composed of subunits called glycosaminoglycans (GAGs). The most common GAG is chondroitin-sulfate (of which there are 2 subtypes, the chondroitin-4-sulfate and the chondroitin-6-sulfate), then the keratan-sulphate (or keratin-sulfate) and the dermatan-sulphate. Chondroitin-4-sulphate is the most abundant and decreases over the years; chondroitin-6-sulfate remains constant; and keratan-sulphate increases with age. The PGs have an average lifespan of three months and have a great capacity for retain water which gives elasticity to the tissue. They are attached to collagen and are responsible for the "porous" structure of cartilage (See Fig.1)Extracellular Glycoproteins: (anchorite CII, fibronectin, laminin, integrin). They serve a binding function between the ECM and chondrocytes. The most important, integrin, interacts with cell receptors and regulates the migration, proliferation and differentiation of the chondrocytes (Sopena Juncosa et al., 2000).

At present it is believed that collagen has a different orientation in articular cartilage in relation to its depth, as described by Benninghoff in 1925 in the Arcade model (Wilson, 2005): Packets of primary fibrils extend perpendicular to the sub-chondral bone; the fibrils are separated near the joint surface presented by the curve of the horizontal; each packet of the vertical surface is assumed to be subdivided into two different directions in the curvature of the radial direction (See Fig. 2). It was assumed that the orientation of the secondary fibrils is random and that on the uppermost, fibers are distributed horizontally.

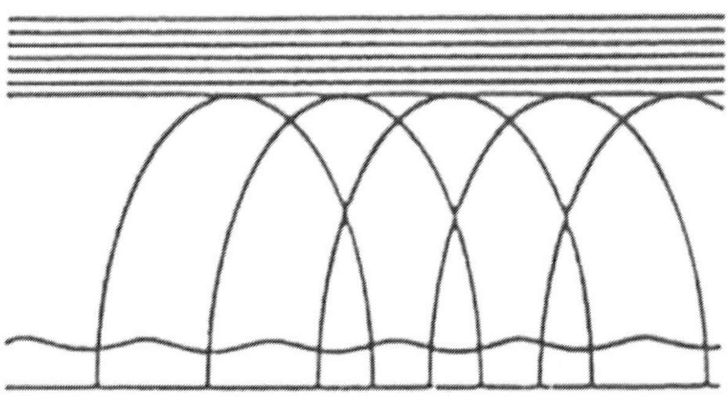

Figure 2. Representation of collagen distribution in articular cartilage.

2.3. Architecture of articular cartilage

Some studies like those reported by Martin, 2002; Sopena et al., 2000; Wilson, 2005 and Meyer & Wiesmann, 2010; mention that with electron microscopy it can be determined that cartilage has a multilayered arrangement with layers of different thickness and composed of fibrils irregularly cross-linked into a plane parallel to the surface. Using the electron microscope, the following layers of articular cartilage were described:

2.3.1. Superficial, tangential or sliding layer

This is adjacent to the joint cavity. The chondrocytes adopt an elongated or ellipsoid shape and they are oriented parallel to the surface. The cells have low activity and poor protein synthesis. This layer possesses few PGs and a high concentration of fine collagen fibers distributed perpendicularly to one another and parallel to the surface in order to withstand the shear forces during joint movement.

2.3.2. Intermediate or transitional layer

Cells adopt a rounded morphology and are larger than those of the previous layer. The chondrocytes are irregularly arranged and show a greater presence of PGs and less collagen with thicker fibers arranged obliquely and randomly in all three planes of space. This layer has high metabolic activity and supports compression forces.

2.3.3. Radial or deep layer

The cells are rounded and have the same characteristics as layer 2 but adopt a columnar arrangement. They present a high protein synthesis. The collagen fibers are thick and they are distributed parallel to each other and perpendicular to the articular surface to provide resistance to compressive forces. The water content is less than in the previous layers and proteoglycans are most abundant.

2.3.4. Calcified layer

This is adjacent to the bone and separated from the previous layer by a basophilic line called tidal or "tidemark", which is a bar wavy tangential arrangement of its fibers and can withstand shear forces. The cells are small and scarce. The matrix is rich in hydroxyapatite crystals. Cartilage anchoring to the sub-chondral bone occurs in this layer (See Fig 3).

The most common tests used for explaining the behavior of articular cartilage under load, expressed in computer models, that include the behavior of swelling or anisotropic properties of the collagen structure for determining the mechanical quality of articular cartilage are: confined compression, the unconfined compression, indentation and swelling (Wilson, 2005). This can be reviewed extensively in the literature.

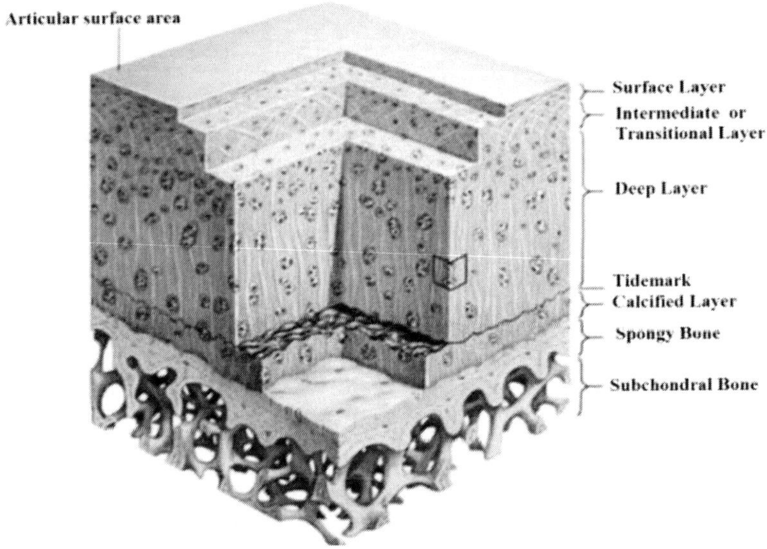

Figure 3. Architectural layout of the articular cartilage according to its various layers. Note the anisotropic distribution of the tissue in relation to the depth thereof. (Sopena-Juncosa et al., 2000)

2.4. Biphasic behavior of articular cartilage

Mechanical properties of articular cartilage are attributed to their complex structure and composition of the extracellular matrix that is comprised of a fluid phase (water containing dissolved ions) and a solid matrix that consists mainly of a fibrous network of collagen type II and aggregates of proteoglycans as well as other type of proteins, lipids, and cells (Wilson et al., 2004).

With the mechanical load, the interstitial fluid is redistributed through the pores of the permeable solid matrix, resulting in predominantly viscoelastic behavior (See Fig 4). This highly viscoelastic behavior of articular cartilage is mainly due to two mechanisms: (a) the frictional drag force of interstitial fluid flow through the porous solid matrix (i.e., the flow-dependent mechanism), and (b) the function of the time-dependent deformability of the solid matrix (i.e. the flux-independent mechanism) (Garzón, 2007).

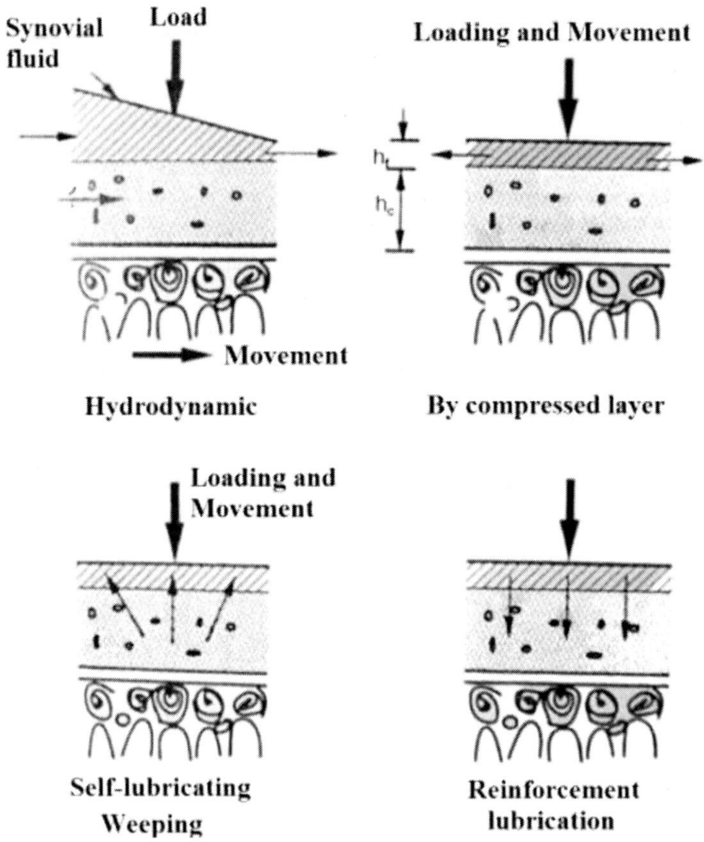

Figure 4. Different forms of lubrication in articular cartilage given by the load applied to the tissue. This lubrication and nutrition takes various forms but mainly by the displacement of the fluid. (Terada et al., 1998).

Mechanical behavior of cartilage is then described by the biphasic or continuous poro-elastic model that describes the mechanical interactions of the different phases (Donzelli & Spilker, 1998).

From the mechanical standpoint, the most important components of articular cartilage are strong and highly organized as a network of collagen together with the load of proteoglycans. Due to the fixed charges of proteoglycans, the cation concentration within the tissue is higher than in the surrounding synovial fluid. This excess of ion particles leads to an osmotic pressure difference, which causes swelling tissue. The fibrillar collagen network resists pressure and swelling. This combination makes cartilage a unique, highly hydrated and pressurized tissue, reinforced by the tension of the collagen network (Wilson, 2005).

2.4.1. Mixtures theory

Articular cartilage can be described by the mixtures theory as a mixture of four elements: a fibrous network (collagen fibres and proteoglycans), a fluid and a positively and negatively charged particle. However, it's important to differentiate between components and phases. Hubertus, 2000; defines a component as a group of particles with the same properties and a phase as a set of miscible components. Thus in theory the four components can be separated into only two phases: a solid and a fluid phase. In this case the fluid is comprised of three components: the liquid, the cations and anions.

Many authors such as Haider & Schugart, 2000; Wilson et al., 2004; Haider & Guilak, 2007; Meng et al., 2002; Wu et al., 1997; Terada et al., 1998; Donzellie et al., 1999 and Donzelli & Spilker, 1998; among others, have conducted their research, viewing cartilage from this biphasic behavior. This has allowed its analysis as a material with a porous-viscoelastic behavior, in an attempt to better understand its response to loads, forces and overload.

2.5. Computational analysis of articular cartilage

Continued use of simulation in medicine has allowed important data to be obtained about the biological, mechanical and chemical behavior of the organs and tissues using mathematical formalization and subsequent numerical simulation of complex biological processes. Various medical problems related to surgery, trauma and rehabilitation have been identified, conceptualized and solved systematically and numerically (Garzón, 2007).

Computer mechano-biology determines the quantitative rules governing actions for cellular expression, differentiation and maintenance of biological and mechanical stimuli, which can be simulated by numerical methods. The procedure for finding such rules is usually through the process of "trial and error" (Van der Meulen & Huiskes, 2002). The computational tests are simulated usually from problems in the contour value by which the mechanical loads on the boundary are transferred to local mechanical variables (stress and strain). On the biological side, these local mechanical or biophysical variables stimulate cell expression to regulate, for example, the composition of the matrix and the expression of molecular substances. Both biological and mechanical parts are combined in a computational model, which considers the application of forces, mechano-transduction, cellular expression, genetics and the transformation of the characteristics of the extracellular matrix. The typical method of numerical implementation of these mechano-biological problems is the finite element method (FEM) (Garzón, 2007).

Finite element computational analysis has been used as an approach to diverse biological processes including the biomechanical behavior of articular cartilage (Wilson et al., 2004; Ateshian et al., 1997; Chan et al., 2004; Wilson, 2005; Donzelli et al, 1999; Donzelli & Spilker, 1998; Almeida & Spilker, 1998; Wu & Herzog, 2000; Levenston et al., 1998). Using the material representation of the continuum phase of cartilage, results have indicated that local intermittent hydrostatic pressure promotes cartilage maintenance (Carter & Wong, 2003).

3. MECHANICAL BEHAVIOR OF ARTICULAR CARTILAGE (AC):

Mechanical properties of articular cartilage (AC) are attributed to its complex structure and to composition of its ECM including a fluid phase (water with dissolved ions), and a solid matrix (collagen type II, aggregates of PGs, proteins, lipids, and cells) (Haider & Guilak, 2007). With the mechanical load, the interstitial fluid is redistributed through the pores of the permeable solid matrix, resulting in predominantly poro-elastic conduct. This behavior of the AC is mainly due to two mechanisms: (a) the frictional force due to drag flow of interstitial fluid through the porous solid matrix (flow-dependent mechanism), and (b) the deformability of the matrix strong function of time (flow-independent mechanism) (Wilson, 2005).

4. MATHEMATICAL MODEL FOR ARTICULAR CARTILAGE

Several authors (Mow, 1980 ; Haider & Schugart, 2006; Wilson et al., 2005; Haider & Guilak, 2007; Meng et al., 2002; Wu et al., 1997; Terada et al., 1998; Donzelli et al., 1999; Donzelli & Spilker, 1998); have conducted their research on cartilage from biphasic behavior that this tissue exhibits. This allowed analysis of the same material as a poro-elastic behavior capable of supporting loads. The mathematical model of the AC as a biphasic material, analyzes the displacement *u (t, x)* of the solid (matrix) and pressure *p (x)* of the fluid displaced by the load, thanks to its characteristic of poro-elasticity. This model is described by the equations (1) and (2):

$$-\nabla.(2\mu_s \underline{\underline{\varepsilon}}(\underline{u}) + \lambda_s \nabla.\underline{u}\,\underline{\underline{I}}) + \nabla p = 0 \text{ en } \Omega \quad (1)$$

$$\frac{\partial}{\partial t}(\nabla.\underline{u}) - \nabla.(k\nabla p) = 0 \text{ en } \Omega \quad (2)$$

Equation (1) is derived from the law of conservation of momentum and corresponds to the linear elasticity equation (term 1a) coupled with a term that represents fluid pressure (term 1b). The term $\underline{\underline{\varepsilon}}(\underline{u})$ corresponds to the strain tensor acting on the Γ surface enclosed by **Ω.**

μ_s y λ_s are the "Lame elastic constants" for the solid, related to Young's modulus and Poisson's ratio (E, ν). For its part, the equation (2) refers to the change of the dilation of the solid matrix (term 2a) due to the mechanical load created by the divergence of the gradient of the pressure of fluid contained in the domain Ω (term 2b) (Frijns, 2000). In this equation, k is a constant representing the permeability of solid module.

4.1. Boundary conditions

Boundary conditions of the model are defined in Γ domain and may be dependent on time. The mathematical expression of these conditions is (3-6):

$$u = g_u^D \text{ en } \Gamma_u^D \quad (3)$$

$$n.(2\mu_s \underline{\underline{\varepsilon}}(\underline{u}) + \lambda_s \nabla.\underline{u}\,\underline{\underline{I}} - p\underline{\underline{I}}) = g_u^N \text{ en } \Gamma_u^N \quad (4)$$

$$p = g_p^D \text{ en } \Gamma_p^D \quad (5)$$

$$-n.(k\nabla p) = g_p^N \text{ en } \Gamma_p^N \quad (6)$$

A widely used method for solving partial differential equations in complex geometries is the finite element method (Garzón, 2007). This method allows implementing the numerical model presented in equations (1) and (2) simply and with low computation cost. The method consists of using a vectorial function *W* or *weighting function* and a scalar function of *q*, which minimizes the terms of the equations (1) and (2). Multiplying (1) by *W* and (2) by *q*, and performing integration by parts in the

Ω domain, we obtain a variational of the form (Frijns, 2000):

$$a(u,W)+b(W,p)=(g_u^N,W) \tag{7}$$

$$\frac{\partial}{\partial t}b(u,q)-c(p,q)=(g_p^N,q) \tag{8}$$

Where:

$$a(u,W)=\int_\Omega (2\mu_s \underline{\underline{\varepsilon}}(\underline{u}):\varepsilon(W)+\lambda_s(\nabla.u)(\nabla.\underline{W}) \tag{9}$$

$$b(W,q)=-\int_\Omega \nabla.Wq \tag{10}$$

$$c(p,q)=\int_\Omega (k\nabla p).\nabla q \tag{11}$$

$$(g_u^N,W)=\int_{\Gamma_u^N} g_u^N.W \tag{12}$$

$$(g_u^N,W)=\int_{\Gamma_u^N} g_u^N.W$$

(1

The solution method is based on dividing the domain Ω (continuum) in which are defined equations (1) and (2) in their integral form in a series of non-intersecting subdomains Ω ^{e} called *finite elements*. The set of finite elements forms a partition of the domain called *discretization* which together are described by equations (14) and (15):

$$a(u_u^h,W_h^h)+b(W_h^h,p_n^h)=(g_u^N,W_h^h) \tag{14}$$

$$b(u_n^h, q^h) - \Delta t\, c(p^h, q_n^h) = b(u_{n-1}, q^h) + \Delta t(g_p^N, q^h) \tag{15}$$

In (15) the n−1n−1 subscript indicates the value of a parameter at a t−1t−1 time, while the subscript nn indicates the value in the next time, $t_n = t_{n-1} + \Delta t$ (Frijns, 2000). Using the Galerkin method, the functions *W* and *q* are approximated by the matrix expression (16) where the *N* vector contains *shape functions standard (* ξ,*η)* which allow interpolation within each element of the domain.

$$q^h = W_h^h = \begin{pmatrix} N_1 \\ N_2 \\ N_3 \end{pmatrix} = \begin{pmatrix} \frac{1}{2}\xi(\xi - 1) \\ (1+\xi)(1-\xi) \\ \frac{1}{2}\xi(\xi + 1) \end{pmatrix} \tag{16}$$

$$W_h^h = \begin{pmatrix} N_1 \\ N_2 \\ N_3 \\ N_4 \end{pmatrix} = \begin{pmatrix} \frac{1}{4}(1-\xi)(1-\eta) \\ \frac{1}{4}(1+\xi)(1-\eta) \\ \frac{1}{4}(1+\xi)(1+\eta) \\ \frac{1}{4}(1-\xi)(1+\eta) \end{pmatrix} \tag{17}$$

In (16), the shape functions correspond to the case of one-dimensional element with three nodes (See Fig. 5a). For two-dimensional case four node elements are used, whose standard shape functions are shown in matrix form (17) and in Figure 5b.

If the integration space (x,y) is changed to (ξ,η) by the Jacobian of the transformation and vector notation is adjusted, the equations (1) and (2) can be reduced to a matrix system of elementary type of the form (18) which corresponds to the algebraic discretization in Ωe domain of an element, with k elementary stiffness matrix, m the unknowns and f the independent term. Joining the result of (18) for total elements in Ω, a general matrix system is obtained defined as (19) where K is the global matrix of stiffness, M is the vector of unknowns and F is the global vector entries.

$$k \cdot m = f \text{ en } \Omega^e \tag{18}$$

$$K \cdot M = F \text{ en } \Omega \tag{19}$$

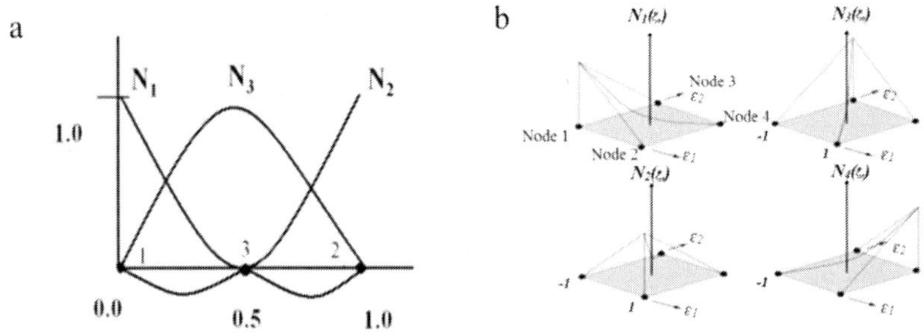

Figure 5. Representation of the shape functions standard. a) Case 1D, 3 nodes per element. b) Case 2D elements with four nodes (Radi, 2007).

4.2. Model implementation

Solution of these equations for both u and p are implemented using a routine programmed in Fortran and a desktop PC with 2.4 GHz AMD processor and 1.0 GB of RAM

4.3. Computer simulation

Conditions for the simulation were based on experiments by Frijns, 2000 and Wu et al., 1999. Simulations were performed for 1D and 2D looking tissue response to the application of: (a) compression, (b) tensile and (c) oscillating or cyclical loads. The simulation time is 45 seconds of load in all cases. The loads applied in 1D and 2D simulations were performed with the parameters shown in Table 1. The tests considered the AC as a continuous and homogeneous material with the chondrocytes being part of the continuum.

Table 1. Parameters for the homogeneous material characteristics (Wu et al., 1999).

	E	ν	K
Matrix	0.40 Mpa	0.1	1.0 $x10^{-2}mmN^{-1}s^{-1}$
Chondrocyte	1.0 kPa	0.2	1.0 $x10^{-5}mmN^{-1}s^{-1}$

4.3.1. Meshing

For the 1D simulation a mesh was performed that represents a fragment of 0.3 mm AC. From this mesh 641 nodes and 320 elements of 3 nodes were obtained. In the 2D simulation a mesh was performed that represents a fragment of articular cartilage 0.09 mm2 (0.3 mm x 0.3 mm). In this case 10,201 nodes were obtained, equivalent to 10,000 elements of 4 nodes.

4.3.2. Boundary conditions

Simulation was performed so as not to allow displacement at the bottom. Load was applied on the upper edge, allowing the fluid outlet only at the bottom of the tissue fragment, as shown in Figure 6a. For 2D, a condition of tissue confinement was simulated, as shown the Figure 6b, so as to present only flow at the bottom. The burden was placed at the top and lateral and bottom movement were restricted, similar conditions to those reported in several experimental studies, including that of Ateshian et al, 1997; Frijns, 2000 and Wu et al, 1999.

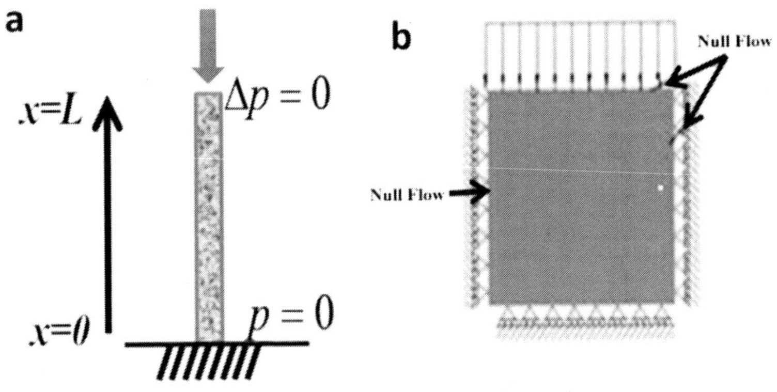

Figure 6. AC scheme for the confinement conditions in (a) 1D and (b) 2D.

4.3.3. Loading conditions

Calculations to simulate the applied load were performed from the data for the AC reported by Wu et al. For the 1D simulations, a load was applied of -2.033 N/m in compression, 2.033 N/m in tension and cyclic loading with frequency equal to 0.1 Hz. In 2D simulations a load was applied on the upper face of cartilage fragment corresponding to the value $-4.0397e^{-6}$ N/m in compression, $4.0397e^{-6}$ N/m in tension. For cyclic loading, it was applied at a frequency equivalent to 0.1 Hz

5. RESULTS OBTAINED

5.1. 1D simulations

5.1.1. Compression loads response

Responses from the evidence to compressive loads can be seen in Figure 7. Figure 7a shows the negative shift of the solid component that increases its value negatively with increasing loading time. The time initially is zero (0) for t = 0.67s and reaching figures of $-15x10^{-4}mmN^{-1}s^{-1}$ at the time of maximum loading, t = 45s for x = 0.3. As the liquid flows, the behavior is similar to a linear elastic behavior, because the only component that supports the load is solid.

Figure 7b shows the decrease in pressure due to the fluid outlet presented by the compression of the tissue. For an initial time t = 0.67s the pressure exerted on the fluid in the tissue is 2 MPa, but as the permanence of the compressive, the pressure decreases and takes values close to 0 MPa, thereby external pressure is balanced, also zero.

5.1.2. Tensil loads response

Figure 8 summarizes the results obtained for tensile strength. Figure 8a shows the positive displacement of the solid component in response to tensile load imposed on tissue. The figure shows how the displacements are positive in presence of tension maintained over time. From an initial value of 0 $mmN^{-1}s^{-1}$ at t = 0.67s, displacement increases up to a maximum of $16x10^{-4}$ $mmN^{-1}s^{-1}$ at a maximum load time t = 45s.

Figure 8b shows the increase of fluid pressure generated by the entry thereof into the tissue due to the pressure difference between the internal and external environment (fluid suction). This pressure difference is caused by displacement of the solid component in the positive direction in response to imposed tensile load. It is shown that this progression of the increase of pressure manifests over time, creating a redistribution of the fluid in the tissue.

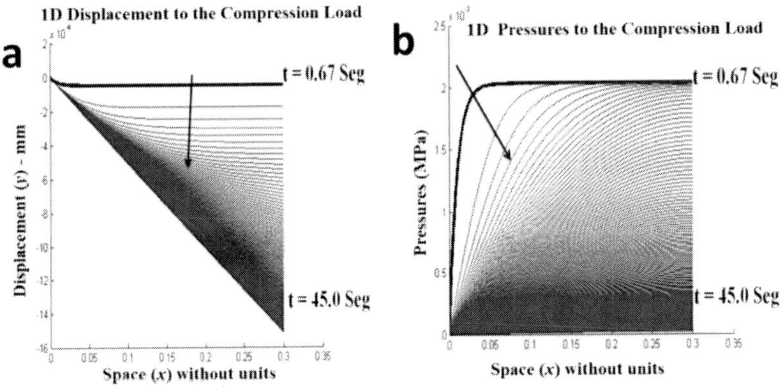

Figure 7. Response of tissue to compression forces. a. Displacement of solid matrix. b. Changes in the fluid pressure in the tissue in presence of the displacement of the same.

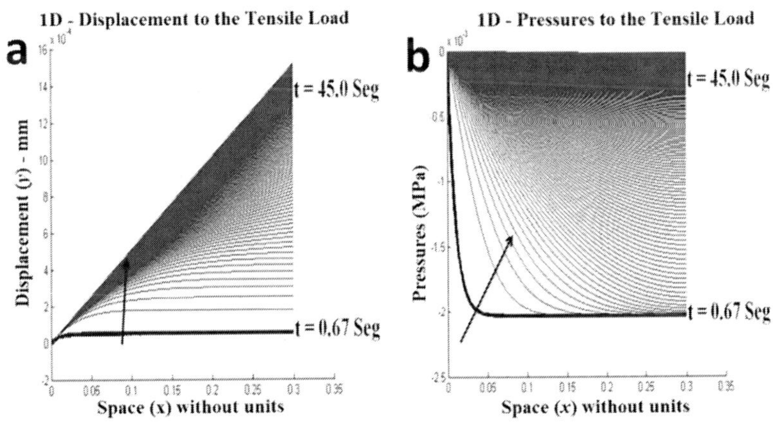

Figure 8. Response of tissue to tensile forces. a. Displacement of solid component tissue upward. b. Change in the fluid pressure generated by the redistribution of fluid in the tissue.

5.1.3. Oscillating loads response

Responses obtained in the simulation of oscillating charges can be seen in Figure 9. Figure 9a shows the alternating displacement presented at each oscillation.

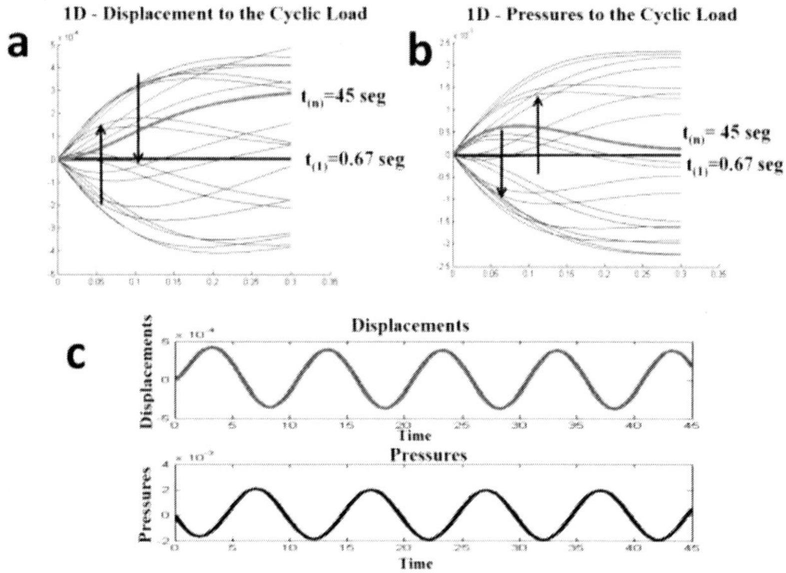

Figure 9. Response to cyclic load. a. Alternate displacement of the solid component. b. Changes in the fluid pressure in oscillate form. c. Delay between the solid deformation and fluid pressure variation.

Initially the movements of the solid are small, close to 0 $mmN^{-1}s^{-1}$ for t = 0.67s. However, they increase over time reaching a value of $6x10^{-4}$ $mmN^{-1}s^{-1}$ or $-6x10^{-4}mmN^{-1}s^{-1}$, at t = 45s, according to the tissue load, tension or compression respectively. Figure 9b shows the development of the process of oscillation in the fluid pressures in response to movement that the solid matrix undergoes upon perceiving the cyclic loading. Initially the pressure lowers and then makes an adjustment that increases, being in inverse phase with the evolution of the deformation of the solid phase of the tissue. I.e., once the solid is deformed in the positive direction, the pressure changes, becoming more negative and vice versa.

The obvious alternating deformation processes of the solid in the face of the application of cyclic loads shows that the displacements are caused by the loads exerted on the tissue and the mobilization of fluid from or into the interior as applied tension or compressive loads respectively. These loads, in turn, generate alternation with respect to each instant of time between the variation of the pressure pattern and the variation of the displacement pattern. However, this action is not in complete phase with displacement. One can appreciate the presence and delay of alternating processes described above, because the equation that represents the displacements is elliptical and corresponds to a displacement equation in space while the equation that represents the pressure corresponds to a parabolic equation and represents a much slower diffusion process than the process of displacement (See Fig. 9c).

5.2. 2D simulations

5.2.1. Compression loads response

Figure 10a corresponds to the displacement of the solid in the y axis for each time instant. The displacements are small in the first moments of the load and it can be seen that as the load increases, the time of displacement increases toward more negative values, demonstrating a greater deformation of the solid phase of the tissue. These displacements are produced by the fluid outlet in response to the maintained compression load and it is observed that the greatest displacement occurs in the upper layers of the tissue responsible for receiving the load directly, while the displacement transference is less at a greater tissue depth.

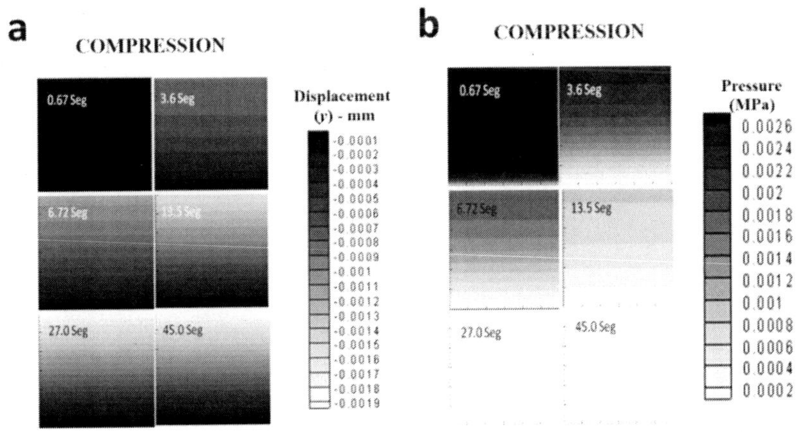

Figure 10. Response to compressive load. a. y displacements during 45 seconds. b. Behavior of the fluid pressure p due to its outflow in presence of the load.

Figure 10b shows the behavior of the fluid pressure at each instant of time. It is observed that the pressure decreases rapidly in all tissue layers reaching values close to 0 MPa in a very short time. This variation of fluid pressure corresponds to the decrease of the same in function of the load exerted over the time and poro-elastic tissue behavior which allows the fluid outlet.

5.2.2. Tensil loads response

Results obtained from the tensile simulation are shown in Figure 11. Figure 11a shows the displacement of the solid in the y axis for different time instants. There are major shifts in the early stages of loading (t = 0.67s) and displacement decreases with increasing load time. It is further noted that with a sustained tensile load at a given time, the greater displacement or elongation occurs in the upper layers of the tissue where the stress is felt in the first instant, which is why the transfer of deformation is smaller at a greater depth of tissue. In this case the fluid pressures tend to increase because the tissue seeks to balance the inside and the outside environment. Then displacement occurs in response to stress and produces a reorganization of the fluid within the tissue, interfering with the variation in the pressure thereof.

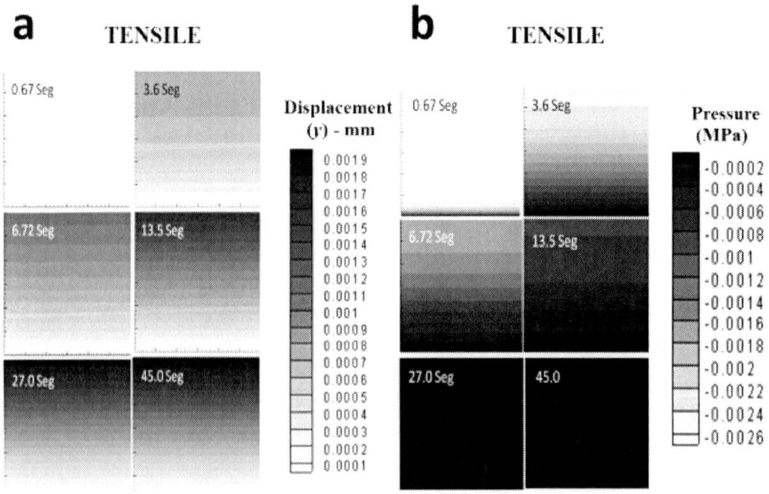

Figure 11. tensile load response. a. y displacements. b.. Behavior of the fluid pressure p due to its inflow in presence of the load.

Figure 11b shows the behavior of the fluid pressure p for different time instants. It is observed that the fluid pressure is increased by small amounts in response to sustained tensile load. This action is due to the compensation of the tissue in response to deformation in elongation by the solid phase of the tissue that requires a redistribution of fluid into the tissue. However, it is evident that at the end of the tensile load time (approximately at t = 27s), the tissue can't undergo greater deformation in elongation and hence the fluid pressure also tends to stabilize at the inside thereof, causing a steady pressure maintained close to zero which balances with the external pressure.

5.2.3. Oscillating loads response

Figure 12a corresponds to the displacement of the solid in the y axis at different time instants. It is noted that the displacements or deformations occur alternately; at the times 1, 3 and 5 (t = 0.67s, t = 27s = 6.57s respectively) the displacements are made positive, behavior similar to that observed during exposure to stress loads. Conversely, at the times 2, 4 and 6 (t = 3.37s, t = 16.5 s and t = 45s respectively) the displacements are negative, consistent with the behavior exhibited by the matrix to compressive loads.

Figure 12b shows the behavior of the fluid pressure p. Similar to what happened with the deformations of the solid phase, the pressure oscillation in response to cyclic loading imposed on the tissue is evident. Note that at times 1, 3 and 5 (t = 0.67s, t = 6.57syt = 27s respectively) the pressure tends to decrease due to the redistribution of fluid in response to perceived stress loads. Because the loads are not maintained, the tissue can't compensate with fluid inlet from the outside which is why the pressure is not increased. Thus, at times 2, 4 and 6 (t = 3.37s, t = 16.5 s and t = 45s respectively), the pressure tends to increase in response to the compression of the tissue. This is due to the rapidity with which compressive and tensile loads are alternated, loads which prevent the behavior of cartilage during the cyclic loading from exhibiting the behavior that corresponds exclusively to the case of tension or compression.

6. FINAL DISCUSSION

There are several theories explaining the behavior of AC in the presence of load conditions, summarized in computational models that include the swelling process and the properties of the anisotropic structure of collagen. The most frequently used tests to determine the mechanical qualities of the AC are the confined compression, the unconfined compression, the indentation and the swelling tests (Wilson, et al., 2005), carried out using numerical approximation tools.

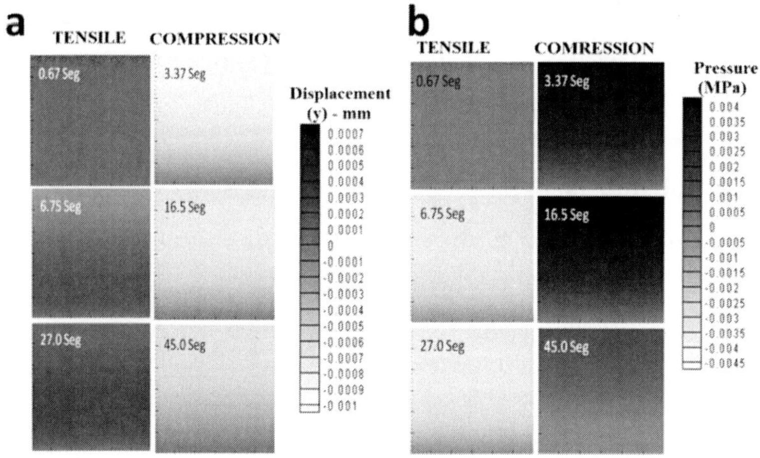

Figure 12. Response to oscillate load. a. y displacements. b. Oscillation of the pressure p within the tissue according to the load.

For purposes of meeting the stated objectives, we simulated a condition of confinement of the tissue that allows the flow at the bottom to restrict lateral and bottom movement. Conditions were similar to those reported in practical experiments as the papers presented by Ateshian et al., 1997; Frijns, 2000 and Wu et al., 1999; among others. The data obtained from the simulations confirm the theory of biphasic articular cartilage, first proposed by Mow et al., 1980; and supported by several authors as Haider et al., 2006; Wilson et al., 2005, Haider & Guilak, 2007; Meng et al., 2002; Wu et al., 1997, Terada et al., 1998, Donzelli et al., 1999 and Donzelli & Spilker, 1998; among others.

Results allow us to conclude that articular cartilage exhibits a displacement response of the solid component (matrix) and a variation in the pressure of fluid component due to the exit or entrance thereof, with decreases in pressure in response to compressive loads and increases at the same tensile loads. The displacement is caused by outflow of fluid in response to the maintained compressive load. However it is important to note that once the tissue reaches its maximum displacement, it behaves as a solid rather than as a poro-elastic material. From this point the fluid can't flow out of the tissue because the pressure is balanced with the external fluid begins to bear part of the load.

If this pressure is maintained for prolonged periods dehydration of the tissue may result and cause changes in normal behavior, making it temporarily or permanently more sensitive to injury. These findings support the conclusion that a load consistently maintained for long periods of time or an excessive load that exceeds the characteristics of the tissue once it has reached the maximum possible displacement, can make one more vulnerable to overuse injuries.

The data reported are the beginning of broader work in the study of cartilaginous tissues which can incorporate the cellular component differentially and cartilage's own biochemistry in the model. The results obtained motivate the efforts that currently seek to simulate the production and destruction of the matrix in the presence of mechanical loads, to simulate the restructuring of the same after an injury, to apply mathematical models in the study of cartilage growth and to study their behavior in vitro and in vivo. These lines of research aim to provide a solid foundation for the development of AC experiments in vivo and in vitro that expands the range of applications of numerical simulation techniques and techniques used in tissue engineering.

ACKNOWLEDGEMENT

The authors wish to thank the Research Division of Bogota (DIB) of the National University of Colombia in the Call for Research for supporting this work under the project "Mathematical Modeling and Simulation of Processes in Mechanical and Biomedical Engineering" and Colciencias 2011 throughout thc project "Model for the definition of term of the Layout of a manufacturing cell through set theory and optimization" who contributed to the financing the chapter.

REFERENCE

1. E. S. Almeida, R. L. Spilker, 1998 Finite element formulations for hyperelasticbiphasic soft tissues transversely isotropic. Computer methods in applied mechanics and engineering. 151 513538
2. T. P. Andriacchi, A. Mûndermann, R. L. Smith, E. J. Alexander, C. O. Dyrby, S. Koo, 2004 A framework for the in vivo pathomechanics of osteoarthritis at the knee. Annals of Biomedical Engineering. 32 447457 .
3. G. A. Ateshian, W. H. Warden, J. J. Kim, R. P. Grelsamer, V. C. Mow, 1997 Finite deformation biphasic material propertieso f bovine articular cartilage from confined compression experiments. J. Bimechanics. 30 Nos. 11/ 12. 11571164 . Published by Elsevier Science
4. F. Boschetti, C. Miotti, F. Massi, M. Colombo, V. Quaglini, G. M. Peretti, R. Pietrabissa, 2002 An Experimental Study on Human Articular Cartilage Permeability. Proceedings of the Second Joint EMBS/BMES Conference. Houston, TX, USA. October 2002. 2326

5. D. R. Carter, M. Wong, Cartilage. Modelling, Mechanobiology, 2003 The Royal society. Philos Trans R Soc Lond B Biol Sci. 2003 September 29; 358 1437 14611471 .
6. B. Chan, P. S. Donzelli, R. L. Spilker, 2000 Mechanical Behavior of Articular Cartilage. Annals of Biomedical Engineering, 28 589597
7. Doblaré Castellano, 3 (3 de noviembre del año 2005). Sobre el Modelado en Biomecánica y Mecano-biología. Discurso de ingreso a la Real Academia de Ciencias Exactas, Físicas, Químicas y Naturales de Zaragoza. http://www.unizar.es/acz/ 02Academicos Numerarios/Discursos/Doblare.pdf
8. P. S. Donzelli, R. L. Spilker, G. A. Ateshian, V. C. Mow, 1999 Mechanical Behavior of Articular Cartilage Journal of Biomechanics. 32 10371047 .
9. P. S. Donzelli, R. L. Spilker, 1998 Mechanical Behavior of Articular Cartilage. Computer methods in applied mechanics and engineering. 153 6379
10. J. H. Frijns, 2000 Mechanical Behavior of Articular Cartilagey of Technology.
11. Garzón, D.A. (Mayo de 2007 Simulación de Procesos de Reacción-Difusión: Aplicación a la Morfogénesis del Tejido Óseo. Tesis doctoral. Centro Politécnico Superior de la Universidad de Zaragoza. Zaragoza.
12. F. Guilak, W. Jones, P. Ting-beall, G. Lee, 1999 The deformation behavior and mechanical properties of condrocitos in articular cartilage. Osteoarthritis and Cartilage. 7, 1: 59-70. January 1999
13. M. A. Haider, R. C. Schugart, 2006 Mechanical Behavior of Articular Cartilage. Journal of Biomechanics. 39 177183 .
14. M. A. Haider, F. Guilak, 2007 Application of a three-dimensional poro-elastic BEM to modelling the biphasic mechanics of cell-matrix interactions in articular cartilage. Computer methods in applied mechanics and engineering. 196 29993010 .
15. C. . Koenig, Noviembre de, 2011 Curso de Histología. Oficina de Educación Médica. Escuela de Medicina. Pontificia Universidad Católica de Chile. Consultado el 25 Noviembre 2011. http//escuela.med.puc.cl
16. R. C. Lawrence, Ch. G. Helmick, F. C. Arnett, 1998 Mechanical Behavior of Articular Cartilageates. Arthritis Rheum; 41 22132218 .
17. M. E. Levenston, E. H. Frank, A. J. Grodzinsky, 1998 Variationally derived 3-field finite element formulations for quasistaticporo-elastic analysis of hydrated biological tissues. Computer methods in applied mechanics and engineering. 156 231246 .
18. C. Martín-Hernández, 2002 Estudio Mecánico, Histológico, e Histo-morfométrico del Regenerado de Cartílago Mechanical Behavior of Articular Cartilage Barcelona.
19. X. N. Meng, M. A. Leroux, T. A. Laursen, L. A. Setton, 2002 Mechanical Behavior of Articular Cartilage. International Journal of Solids and Structures. 39 879895 .
20. U. Meyer, H. P. . Wiesmann, 1. April, 2006 Bone and Cartilage Engineering. (1 edition) ED Springer. 978-3-54025-347-1 Germany. 2527 .
21. V. C. Mow, S. C. Kuei, W. M. Lai, C. G. Armstrong, 2006 Mechanical Behavior of Articular Cartilage Journal of Biomechanical Engineering. 102 7384 . X
22. M. Nordin, V. Frankel, 2004 Biomecánica Básica del sistema Musculoesqueletico. ED McGrawn-Hill Interamericana. 61
23. M. Radi, 1998 Mechanical Behavior of Articular Cartilage. Dissertation. Institute for Microelectronics. Faculty of Electrical Engineering and information technology 1998. Consultado el 25 de Abril de 2012. http://www.iue.tuwien.ac.at/phd/radi
24. J. J. Sopena-Juncosa, J. M. Carrillo-Poveda, M. Rubio-Zaragoza, J. I. Redondo-García, I. Serra-Aguado, I. Soleri-Canet, 2000 Estructura y función del cartílago articular. Portada: Armas Frente a la Patología Articular.
25. K. Terada, T. Ito, N. Kikuchi, 1998 Mechanical Behavior of Articular Cartilage. Computer methods in applied mechanics and engineering. 153 223257 .
26. M. Van der Meulen, R. Huiskes, 2002 Why Mecanobiology? A survey article. Journal of Biomechanics. 35 4 Pag: 401-414.
27. W. Wilson, C. C. Van Donkelaar, B. Van Rietbergen, K. Itoa, R. Huiskes, 2004 Mechanical Behavior of Articular Cartilage Journal of Biomechanics. 37 357366 .
28. W. Wilson, C. C. Van Donkelaar, Van Rietbergen, R. Huiskes, 2005 Mechanical Behavior of Articular Cartilage Journal of Biomechanics. 38 11951204 .

29. W. Wilson, 2005 Mechanical Behavior of Articular Cartilage. Tesis Doctoral. Technische Universite it Eindhoven. Eindhoven.
30. J. Z. Wu, W. Herzog, M. Epstein, 1997 An improved solution for the contact of two biphasic cartilage layers. Journal of Biomechanics. 30 4 371 -375
31. J. Z. Wu, W. Herzog, M. Epstein, 1999 Mechanical Behavior of Articular Cartilage Journal of Biomechanics. 32 563572 .
32. J. Z. Wu, W. Herzog, 2000 Mechanical Behavior of Articular Cartilage Annals of Biomedical Engineering. 28 318330
33. J. P. Wu, T. B. . Kirk, November, 2001 A Study of the Shape Change of the Sheep Chondrocytes with Application of Compression to Cartilage. Seventh Australian and New Zealand Intelligent Information Systems Conference, 1821 . Perth, Western Australia

CHAPTER 16

Synthesis and Characterization of Types A and B Gelatin Methacryloyl for Bioink Applications

Bae Hoon Lee [1,*,†], Nathaniel Lum [1,†], Li Yuan Seow [1,†], Pei Qi Lim [1] and Lay Poh Tan [1,2,*]

[1]*School of Materials Science and Engineering, Nanyang Technological University, 50 Nanyang Avenue, Singapore 639798, Singapore*

[2]*Singapore Centre for 3D Printing (SC3DP), Singapore 639798, Singapore*

ABSTRACT

Gelatin methacryloyl (GelMA) has been increasingly considered as an important bioink material due to its tailorable mechanical properties, good biocompatibility, and ability to be photopolymerized in situ as well as printability. GelMA can be classified into two types: type A GelMA (a product from acid treatment) and type B GelMA (a product from alkali treatment). In current literature, there is little research on the comparison of type A GelMA and type B GelMA in terms of synthesis, rheological properties, and printability for bioink applications. Here, we report the synthesis, rheological properties, and printability of types A and B GelMA. Types A and B GelMA samples with different degrees of substitution (DS) were prepared in a controllable manner by a time-lapse loading method of methacrylic anhydride (MAA) and different feed ratios of MAA to gelatin. Type B GelMA tended to have a slightly higher DS compared to type A GelMA, especially in a lower feed ratio of MAA to gelatin. All the type A and type B GelMA solutions with different DS exhibited shear thinning behaviours at 37 °C. However, only GelMA with a high DS had an easy-to-extrude feature at room temperature. The cell-laden printed constructs of types A and B GelMA at 20% *w/v* showed around 75% cell viability.

Keywords: gelatin; methacryloyl; bioink; type A and type B; 3D bioprinting

1. INTRODUCTION

Three-dimensional (3D) printing is the new cutting edge technology that has been developed rapidly for various applications such as bioprinting [1,2,3]. As 3D printing technology evolves, an increasing number of materials such as acrylonitrile butadiene styrene, photo-curable resins, and even stainless steel have been developed and used along for manufacturing all forms of complex delicate building blocks. This development of 3D printing has gained much attention from various industries because of its ability to fabricate high-precision products and reduce the product design cycle for various industries [4,5]. 3D bioprinting is another powerful tool from 3D printing technology for tissue engineering and regenerative medicine applications [6,7,8,9]. A common method for bioprinting is through an extrusion technique, where a mixture of viscous cell-supportive materials and biologically active components including cells is dispensed onto a substrate [10,11,12,13,14].

3D bioprinting is basically incorporating a 3D printing technique into tissue engineering application. Currently, there are numerous studies that focus on 3D bioprinting of cells and scaffold materials together to create 3D artificial constructs for organ transplantation and drug screening [15,16]. 3D bioprinting is not only useful for 3D cell interaction studies, but also it potentially helps in regeneration therapy. By using regeneration therapy, an organ can be regenerated and the chance of patients receiving an organ transplant will increase.

In parallel with extensive exploration in the field of 3D bioprinting, the development of a suitable bioink is another important research area needed to meet all requirements for printing different types of artificial tissues and organs [17,18,19]. Basically, a mixture of materials and cells is known as a bioink in the 3D bioprinting industry. An ink material can become a scaffold after a certain treatment such as heat, ions, enzymes and light. Examples of typical bioink materials are alginate [20], gelatin

[21], and gelatin methacryloyl [16,22,23]. Ideal bioinks have some important characteristics such as low cost, printability, cell compatibility and cell-specific functionality [10]. A bioink should allow cell encapsulation, should be printable with a good resolution, and should ideally guide cells in the construct towards specific functions or behaviour.

Gelatin methacryloyl (GelMA) is an attractive photocurable material which is made from chemically modified gelatin with methacrylic anhydride (MAA) [24]. GelMA synthesis was first introduced by Van Den Bulcke et al. [25]. Recently, we developed an efficient and controllable synthesis of type A GelMA for mechanically stiff hydrogels using a sequential time-lapse loading method of MAA as well as one-pot GelMA synthesis [26,27]. GelMA is able to form a stable chemical gel with a water soluble photoinitiator upon exposure to a long wavelength ultraviolet (UV). The resultant hydrogel retains its physical form in body temperature and maintains its structural stability. This is in contrast to pure gelatin which is soluble at body temperature [25]. GelMA has controllable mechanical properties depending on its degree of methacryloylation and its concentration. In addition, GelMA retains its intrinsic biological properties such as enzyme-degradability and cell-adhesion promotion [23,24,28]. In these respects, GelMA could be a good candidate for bioink applications. The research conducted by Billiet et al. in 2014 showed that the 3D printed GelMA scaffold had excellent cell viability and scaffold stability for 14 days, indicating that GelMA could have great potential as a bioink for 3D bioprinting [16]. Most of the existing research studies carried out on GelMA focused on rheological studies [21], printing parameters [16] and cell viability [16,21,22].

However, there has been little research on comparing type A GelMA with type B GelMA as to synthesis, rheological properties, and printability for bioink applications. In this report, we synthesized type A and type B GelMA with different degrees of substitution, and measured their rheological properties, and discussed feasibility of type A and type B GelMA as a bioink material.

2. EXPERIMENTAL SECTION

2.1. Types A and B GelMA Synthesis

Gelatin Methacryloyl (GelMA) was synthesized in a buffer (0.1 M CB) of sodium carbonate (3.18 g) and sodium bicarbonate (5.86 g) in 1 L of distilled water. Gelatin (20 g) was added to 200 mL of the CB solution.

Two types of gelatin were used-type A (Bloom 175 g, porcine skin, Sigma-Aldrich, St. Louis, MO, USA) and type B (Bloom 225 g, bovine skin, Sigma-Aldrich, St. Louis, MO, USA). The gelatin solutions were adjusted to pH 9 by addition of 5 N NaOH to promote the reaction of the amino groups on gelatin. Methacrylate anhydride (MAA, 94%, Sigma-Aldrich, St. Louis, MO, USA) was added to the gelatin solutions at various feed ratios of 0.1:1, 0.05:1, 0.025:1, and 0.0125:1 (mL of MAA to g of gelatin), and the addition of MAA was divided into 6 separate times with an average of 11 min between each addition. As the reaction proceeded, the pH decreased. As such, the pH of the reaction solutions was adjusted to pH 9 after each addition for optimal reaction of gelatin with MAA. After around 1.5 h of the reaction, 1 N HCl was added to the solutions to be adjusted to the physiological pH of 7.4. The solutions were then filtered with 70 mm paper filter, followed by an antibacterial filter (0.2 μm). Then, the solutions were dialyzed for about 7 h through Minimate™ Tangential Flow Filtration system (Pall Corporation, Port Washington, WA, USA) which filtered out small molecules of less than 10 kDa, unreacted MAA, and methacrylic acid byproduct. After dialysis, the solutions were placed at −70 °C for 3–5 h, and finally lyophilized to obtain the dry GelMA products.

2.2. Determination of GelMA DS

To determine the degree of substitution, TNBS (2,4,6-Trinitrobenzenesulfonic acid, Sigma-Aldrich, St. Louis, MO, USA) method was used. GelMA and pure gelatin (type A and type B) were diluted in 0.1 M sodium bicarbonate buffer to 1.6 $mg \cdot mL^{-1}$. Then, 500 μL of 0.2% TNBS solution was added to 500 μL of each sample and placed in 37 °C for 3 h. Subsequently, 250 μL of 1 M hydrochloric acid and 500 μL of 10% *w*/*v* sodium dodecyl sulphate were added to stop the reaction. The absorbance of the samples was then measured at 335 nm. The molar concentration of free primary amino groups was compared to glycine standard solutions at concentrations of 0, 0.8, 8, 16, 32, and 64 $\mu g \cdot mL^{-1}$.

In order to verify the methacryloylation, ^1H NMR measurement was conducted. Types A and B GelMA samples were separately dissolved at around 50 $mg \cdot mL^{-1}$ in deuterium oxide (Merck, Kenilworth, NJ, USA) at 37 °C, and the 1H chemical shifts of each sample were observed.

2.3. Viscoelastic Properties of Types A and B GelMA

Characterization of the mechanical properties of the GelMA hydrogels was carried out via Rheometer (Anton Paar Physica, Ostfildern, Germany) MCR 501 testing using an Anton Paar CP25/TG 25-mm diameter cone-plate geometry with one-degree angle. The GelMA was photo-crosslinked in situ, and time sweep and frequency sweep tests were performed. The parameters for the time sweep test were set at frequency of 1 Hz for the entire duration of the test, and shear strain of 2% for about 6 min. The samples were irradiated with UV for 60 s after the time sweep test started. For the frequency sweep test of the cured GelMA, the parameters were set at 0.5% shear strain within the linear viscoelastic region and the frequency range of 0.1 to 10 Hz. About 70 μL of GelMA solutions (10%, 20%, and 30% *w*/*v*) containing 0.1% Irgacure 2959 (2-Hydroxy-4′-(2-hydroxyethoxy)-2-methylpropiophenone, Sigma-Aldrich, St. Louis, MO, USA) was used for each test and the temperature was set at 37 °C.

To investigate the effect of shear rate on shear stress and viscosity of GelMA at 37 °C and 25 °C, a rheology test was conducted under shear rate from 1 to 1000 s^{-1}. For testing the effect of temperature on the viscosity of GelMA, the shear rate was kept constant at 50 s^{-1} and the viscosity of the GelMA samples was measured from 40 °C to 20 °C at the speed of 2 $°C·min^{-1}$. The storage and loss modulus profiles of the GelMA samples were obtained at 0.5% shear strain and a frequency range of 0.1 to 10 Hz, and both at 37 °C and 25 °C.

2.4. Printability of Types A and B GelMA

Type A GelMA 2.2 and type B GelMA 2.2 with 20% *w*/*v* were tested for printability in Desktop Robot 2300N (SAN-EI-TECH Ltd., Chiba, Japan). Both samples were transferred and left to cool to room temperature (25 °C) in syringe before printing. The extrusion pressure was varied from 0.1 to 0.4 MPa in order to determine an optimal pressure for obtaining a good resolution print for both types A and B GelMA solutions. A conical needle size with gauge size 25 (inner diameter = 0.25 mm) was used. The printing temperature was kept at room temperature (25 °C) and printing speed used in the setup is 10 $mm·s^{-1}$. A pressure of 0.15 MPa was finally selected for type A GelMA and type B GelMA printing. For actual printing, a 3D grid pattern with the dimension of 2 cm × 2 cm with line spacing of 4 mm was printed using both types A and B GelMA with curing or without curing.

2.5. 3D Printing of Cell-Laden Constructs of Types A and B GelMA

Human hepatocarcinoma cells (Huh-7.5, Apath) were cultured in Dulbecco's Modified Eagle's Medium (DMEM, Hyclone) supplemented with 10% fetal bovine serum (FBS, Hyclone) and 1%

antibiotic/antimycotic (ABAM, Life Technologies, Gaithersburg, MD, USA) in a humidified atmosphere with 5% CO_2 at 37 °C. Types A and B GelMA samples at 20% were mixed with 2 × 10^6 cells in PBS containing 0.1% Irgacure 2959 and used for 3D printing of the 3D grid pattern with the dimension of 2 cm × 2 cm with line spacing of 4 mm. The cell-laden printed constructs with five layers were cured at 1–2 $mW \cdot cm^{-2}$ for 5 min. The cell viability on photopolymerized cell-laden print constructs of types A and B GelMA was measured using the LIVE/DEAD® Cell Viability/Cytotoxicity kit (Life Technologies, Gaithersburg, MD, USA) 4 h after printing. In brief, 4 μM calcein-AM and 8 μM ethidium homodimer-1 (EthD-1) were prepared in media and added to the cell-laden print constructs and incubated at 37 °C for 1 h. Live cells stained by calcein-AM were imaged as green, and dead cells stained by EthD-1 were imaged as red by microscope. Live and dead cells (n = 5 images, the average total number of cells = 387) were counted using ImageJ.

2.6. Statistic Analysis

Statistical analysis was performed using the Microsoft Excel statistical analysis software package. Standard deviation was calculated and presented for each treatment group (mean ± SD). Significant difference among the experimental groups was determined by a two-tailed student's t-test. p values < 0.05 were considered to be statistically significant.

3. RESULTS

3.1. Type A and Type B GelMA Synthesis

Both type A and type B gelatin samples were reacted with MAA at various feed ratios in a sequential addition method similar to reported literature [26]. The degree of substitution (DS) of methacryloyl groups on both types A and B gelatin was determined by 2,4,6-Trinitrobenzenesulfonic acid (TNBS) method. Similar results and trends were found for both type A GelMA and type B GelMA as presented in Figure 1 and Figure 2. Both Types A and B GelMA showed increasing degrees of substitution with increasing feed ratios of MAA to gelatin. The increase in degree of substitution was approximately proportional to the feed ratio up to a feed ratio of MAA (0.1 mL) to gelatin (1 g). At the highest feed ratio of MAA (0.1 mL) to gelatin (1 g), the reaction reached almost complete methacryloylation of lysine groups. The degree of substitution of type B

GelMA was higher than that of type A GelMA especially at a lower feed ratio of MAA to gelatin.

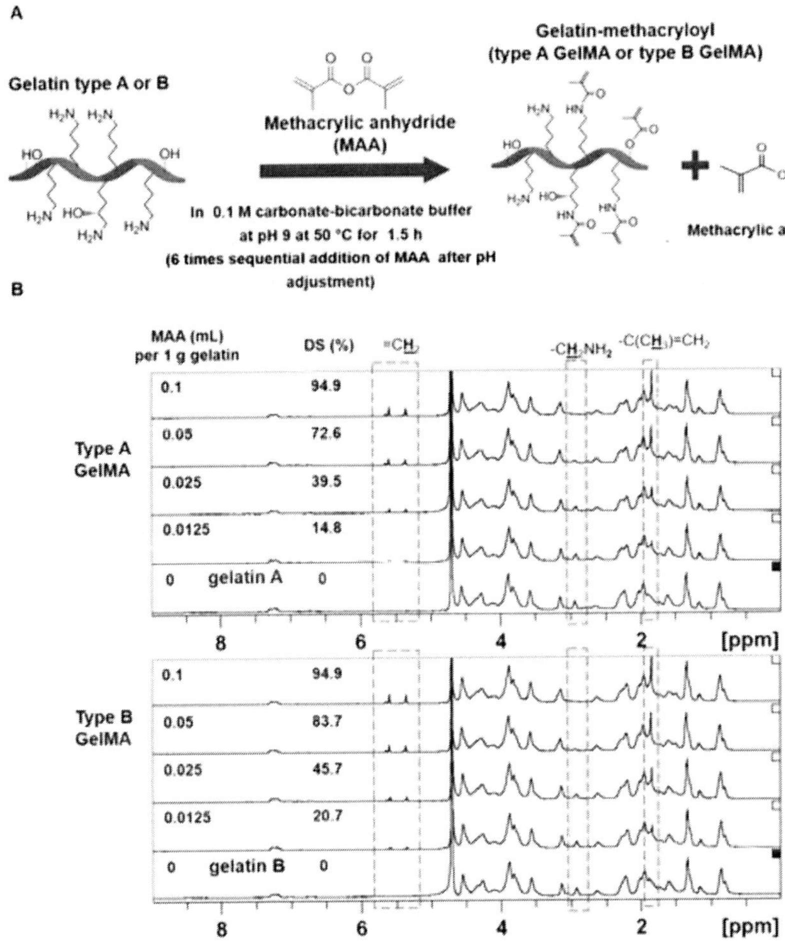

Figure 1. Preparation of types A and B GelMA. (**A**) Schematic illustration of GelMA reaction; (**B**) ^{1}H NMR verification of types A and B GelMA. Peaks correspond to acrylic protons (2H) of methacryloylated grafts of lysine groups and those of hydroxyl lysine groups, methylene protons (2H) of unreacted lysine groups, and methyl protons (3H) of methacryloylated grafts.

NMR characterization was conducted to verify that gelatin methacryloyl (GelMA) had been synthesized from the constituent gelatin and MAA. For the NMR analysis, the phenylalanine peaks (7.1–7.4 ppm) were taken as the standard among all the different samples as it was not modified by reaction with MAA. Both type A and type B GelMA samples formed new peaks from the reaction at around 5.3 and 5.5 ppm which were attributed to acrylic protons (2H) of methacryloylated grafts. The intensity of the peaks increased with the degree of methacryloylation, showing that the higher feed ratio of MAA to gelatin could produce GelMA with the higher degree of substitution. There was also a peak at around 1.8 ppm from the methyl group (3H) of methacryloylated grafts. This peak also increased with the degree of substitution, which indicates increasing methacryloylation. There was also a decreasing trend observed for the lysine methylene peak (2H) from gelatin at 2.8–3.0 ppm. This peak was the highest in pure gelatin and gradually decreased as the degree of substitution increased. As lysine is the bonding site for methacryloyl conjugation, it indicates that more lysine groups reacted with MAA and there was increased methacryloylation with lysine groups.

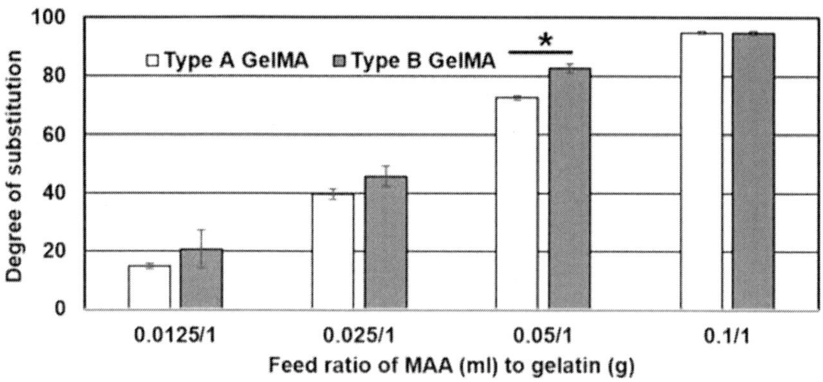

Figure 2. Comparison of the effect of a feed ratio of MAA/gelatin on the degree of substitution of types A and B GelMA. Error bar stands for standard deviation of means (* $p < 0.01$, $n = 3$).

3.2. In Situ Photopolymerizing Properties of Types A and B GelMA

The samples used for in situ photopolymerizing properties of GelMA were GelMA 0.2, GelMA 0.5, GelMA 1.1, and GelMA 2.2 from Type A and Type B as seen in Table S1. First, the time sweep test of the storage modulus versus time was carried out to determine the gelation behaviour during crosslinking. As presented in Figure 3, the time sweep results showed that types A and B GelMA solutions (10%, 20%, and 30%)

were photocrosslinked rapidly, within 20–30 s after the exposure to UV irradiation. The storage moduli of the formed GelMA hydrogels reached a plateau 1 min after UV exposure and showed a clear trend that the storage modulus increased with an increase in degree of substitution and hydrogel concentration. For type A GelMA, GelMA 2.2 (94.9% DS), GelMA 1.1 (72.6% DS), GelMA 0.5 (39.5% DS), and GelMA 0.2 (14.8% DS) at 30% *w/v* exhibited 67.6, 26.7, 11.3, and 1.9 kPa, respectively. The storage moduli of GelMA 2.2 at 30%, GelMA 2.2 at 20%, GelMA 2.2 at 10% were 67.6, 34.7, and 9.1 kPa, respectively. Type B GelMA exhibited trends similar to that of type A GelMA. However, type B GelMA hydrogel showed a higher storage modulus than type A GelMA hydrogel at the same concentration. Type A GelMA 2.2 (94.9% DS) at 30% exhibited 67.6 kPa storage modulus whereas type B GelMA 2.2 (94.9% DS) at 30% exhibited 147.3 kPa storage modulus ($p < 0.01$, $n = 3$).

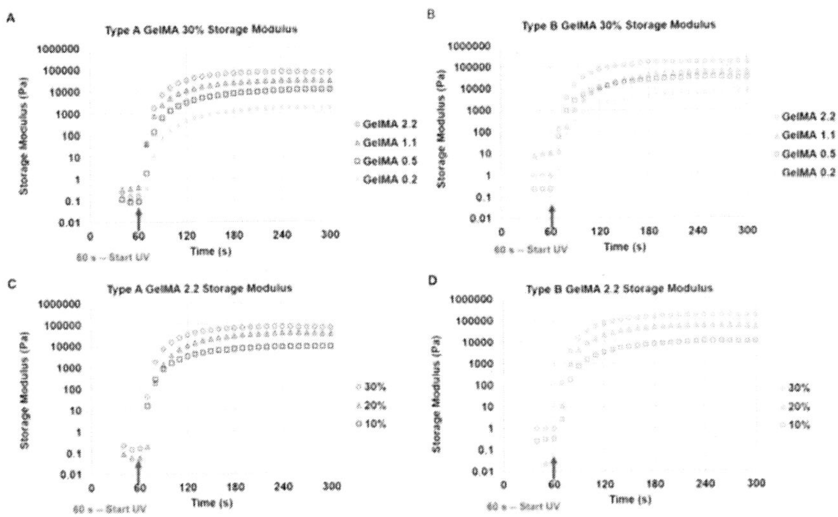

Figure 3. Rheological analysis of time-dependent storage modulus on UV irradiated GelMA with different DS and different concentrations at 37 °C. (**A**) 30% type A GelMA with different DS; (**B**) 30% type B GelMA with different DS; (**C**) Type A GelMA 2.2 with different concentrations (30%, 20%, and 10%); (**D**) Type B GelMA 2.2 with different concentrations (30%, 20%, and 10%).

3.3. Viscoelastictic Properties of Types A and B GelMA

We investigated the effect of shear rate on the shear stress and viscosity of types A and B GelMA in order to predict which DS and concentration of GelMA can be extruded. Figure 4 shows the shear stress and temperature sweep viscosity profiles of 30% *w/v* GelMA from type A and type B. The GelMA solutions at 37 °C from both types A and B behaved like shear thinning fluids. This can be seen as the viscosity of both types GelMA, as indicated by the slope of stress versus shear rate, decreased with increasing the shear rate. In types A and B GelMA solutions, GelMA with a higher DS exhibited lower shear stress and lower viscosity. Comparing type A GelMA and type B GelMA, it was observed that the overall shear stress and viscosity of type A GelMA were lower than type B GelMA. In a temperature sweep testing from 40 °C to 20 °C, the viscosity of all the GelMA solutions at 30% started to increase abruptly below 30 °C as seen in Figure 4C,D. The viscosity of the original gelatin solution started to increase at higher temperatures than GelMA solutions. The onset temperature of the abrupt increase in viscosity lowered with increasing the DS of the GelMA. Type A GelMA samples showed a more distinct viscosity profile as a function of temperature as compared to type B GelMA samples.

At 25 °C, GelMA at 30% *w/v* formed a physical gel and no longer exhibited shear thinning behaviour. Thus, we tested types A and B GelMA with 20% *w/v* on stress versus shear rate at 25 °C. The results are presented in Figure 5. Type A GelMA 2.2, type B GelMA 2.2 and type B GelMA 1.1 exhibited comparable pseudoplastic behaviours across the shear rate (1–1000 s^{-1}) as the viscosity reduced with increasing the shear rate. Types A and B GelMA 2.2 solutions were selected for further investigation for 3D printing.

A frequency sweep test was conducted to investigate the viscoelasticity of types A and B GelMA 2.2 solutions at 20% *w/v* at 25 °C as seen in Figure 6. Uncured types A and B GelMA 2.2 solutions exhibited a higher value of storage modulus than that of loss modulus at 25 °C. The storage modulus of type A GelMA 2.2 was higher than that of type B GelMA 2.2 across the frequency range of 0.1–1 Hz. After exposure to UV for 3 min, cured types A and B GelMA 2.2 samples showed more stability and higher storage moduli than uncured types A and B GelMA 2.2. After curing, both types A and B GelMA 2.2 showed structural stability as the G′ values were much higher than the G″ values across the frequency range. Type B GelMA 2.2 exhibited a higher storage modulus than type A GelMA 2.2; the storage modulus of type B GelMA 2.2 is 53.1 ± 0.7 kPa, whereas that of type A GelMA 2.2 is 34.7 ± 2.4 kPa.

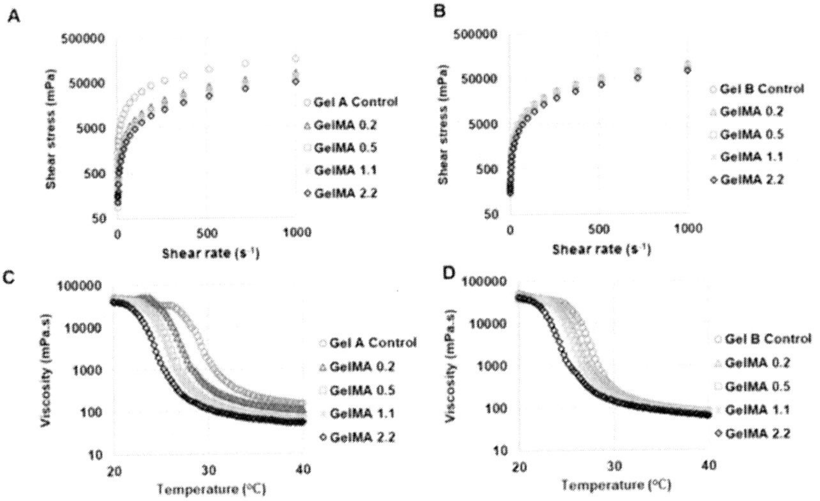

Figure 4. Rheological properties of types A and B GelMA. (**A**) Shear stress profile of 30% *w*/*v* type A GelMA at 37 °C ($n = 3$); (**B**) Shear stress profile of 30% *w*/*v* type B GelMA at 37 °C ($n = 3$); (**C**) Viscosity profile of 30% *w*/*v* type A GelMA as a function of temperature ($n = 3$); (**D**) Viscosity profile of 30% *w*/*v* type B GelMA as a function of temperature ($n = 3$).

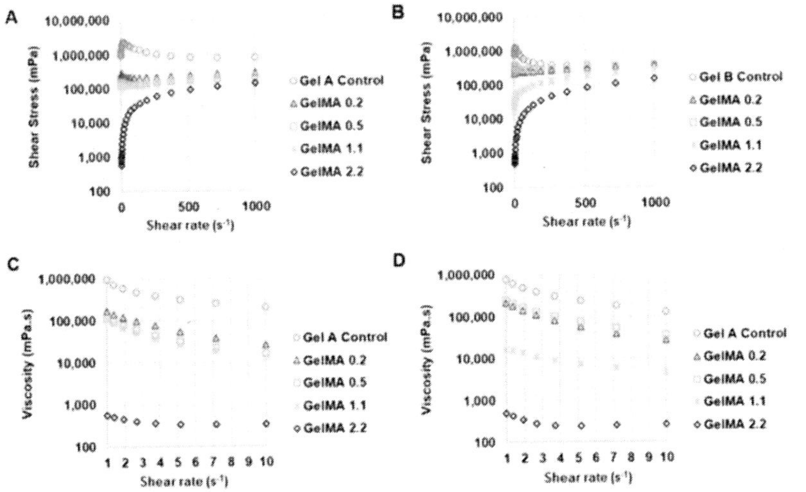

Figure 5. Shear thinning behaviour of types A and B GelMA at room temperature. (**A**) Overall shear stress profile of 20% *w*/*v* type A GelMA at 25 °C ($n = 3$); (**B**) Overall shear stress profile of 20% *w*/*v* type B GelMA at 25 °C ($n = 3$); (**C**) Viscosity profile of 20% *w*/*v* type A GelMA at 25 °C ($n = 3$); (**D**) Viscosity profile of 20% *w*/*v* type B GelMA at 25 °C ($n = 3$).

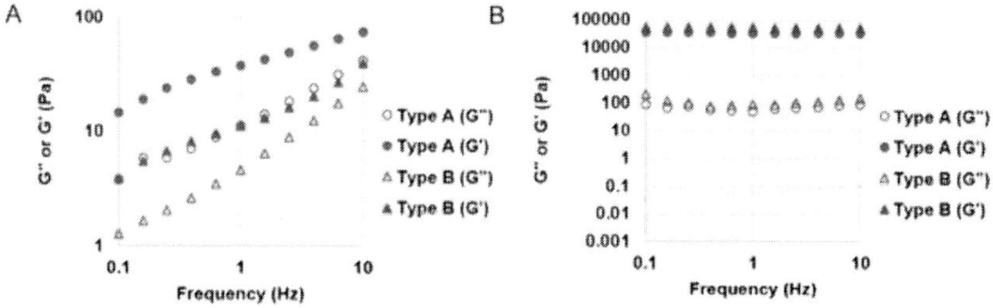

Figure 6. Viscoelastic properties of GelMA before and after curing. (**A**) Storage (G′) and loss modulus (G″) of 20% *w*/*v* type A GelMA 2.2 and type B GelMA at 25 °C (*n* = 3). G′ values were higher than G″ values across the frequency range for both of types A and B GelMA 2.2, which could provide initial shape fidelity after extrusion; (**B**) Storage (G′) and loss modulus (G″) of cured 20% *w*/*v* types A and B GelMA 2.2 at 37 °C (*n* = 3). After curing, both types A and B GelMA 2.2 could show structural stability for G′ values are much higher than G″ values across the frequency range.

3.4. Printability of Types A and B GelMA

Based on the rheology profiles obtained from the rheology testing, suitable samples (20% *w*/*v* types A and B GelMA 2.2) were selected for 3D extrusion style printing owing to their shear thinning behaviours, good shape fidelity at room temperature, and strong structural stability on curing. A trial was conducted to ensure that a proper fine line can be drawn during printing. Figure 7 showed that 0.10 MPa pressure produced discontinuous lines for both types A and B GelMA 2.2. The robot produced continuous lines of GelMA 2.2 above 0.15 MPa. However, the lines of type A GelMA were thinner than those of type B GelMA 2.2. Thus, 0.15 MPa was selected as the pressure for type A GelMA 2.2 and type B GelMA 2.2 extrusion at 20% *w*/*v*.

For the actual printing, a 3D grid pattern was designed in Microsoft Excel with the dimension of 2 cm × 2 cm with 4 mm line spacing. The 3D printing of 20% *w*/*v* type A GelMA 2.2 was compared between an uncured construct and a cured construct. Each of them was printed in five layers. Figure 8 presented the printing results of uncured and cured 20% *w*/*v* types A and B GelMA 2.2. The uncured constructs of types A and B GelMA displayed a defined grid pattern similar to the cured constructs of types A and B GelMA, respectively, as seen from the top view. In addition, type A GelMA produced a thinner construct compared to type B GelMA 2.2 as the extruded line (0.72 ± 0.10 mm) from type A GelMA was thinner from that (1.09 ± 0.21 mm) from type A GelMA under the same conditions.

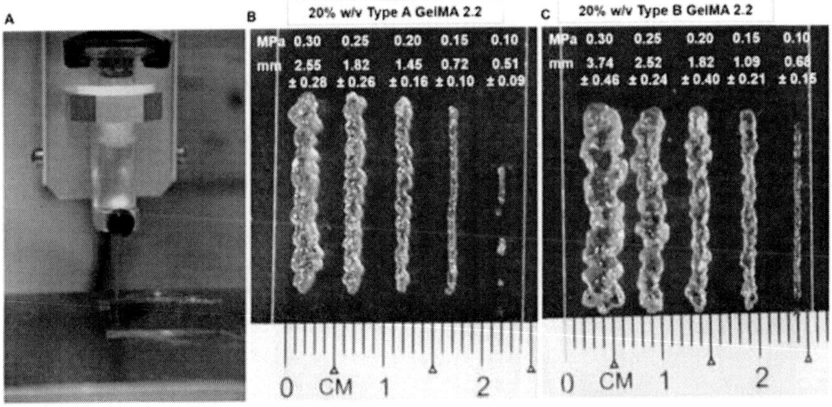

Figure 7. The effect of extrusion pressure on the printed line to find an optimal pressure for producing the best printed line. (**A**) Demonstration of the 3D extrusion style printing; (**B**) Type A GelMA 2.2 at 20% *w/v*; (**C**) Type B GelMA 2.2 at 20% *w/v*. * discontinuous lines.

3.5. 3D Printing of Cell-Laden Constructs of Types A and B GelMA

In order to evaluate the cell viability of cell-laden printed constructs of types A an B GelMA, the live/dead assay was conducted using calcein-AM (green fluorescent staining) and ethidium homodimer-1 (red fluorescent staining). Huh-7.5 cell-laden printed constructs with five layers from types A and B GelMA at 20% containing 0.1% Irgacure were photocrosslinked at 1–2 mW·cm^{-2} of 365 nm for 5 min. Both types A and B GelMA printed constructs exhibited good shape fidelity and maintained their form when handled with a pair of forceps as seen in Figure 9A. Live/dead staining images showed that Huh-7.5 cells in the 3D printed constructs of types A and B GelMA were considerably viable (around 75%) and their viability was not significantly different as presented in Figure 9B.

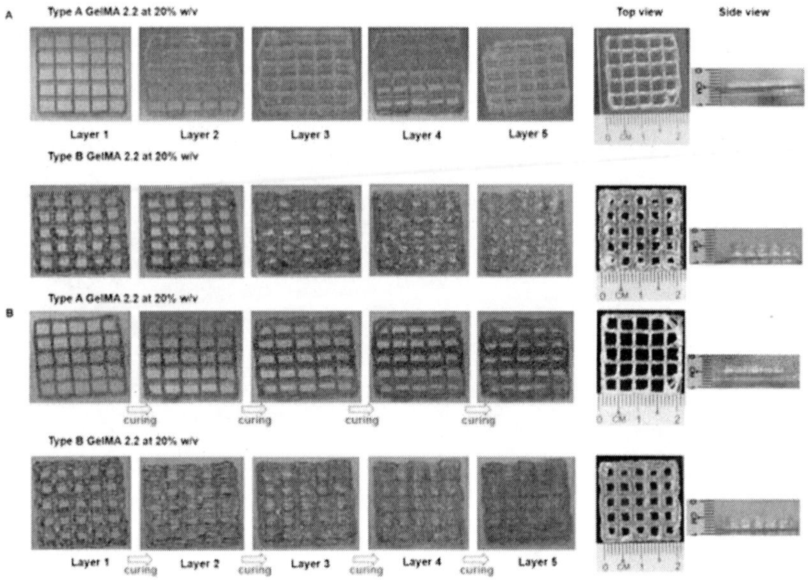

Figure 8. 3D printing of 20% *w/v* GelMA containing 0.1% I2959 with five layers. (**A**) Types A and B GelMA 2.2 without curing; (**B**) Types A and B GelMA 2.2 with layer-by-layer curing by 365 nm U.V.

4. DISCUSSION

So far, most of the GelMA studies have focused on one type of GelMA, either type A or type B for their specific applications [16]. In order to compare type A GelMA with type B GelMA, type A and type B GelMA with a wide range of DS (15%–95%) were synthesized in a well controllable manner by the sequential MAA addition method using various feed ratios of MAA (mL) to gelatin (g) (0.1/1–0.0125/1). Type B GelMA exhibited higher degrees of substitution across the entire range as compared to type A GelMA, except at the highest feed ratio where the degree of substitution is close to the maximum. This may be due to the difference in IEP (isoelectric point) of type A and type B gelatin. Type B gelatin has a lower IEP of pH 4–6, compared to an IEP of pH 7–9 for type A gelatin. As the reaction of gelatin with MAA was maintained at pH 9, type B gelatin had more free amine group exposed owing to its lower IEP in the same reaction buffer. A larger number of freely available amino groups exposed in type B gelatin would be able to react with MAA and hence would lead to a higher degree of methacryloylation as compared to type A gelatin.

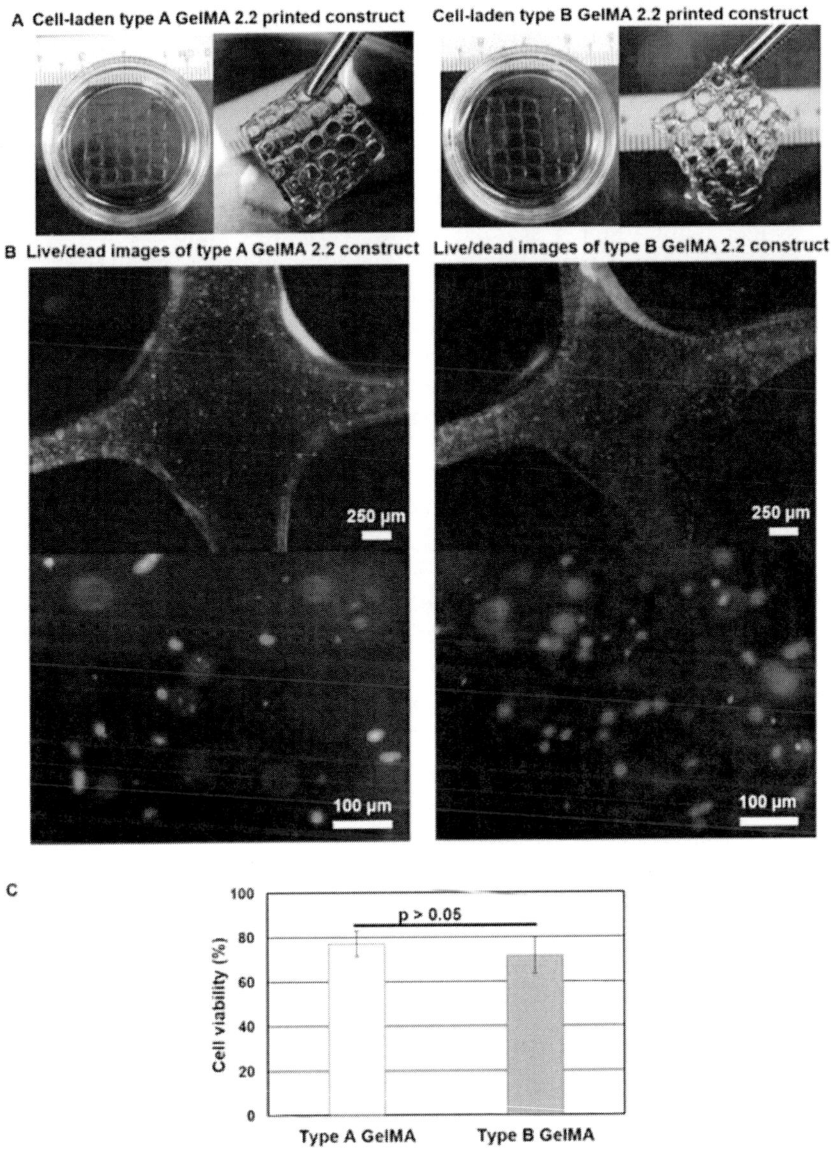

Figure 9. 3D printing of Huh-7.5 cell-laden constructs of types A and B GelMA. (**A**) Gross pictures of cell-laden constructs of types A and B GelMA; (**B**) Live and dead staining images of cells in the cell-laden constructs of types A and B GelMA (Top: 2×; Bottom: 10×); (**C**) The viability of cells in the cell-laden constructs of types A and B GelMA.

Both types A and B GelMA solutions were crosslinked within a few minutes to maintain their stable storage moduli after the exposure to a long wavelength UV (365 nm). The storage moduli of both types GelMA hydrogels were constant across the range of frequencies. The GelMA hydrogels also showed very good reproducibility, with all the storage moduli maintaining a relatively narrow standard deviation. The GelMA hydrogels with a higher storage modulus are more rigid and are able to withstand higher deformation stresses. Additionally, because the storage modulus is constant across the range of frequencies, the GelMA hydrogels are expected to maintain their elastic properties for a longer time period. The stiffness of GelMA hydrogels depends on their degrees of substitution (DS) and concentrations. A higher DS and concentration of GelMA solutions gave rise to hydrogels that have a higher storage modulus owing to the increased availability of crosslinkable methacryloyl groups (higher crosslinking density). Type B GelMA hydrogels exhibited a higher storage modulus than type A GelMA hydrogels possibly on the account of a slightly higher DS and a higher bloom number: Type A gelatin has a gel strength of 175 g bloom, whereas type B gelatin has a gel strength of 225 g bloom. Our GelMA hydrogels with various DS exhibit a wide range of stiffness (from 0.1 kPa to 150 kPa) which can be advantageous for diverse bioink applications. In situ photocrosslinkable properties of GelMA can secure a prolonged cell culture as well as structural stability of its scaffold.

The viscosity of GelMA solutions was varied as a function of temperature. GelMA with a higher DS tended to form a physical gel at lower temperatures. Gelatin formed a physical gel at higher temperature than GelMA. This could be due to the effect of the methacryloylation of GelMA, which may hinder the formation of physical bonds and helix structures in GelMA [15]. This phenomenon appeared to be more pronounced in type A GelMA than type B GelMA. Type A gelatin is less denatured and hydrolysed than type B gelatin in terms of IEP and the composition of amino acids [26]. Type A gelatin has more similarity of amino acid composition and IEP to collagen than type B gelatin. Therefore, the effect of the methacryloylation of type A gelatin on temperature-driven physical gelation was possibly more significant and obvious than that of type B GelMA. The viscosity of GelMA solutions decreased with increasing the shear rate at 37 °C, showing that GelMA solutions are pseudo-plastic (shear thinning) at 37 °C. However, GelMA at 30% *w/v* formed a strong physical gel at 25 °C, which rendered it unsuitable for extrusion-based 3D printing at room temperature, whereas type A and type B GelMA 2.2 at 20% *w/v* showed shear thinning properties under shear strain at 25 °C and extrusion through a syringe was possible.

Based on the material property studies, we chose 20% *w/v* type A GelMA 2.2 and type B GelMA 2.2 for further optimization of bioprinting parameters. Both types GelMA met two of the requirements of a bio-ink material which are shear thinning properties and shape fidelity. At the shear stress versus shear rate test, both types A and B GelMA solutions showed shear thinning behaviours at 25 °C as their viscosity reduced with increasing the shear rate. For shape fidelity, both types of GelMA exhibited higher storage moduli (G′) than their loss moduli (G″) at 25 °C, which would probably allow the GelMA ink solutions to retain their shapes after printing. As the storage modulus of type A GelMA was higher than that of type B GelMA 2.2 at room temperature, the 3D printing construct of type A GelMA had a better resolution than that of type B GelMA under the same conditions. Compared to type B gelatin, type A gelatin is more similar to the original collagen in terms of amino acid components, IEP and capability of helix structure formation. In this respect, type A GelMA can maintain better physical gelation properties at room temperature than type B GelMA, leading to good shape fidelity. Both types A and B GelMA printing constructs presented structural stability after layer-by-layer curing.

Cell-laden printed constructs of types A and B GelMA at 20% showed good shape fidelity in culture media and could be easily handled with a pair of forceps. Both of them exhibited good viability (around 75%). In summary, we synthesized both types of GelMA, compared their viscoelastic properties, and investigated their feasibility as bioinks for 3D printing. Even though increasing the concentration of bioinks to obtain shape fidelity is a common strategy, the high concentration of a printing ink may compromise cell viability during cell culture. As the printability of GelMA is considerably sensitive to the temperature change, a lower concentration of GelMA can be used for 3D printing once 3D printers are equipped with a more precise temperature control system from low (5 °C) to high (37 °C). In addition, the employment of viscosity enhancers such as Ca^{++}, Mg^{++}, and other small molecules could reduce the concentration of GelMA for 3D printing, and that of various bioactive molecules could provide GelMA with more attractive features as bioink materials [29].

5. CONCLUSIONS

The aim of this work was to synthesize GelMA from different types of gelatin precursors (type A and type B) and compare them in terms of degree of substitution, mechanical properties, and feasibility for bioink applications. Both types of gelatin reacted with MAA to form gelatin methacryloyl (GelMA) in a similar pattern; the degree of methacryloylation of type B GelMA was slightly higher than that of type A

GelMA at the same feed ratio of MAA to gelatin. In terms of mechanical properties, an increase in storage modulus could be brought about by increasing the degree of substitution of GelMA, GelMA concentration, and by using GelMA with higher bloom strength. For printability of GelMA as a bioink, both types A and B GelMA solutions with the high DS exhibited shear thinning properties, shape fidelity at 25 °C, and structural stability on curing. Type A GelMA produced the printed construct with a better resolution that type B GelMA at the same extrusion pressure. Cell-laden GelMA printed constructs had an easy-to-handle feature in culture media. Huh-7 cells in both types A and B GelMA printed constructs displayed relatively high viability even at 20% concentration.

ACKNOWLEDGMENTS

This research is financially supported by the National Research Foundation (NRF) Competitive Research Programme (NRF-CRP10-2012-07) grant and is partially supported by Final Year Project Programme in MSE and Singapore Centre for 3D Printing (SC3DP). Bae Hoon Lee, Li Yuan Seow, and Nathaniel Lum appreciated Hitomi Shirahama's generous help in NMR measurement.

AUTHOR CONTRIBUTIONS

Bae Hoon Lee, Nathaniel Lum, Li Yuan Seow, and Lay Poh Tan conceived and designed the experiments; Bae Hoon Lee, Nathaniel Lum, Li Yuan Seow, and Pei Qi Lim performed the experiments and analyzed the data; Bae Hoon Lee, Nathaniel Lum, Li Yuan Seow, Pei Qi Lim, and Lay Poh Tan wrote the paper.

CONFLICTS OF INTEREST

The authors declare no conflict of interest.

REFERENCES

1. Patra, S.; Young, V. A review of 3D printing techniques and the future in biofabrication of bioprinted tissue. Cell Biochem. Biophys. 2016, 74, 93–98.

2. Chimene, D.; Lennox, K.K.; Kaunas, R.R.; Gaharwar, A.K. Advanced bioinks for 3D printing: A materials science perspective. Ann. Biomed. Eng. 2016, 44, 2090–2102.
3. Khoo, Z.X.; Teoh, J.E.M.; Liu, Y.; Chua, C.K.; Yang, S.; An, J.; Leong, K.F.; Yeong, W.Y. 3D printing of smart materials: A review on recent progresses in 4d printing. Virtual Phys. Prototyp. 2015, 10, 103–122.
4. Liu, W.; Xu, S.Q. 3D Printing Technology and Its Applications; World Scientific Publ Co Pte Ltd.: Singapore, 2016; pp. 506–514.
5. Zhang, J.C.; Yu, Z.H. Overview of 3D printing technologies for reverse engineering product design. Autom. Control Comput. Sci. 2016, 50, 91–97.
6. Wu, G.H.; Hsu, S.H. Review: Polymeric-based 3D printing for tissue engineering. J. Med. Biol. Eng. 2015, 35, 285–292.
7. Ng, W.L.; Yeong, W.Y.; Naing, M.W. Polyelectrolyte gelatin-chitosan hydrogel optimized for 3D bioprinting in skin tissue engineering. Int. J. Bioprint. 2016, 2.
8. Bhuthalingam, R.; Lim, P.Q.; Irvine, S.; Agrawal, A.; Mhaisalkar, P.; An, J.; Chua, C.K.; Venkatraman, S. A novel 3D printing method for cell alignment and differentiation. Int. J. Bioprint. 2015, 1.
9. Lee, J.M.; Yeong, W.Y. A preliminary model of time-pressure dispensing system for bioprinting based on printing and material parameters. Virtual Phys. Prototyp. 2014, 10, 3–8.
10. Panwar, A.; Tan, L.P. Current status of bioinks for micro-extrusion-based 3D bioprinting. Molecules 2016, 21, 685.
11. Ozbolat, I.T.; Hospodiuk, M. Current advances and future perspectives in extrusion-based bioprinting. Biomaterials2016, 76, 321–343.
12. Zhao, Y.; Li, Y.; Mao, S.; Sun, W.; Yao, R. The influence of printing parameters on cell survival rate and printability in microextrusion-based 3D cell printing technology. Biofabrication 2015, 7, 045002.
13. Chung, J.H.Y.; Naficy, S.; Yue, Z.L.; Kapsa, R.; Quigley, A.; Moulton, S.E.; Wallace, G.G. Bio-ink properties and printability for extrusion printing living cells. Biomater. Sci. 2013, 1, 763–773.
14. Suntornnond, R.; Tan, E.; An, J.; Chua, C. A mathematical model on the resolution of extrusion bioprinting for the development of new bioinks. Materials 2016, 9, 756.
15. Du, M.C.; Chen, B.; Meng, Q.Y.; Liu, S.M.; Zheng, X.F.; Zhang, C.; Wang, H.R.; Li, H.Y.; Wang, N.; Dai, J.W. 3D bioprinting of BMSC-laden methacrylamide gelatin scaffolds with CBD-BMP2-collagen microfibers. Biofabrication2015, 7, 044104.
16. Billiet, T.; Gevaert, E.; De Schryver, T.; Cornelissen, M.; Dubruel, P. The 3D printing of gelatin methacrylamide cell-laden tissue-engineered constructs with high cell viability. Biomaterials 2014, 35, 49–62.
17. Rutz, A.L.; Hyland, K.E.; Jakus, A.E.; Burghardt, W.R.; Shah, R.N. A multimaterial bioink method for 3D printing tunable, cell-compatible hydrogels. Adv. Mater. 2015, 27, 1607–1614.
18. Pati, F.; Jang, J.; Ha, D.H.; Won Kim, S.; Rhie, J.W.; Shim, J.H.; Kim, D.H.; Cho, D.W. Printing three-dimensional tissue analogues with decellularized extracellular matrix bioink. Nat. Commun. 2014, 5, 3935.
19. Yu, Y.; Moncal, K.K.; Li, J.; Peng, W.; Rivero, I.; Martin, J.A.; Ozbolat, I.T. Three-dimensional bioprinting using self-assembling scalable scaffold-free "tissue strands" as a new bioink. Sci. Rep. 2016, 6, 28714.
20. Markstedt, K.; Mantas, A.; Tournier, I.; Avila, H.M.; Hagg, D.; Gatenholm, P. 3D bioprinting human chondrocytes with nanocellulose-alginate bioink for cartilage tissue engineering applications. Biomacromolecules 2015, 16, 1489–1496.
21. Irvine, S.A.; Agrawal, A.; Lee, B.H.; Chua, H.Y.; Low, K.Y.; Lau, B.C.; Machluf, M.; Venkatraman, S. Printing cell-laden gelatin constructs by free-form fabrication and enzymatic protein crosslinking. Biomed. Microdevices 2015, 17, 1–8.
22. Nichol, J.W.; Koshy, S.T.; Bae, H.; Hwang, C.M.; Yamanlar, S.; Khademhosseini, A. Cell-laden microengineered gelatin methacrylate hydrogels. Biomaterials 2010, 31, 5536–5544.
23. Loessner, D.; Meinert, C.; Kaemmerer, E.; Martine, L.C.; Yue, K.; Levett, P.A.; Klein, T.J.; Melchels, F.P.W.; Khademhosseini, A.; Hutmacher, D.W. Functionalization, preparation

and use of cell-laden gelatin methacryloyl-based hydrogels as modular tissue culture platforms. Nat. Protoc. 2016, 11, 727–746.

24. Yue, K.; Trujillo-de Santiago, G.; Alvarez, M.M.; Tamayol, A.; Annabi, N.; Khademhosseini, A. Synthesis, properties, and biomedical applications of gelatin methacryloyl (GeLMA) hydrogels. Biomaterials 2015, 73, 254–271.
25. Van den Bulcke, A.I.; Bogdanov, B.; De Rooze, N.; Schacht, E.H.; Cornelissen, M.; Berghmans, H. Structural and rheological properties of methacrylamide modified gelatin hydrogels. Biomacromolecules 2000, 1, 31–38.
26. Lee, B.H.; Shirahama, H.; Cho, N.J.; Tan, L.P. Efficient and controllable synthesis of highly substituted gelatin methacrylamide for mechanically stiff hydrogels. RSC Adv. 2015, 5, 106094–106097.
27. Shirahama, H.; Lee, B.H.; Tan, L.P.; Cho, N.J. Precise tuning of facile one-pot gelatin methacryloyl (GeLMA) synthesis. Sci. Rep. 2016, 6, 31036.
28. Klotz, B.J.; Gawlitta, D.; Rosenberg, A.J.W.P.; Malda, J.; Melchels, F.P.W. Gelatin-methacryloyl hydrogels: Towards biofabrication-based tissue repair. Trends Biotechnol. 2016, 34, 394–407.
29. Colosi, C.; Shin, S.R.; Manoharan, V.; Massa, S.; Costantini, M.; Barbetta, A.; Dokmeci, M.R.; Dentini, M.; Khademhosseini, A. Microfluidic bioprinting of heterogeneous 3D tissue constructs using low-viscosity bioink. Adv. Mater. 2016, 28, 677–684.

CHAPTER 17

Synthesis, Characterization, and Study of PLGA Copolymer in Vitro Degradation

Anamaria Teodora Coêlho Rios Silva, Barbara Camilla Oliveira Cardoso, Maria Elisa Scarpelli Ribeiro e Silva, Roberto Fernando Souza Freitas, Ricardo Geraldo Sousa*

Polymer Science and Technology Laboratory, Chemical Engineering Department, School of Engineering, Federal University of Minas Gerais, Belo Horizonte, Brazil

ABSTRACT

The poly(lactic-co-glycolic acid), known as PLGA, is one of the main bioreabsorbable polymers used in the field of medicine today. This copolymer is widely applied in sutures, devices geared toward the controlled release of medication, and the guided regeneration of bone tissue as it presents a short degradation time. This work aimed to synthesize the 82/18 PLGA (expressed by the mass ratio of D,L-lactide and glycolide, respectively), to characterize and study the in vitro degradation in the form of rods in phosphate buffer solution (PBS). The copolymer was synthesized by opening the cyclic dimer rings of the monomers D,L-lactide and glycolide, in the presence of the tin octanoate initiator and of the lauryl alcohol co-initiator. The characterization of the copolymer and the follow-up of its in vitro degradation were studied using: Differential Scanning Calorimetry (DSC), Thermogravimetry (TG), Infrared Molecular Absorption Spectroscopy with Fourier Transform (FTIR), Rheometry, and Scanning Electron Microscopy (SEM). Through these characterization techniques, it was possible to obtain the glass transition temperature, thermal stability, chemical composition, morphology, and molar mass of both the synthesized and the degraded copolymer. The molar mass of the synthesized copolymer was, approximately, 10^6 g·mol^{-1}. The degradation rate of PLGA significantly increased from the 19th to the 28th day in PBS. After 28 days in PBS, the glass transition temperature and the molar mass reduced from 45°C to 17°C and from 1.5×10^6 g·mol^{-1} to 7.5×10^4g·mol^{-1}, respectively. The pH of the medium has a significant influence on the copolymer degradation profile. When it diminishes, it accelerates the degradation process, resulting in smaller PLGA polymer chains. This pH dependent degradation can be useful for drug release systems.

Keywords: Synthesis, Characterization, In Vitro Degradation, PLGA

1. INTRODUCTION

Many types of surgically implantable devices, which serve a temporary function within the body, can be obtained through polymers that, upon undergoing degradation, generate products that are fully metabolized in the physiological medium [1] - [6] .

One major advantage of these polymers, which are used as bioreabsorbable materials, is that, before being eliminated, they fulfill the initial purpose of tissue (or organ) recovery, or release of a drug [3] [7] . Although the first applications of these polymers have been as materials used for surgical sutures, other applications, such as implants to repair fractures and drug release devices are already commonplace in medical procedures [8] - [10] . Great effort has been made to improve the profiles of drug release systems, aimed at diminishing the effects of therapy and treatment time [11] [12] .

Due to their diverse applications and functionalities, especially in drug release therapies, the polymers are among the most widely used excipients when obtaining pharmaceutical structures [3] [13] . During drug release, the polymer is used as an active protection agent [14] - [16] .

The first copolymer (Vicril®), made up of units from lactic and glycolic acid, the poly(lactic co-glycolic acid), known as PLGA, was commercialized in the 1970's. The great interest in this copolymer stems from the fact that its mechanical and degradation time properties can be controlled through monomer ratios [10] [17] -[22] . Figure 1 presents the chemical structure of the lactide and glycolide dimers and the reaction of copolymerization, with the polymer structures.

These copolymers can present linear and saturated chains, a constant rate of biodegradation, mechanical resistance, crystallinity, hydrophobicity, regular geometry of individual chains, and thermoplastic behavior. They possess an asymmetric center, presented by the lactic acid (PLA) in a racemic mixture (D,L-PLA) or by one of its forms—levogyrous (L-PLA) or dextrogyrous (D-PLA)—alternately connected to the glycolic acid polymer (PGA) [18] [22] [23] .

The biodegradation of PLGA occurs through bulk erosion (cleavage of polymer chains by hydrolysis), resulting in free lactic and glycolic acids. These acids, as they are natural metabolites from the body, are eliminated by the Kreb's cycle in the form of carbon dioxide (CO_2) and water (H_2O) [24] -[27] , as shown in Figure 2.

The main advantage of PLGA over other bioreabsorbable polymers is that this copolymer requires a shorter time for its complete degradation, implying a lower probability of adverse reactions, which often results from the crystalline fragments released by polymers whose degradation time is excessively long. As its polymer chain possesses the mers from glycolic acid, which presents a weaker barrier against attacks from H_2O molecules when compared to lactic acid, the chemical structure of PLGA is more susceptible to hydrolysis reaction, when compared to PLLA, whose polymer chain is made up of meres only from lactic acid [29] - [33] .

Polymers derived from lactic and glycolic acids have received great attention in the research on alternative biodegradable polymers, as they have already received approval from the Food and Drug Administration (FDA) for use as drug release systems. A wide range of studies has been published demonstrating their low toxicity [3] [9] [34] .

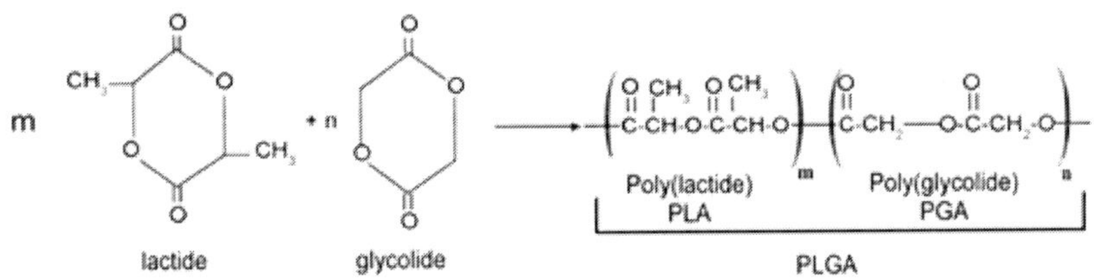

Figure 1. Chemical structure of the dimers, polymers, and copolymerization reaction.

For the diverse application of PLGA, it is necessary to follow up on the degradation process. Many studies have developed methods for monitoring the stages of copolymer degradation [19] . The present work aimed to synthesize, characterize, and study the PLGA copolymer in vitro degradation in the form of a rod placed in phosphate buffer solution (PBS) (pH 7.4). The following characterization techniques were applied: Infrared Molecular Absorption Spectroscopy with Fourier Transform (FTIR),

Differential Scanning Calorimetry (DSC), Thermogravimetry (TG), Scanning Electron Microscopy (SEM), and Rheometry analysis.

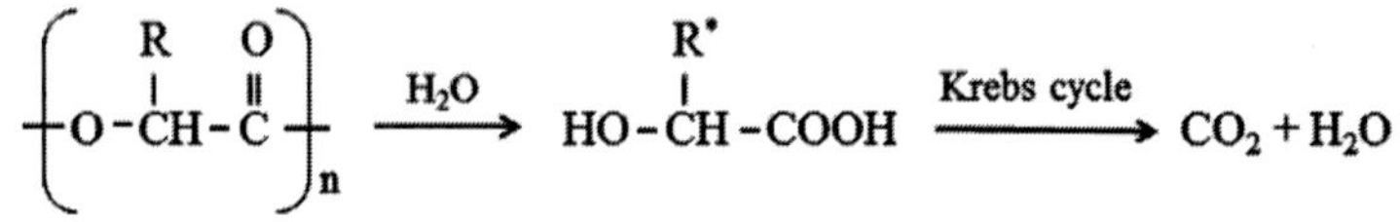

R = H (glycolic acid) or R = CH_3 (lactic acid). *Degradation products: 1) lactic acid (if the polymer is PLA); 2) glycolic acid (PGA polymer source); 3) both (source PLGA polymer).

Figure 2. Reaction of PGA/PLA/PLGA degradation [28] .

2. MATERIALS AND METHODS

2.1. PLGA Copolymer Synthesis

The copolymer was synthesized by opening the cyclic dimer rings of the monomers D,L-lactide and glycolide, in the presence of a tin octanoate initiator (0.02% over the total mass of monomers) and of the lauryl alcohol co- initiator (0.01% over the total mass of monomers). The polymerization occurred in a glass flask-ampoule, where D,L-lactide and glycolide were added in the mass proportions of 82/18, respectively. After adding the initiator and co-initiator, the flask was closed, vacuumed sealed in the ampoule, and immersed in a silicone oil bath at 175°C. After two hours of polymerization, the flask was removed from the bath and dichloromethane was added to dissolve the PLGA. Next, anhydrous methanol was added for PLGA precipitation. The copolymer was then kept in a heated vacuum oven at 60°C, for 24 hours. The efficient vacuum control, the absence of humidity in the reactional medium, coupled with adequate stirring, were keys to achieving a successful synthesis [34] . The synthesized PLGA copolymer was then placed in hermetically sealed flasks in a refrigerator.

The tin octanoate initiator was used in the PLGA copolymerization reaction, for being quite reactive, of great use in the food and pharmaceuticals industries, and recognized by the FDA for this very use. The co-initiator, lauryl alcohol, and the synthesis route for the copolymerization of the monomers of cyclic diesters from lactic and glycolic acids, through the opening in the rings of the lactide and glycolide, were chosen aiming to produce copolymers with a greater molar mass for their use in drug release systems.

2.2. Study of in Vitro Hydrolytic Degradation

The PLGA rods were molded using a spatula and a non-stick surface over a steel plate heated to ±80°C. The average mass of the produced rods was 203 ± 4 mg. The PLGA copolymer rods were immersed in closed test tubes, containing phosphate buffer solution (PBS), in a metabolic bath at 37°C, under stirring, at 30 rpm. At determined period of times: 0, 6, 12, 19, 28, 40, and 55 days, the rods were removed, rinsed in distilled water (80 mL), dried in a heated vacuum oven at 60°C for 48 hours, and characterized by DSC, TG, rheometry, FTIR, and SEM. After having removed the rods from the medium, the pH of the solutions was measured, by using a Denver Instrument, model 220, pH meter. All analyses were carried out in triplicate and the analysis conditions were as follows.

2.3. Fourier Transform Infrared Spectroscopy (FTIR)

The absorption spectra in the infrared region for the samples, within the range of 4000 cm^{-1} to 650 cm^{-1}, were obtained in a Nicolet 6700 FTIR spectrometer (Thermo Fisher SCIENTIFIC), in the Attenuated Total Reflection—ATR mode, with 64 scans and a resolution of 4 cm^{-1}. The spectra were obtained at room temperature (±20°C) with the sample added directly into the device, with no prior treatment.

2.4. Thermogravimetry (TG)

The TG analysis was performed to verify the mass variation with increasing temperature, the thermal stability, and the maximum degradation temperature of the samples. Using a Shimadzu model TGA-50, the test was conducted at a heating ratio of 10°C·min^{-1}, from room temperature (±20°C) up to 400°C, in a platinum sample holder and in a nitrogen atmosphere (N_2), at a flow rate of 50 mL·min^{-1}. The mass of the analyzed samples was 5 ± 1 mg.

2.5. Differential Scanning Calorimetry (DSC)

The DSC equipment (Shimadzu model DSC-60) used in this study was programmed to heat the samples in two distinct runs. The first run was performed from room temperature (±20°C) up to 150°C, at a heating ratio of 20°C·min^{-1}, intended to erase the thermal history of the material. The second run was performed from −50°C (cooled with liquid N_2) up to 200°C, at a heating ratio of 20°C·min^{-1}. An aluminum sealed sample holder was used, with a hole punctured on the cover. Nitrogen was used as the flow gas, at a rate of 50 mL·min^{-1}. The mass of the analyzed samples was 10 ± 1 mg.

2.6. Rheometry Analysis

The rheometric tests were conducted in a rotational rheometer, model AR-G2 (TA Instruments), using an 8 mm plate to plate geometry, heated by the equipment environmental test chamber (ETC). All rheological tests were conducted at room temperature. Through a strain sweep test, at a constant frequency of 1 Hz, curves for the elastic (G') and viscous (G") moduli were obtained as a function of a wide strain range (0.01% to 100%), in order to determine the strain value to be applied in the rheological tests (region of linear viscoelasticity). Frequency sweep test was carried out in order to determine the average molar mass variation of the samples. The frequency varied from 0.8 rad·s^{-1} to 400 rad·s^{-1} (0.1 Hz to 64 Hz), with a strain of 5%, which falls into the linear viscoelasticity range. The tests were conducted at a temperature of 80°C (melt state).

2.7 Scanning Electron Microscopy (SEM)

A scanning electron microscope model TM3000 tabletop microscope (Hitachi), at a voltage of 5.0 kV and at a working distance between 3 mm and 10 mm, was used for the SEM tests, and the obtained images were processed using the TM3000 software. All samples were analyzed at room temperature (±20°C). There was no need for sample coating with gold, as this analysis was conducted under low vacuum.

3. RESULTS AND DISCUSSION

PLGA was synthesized with a yield close to 90% and the average molar mass of the samples was approximately 10^6 g·mol^{-1}. The studied samples were identified according to the in vitro time of degradation and are presented in Table 1.

The FTIR spectrum of the synthesized PLGA sample can be seen in Figure 3. In Figure 4, it's shown the FTIR spectra of the synthesized PLGA sample and of all the copolymer samples removed from the phosphate buffer solution at the pre-determined time intervals, during the in vitro degradation study.

In the synthesized PLGA sample spectrum, it is possible to observe a thick band in the range between 1760 cm^{-1} and 1750 cm^{-1}, a characteristic range of carbonyl (C=O), present in the two monomers. Also, a grouping band (C-O), between 1300 cm^{-1} and 1150 cm^{-1}, can be observed, characteristic of ester groups. The absorption bands that are characteristic of the functional groups present in the PLGA copolymer can be observed in the spectrum and are presented in Table 2. The FTIR spectra, shown in Figure 4, are quite similar, which indicates that there was no significant modification on the chemical groups of PLGA, due to degradation, during the investigated period.

Table 1. Identification of the studied samples according to the in vitro time of degradation.

Identification	Time of degradation (days)
PLGA	0
PLGA sample 1	6
PLGA sample 2	12
PLGA sample 3	19
PLGA sample 4	28
PLGA sample 5	40
PLGA sample 6	55

Table 2. Absorption bands concerning the PLGA and it attributions (adapted from [34]).

Absorption bands (cm^{-1})	Attributions
3000 - 2700	CH. CH_3 e CH_2 (asymmetric stretching)
1760 - 1750	C=O (stretch)
1500 - 1250	CH_3 e CH_2 (symmetric angular deformation)
1350 - 1150	CH_2 e CH (asymmetric angular deformation)
1300 - 1150	C-O (ester) (stretch)

In Figure 5, it's shown the TG curves for the synthesized PLGA sample and for all the copolymer samples during the degradation study in PBS.

In the TG curves, it is possible to observe that the values of T_{onset} (onset temperature of the thermal event, measured by the software), T_i (initial temperature of the thermal event) and of $T_{deg.max.}$ (temperature where the decomposition rate of the thermal event reaches its maximum level?this temperature was obtained by the TG curve) diminished as the time of degradation increased. This is most likely due to the fact that, as the PLGA copolymer begins to degrade over time, it loses thermal stability. The profiles of the PLGA decomposition curves, samples 5 and 6, correspond to the degradation times of 40 and 55 days, and proved to be quite similar, indicating that these samples are, apparently, at the same stage of degradation. Moreover, these two samples present a stage of degradation that is more pronounced than the others, given that their T_{onset}, T_i, and

$T_{deg.max.}$ temperatures are lower than those observed for the other samples. The values of these temperatures are summarized in Table 3.

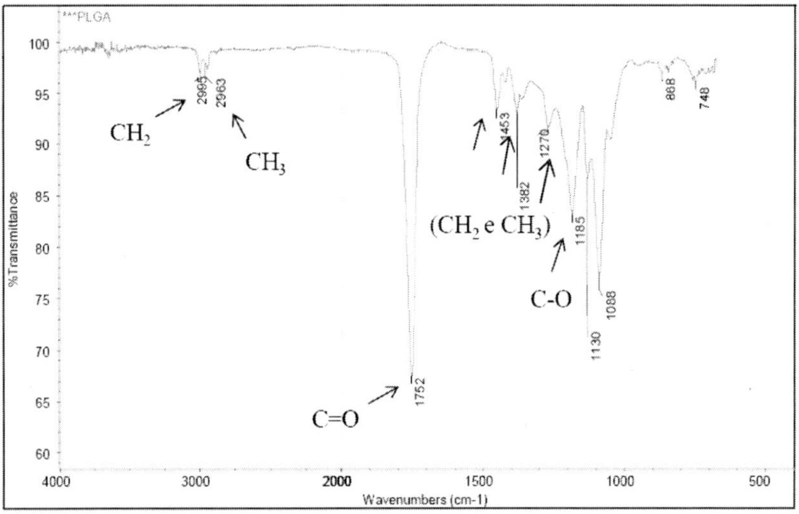

Figure 3. FTIR spectrum of the synthesized PLGA sample.

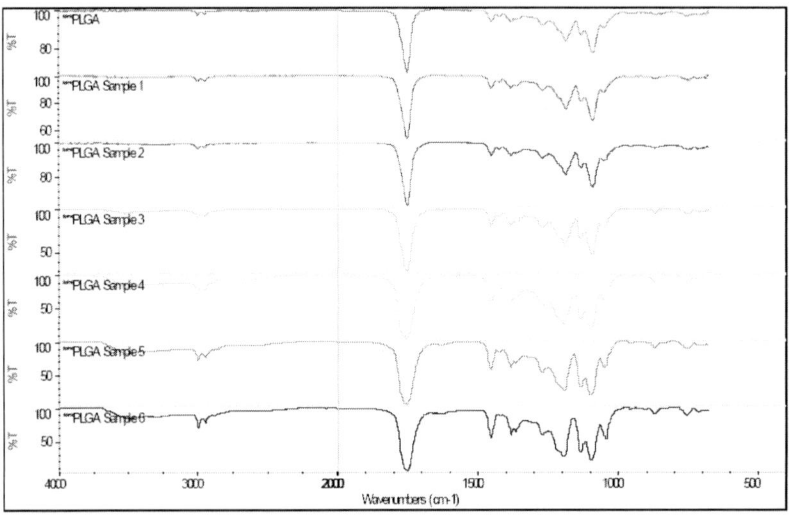

Figure 4. FTIR Spectra of synthesized PLGA and of PLGA samples degraded in vitro in PBS.

The obtained results from thermogravimetry, within the studied temperature interval, indicate that the PLGA samples decomposed almost completely and in a single stage.

The decomposition of the polymer chains occurs mainly between 180°C and 340°C. Samples 4, 5 and 6 presented a slight loss in mass when they reached 100°C, which may be related to the presence of moisture in these samples or, most probably, to the beginning of polymer degradation, since those are the samples which were kept longer at PBS (greater time of degradation). It is important to register that the PLGA samples studied are thermally stable until approximately 100°C, which means that, with respect to such aspect, they are adequate for being used in drug release systems.

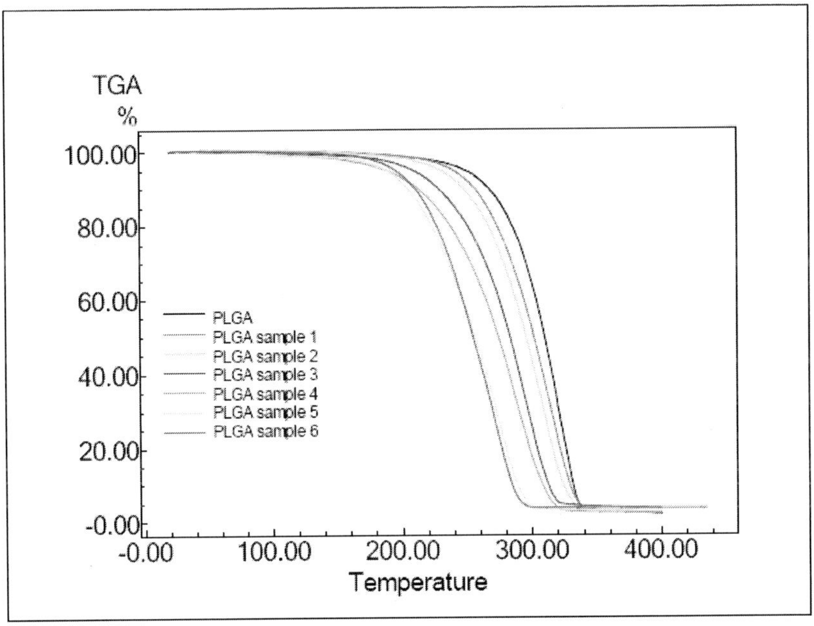

Figure 5. TG curves of synthesized PLGA and of PLGA samples degraded in vitro in PBS.

The DSC curves for both the PLGA and the PLGA samples tested in vitro in PBS are presented in Figure 6. The glass transition temperature values (T_gs), obtained through these curves, are shown in Table 4. The synthesized PLGA sample presented a glass transition temperature (T_g) of 45°C, which is in agreement with the values (between 40°C and 60°C) found in literature [29].

Table 3. Results of the TG analyses for the PLGA in vitro degradation study.

Samples	Time of degradation (days)	TG			
		T_{onset} (°C)	T_i (°C)	$T_{deg\,máx}$ (°C)	% mass loss
PLGA	0	270 ± 3.2	154 ± 1.8	309 ± 5.4	98 ± 0.4
PLGA sample 1	6	276 ± 7.7	159 ± 2.2	315 ± 6.8	98 ± 0.0
PLGA sample 2	12	260 ± 7.0	150 ± 4.5	298 ± 11.0	98 ± 0.0
PLGA sample 3	19	257 ± 8.0	132 ± 11.0	303 ± 12.0	98 ± 0.9
PLGA sample 4	28	257 ± 20.9	125 ± 5.2	302 ± 25.9	98 ± 1.2
PLGA sample 5	40	230 ± 3.5	124 ± 18.0	269 ± 7.0	98 ± 1.5
PLGA sample 6	55	232 ± 6.9	129 ± 3.5	266 ± 4.5	97 ± 0.5

During the in vitro degradation process, there was a significant decrease of the glass transition temperature, which changed from 45°C, at time 0, to 17°C, after 28 days of the degradation process. From 28 to 55 days, the T_g stabilized at around 18.5°C ± 1.5°C. The slow start of the degradation process, observed through the relatively small decrease in T_g during the first 19 days, may well be associated with the plasticizer effect caused by the absorption of H_2O by the PLGA rods [35] . The increase in the copolymer degradation rate, shown by the sharp fall in T_g, may be due to the reduction in the pH of the medium where the PLGA undergoes the degradation process (Table 4). As of 28 days of degradation, the pH level of the medium, where the PLGA samples can be found, undergo an abrupt reduction, going from 6.9 to approximately 2.4. With the PLGA degradation, the tendency of the medium's pH level is to diminish due to the release of lactic and glycolic acid. During this reduction in pH, the T_g of the PLGA samples undergoes a pronounced decrease, thus demonstrating the increase in their degradation.

Through the strain sweep test, the linear viscoelasticity range could be determined as being between 0.2% and 10% of the strain. The strain value chosen for the frequency sweep test was 5%. The obtained result by the frequency sweep rheological test, for the synthesized PLGA sample, i.e., before the in vitro degradation study, is presented in Figure 7. The curves generated by the software show G' (elastic module, in red) and G" (viscous module, in blue) as a function of angular frequency.

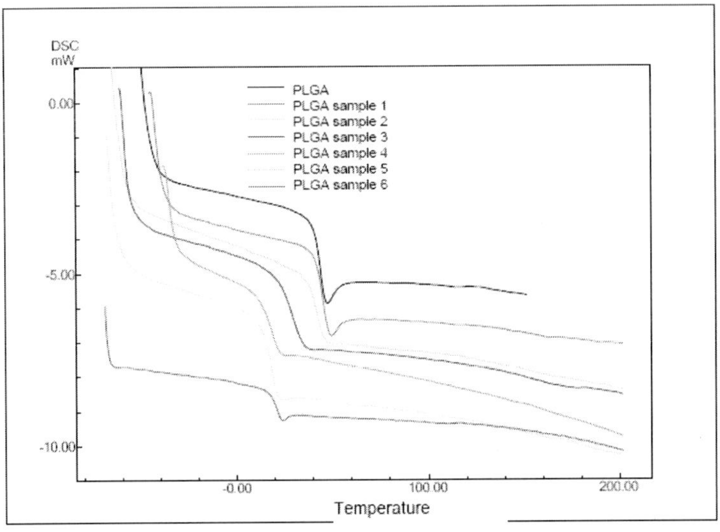

Figure 6. DSC curves of synthesized PLGA and of PLGA samples degraded in vitro in PBS.

Table 4. Results of T_gs and pHs for the PLGA in vitro degradation study.

Samples	Time of degradation (days)	T_g (°C)	pH
PLGA	0	45 ± 0.9	-
PLGA sample 1	6	44 ± 0.5	7.4 ± 0.12
PLGA sample 2	12	42 ± 0.8	7.4 ± 0
PLGA sample 3	19	31 ± 1.2	7.4 ± 0
PLGA sample 4	28	17 ± 0.5	6.9 ± 0
PLGA sample 5	40	18 ± 0.5	2.5 ± 0.05
PLGA sample 6	55	20 ± 0.5	2.3 ± 0

Through the curves shown Figure 7, one can observe the predominance of the viscous behavior over the elastic one (G" > G') up to the angular frequency of approximately 10 rad·s^{-1}, with the inversion of these behaviors occurring when this frequency is reached. The curves of the elastic and viscous moduli, obtained for the PLGA samples that underwent in vitro degradation in PBS, are similar to those presented in Figure 7. The average molar mass of each sample was estimated based on the results from the frequency sweep test and in models that describe the relation of rheological data and the relaxation times with molar mass distribution [36] -[38] . The results of the average molar mass of the PLGA samples, in the in vitro degradation study, obtained indirectly through rheometry, are shown in Figure 8. Table 5 shows the average molar mass obtained by the frequency sweep tests for the PLGA samples that had been degraded in vitro in PBS.

Figure 8 shows that the synthesized PLGA copolymer, with an average molar mass in the order of 10^6 g·mol^{-1}, during the in vitro degradation process in PBS, presented a reduction in its average molar mass to around 10^4 g·mol^{-1}, within 28 days (PLGA sample 4), which corresponds to a decrease of approximately 95%. That is, the average molar mass of the PLGA sample after 28 days in PBS is approximately 5% of the average molar mass of the synthesized PLGA sample. Before reaching this value, a sharp decrease in the PLGA's molar mass, in direct proportion with degradation time, could be observed, particularly between the 19th and the 28th day in PBS.

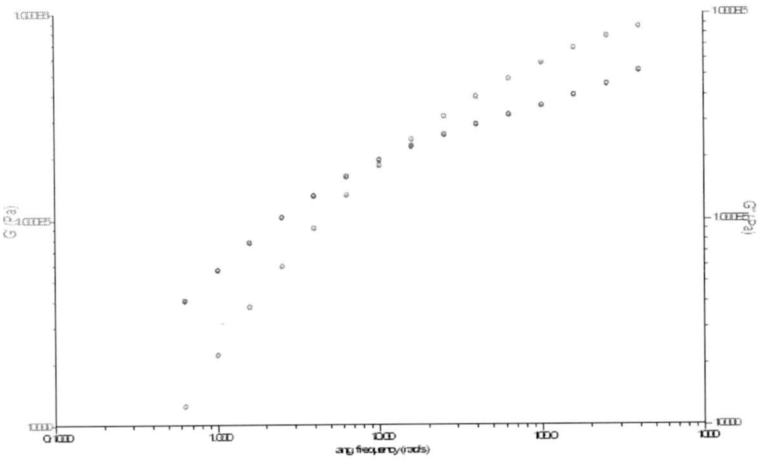

Figure 7. Curves for the elastic and viscous moduli as a function of angular frequency for synthesized PLGA sample.

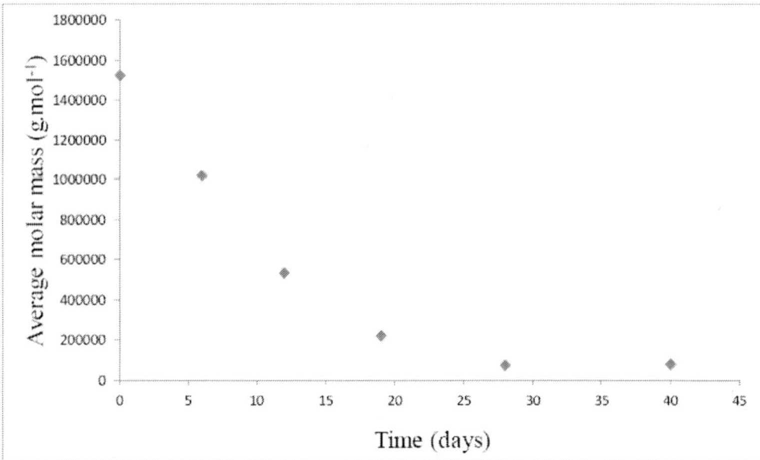

Figure 8. Average molar mass for the PLGA samples with respect to degradation time.

Due to the PLGA sample's stage of degradation after 55 days, it was impossible to perform a rheometric analysis to obtain the average molar mass of this sample, given that the sample size was insufficient to perform an effective rheometric test.

In Figures 9(a), it can be seen the micrograph obtained for the synthesized PLGA sample. In Figure 9(b), Figure 9(c), and Figure 9(d), the micrographs for the PLGA samples 1, 2 and 3, respectively, are shown. The morphology of the surface of the PLGA rods was followed up during the degradation process until the 19th day. In the subsequent samples, the SEM analysis of the sample surfaces became unfeasible, due to the samples' stage of degradation. These samples, upon being inserted into the equipment, had their degradation processes accelerated due to the presence of the electron beam, rendering it impossible to obtain the micrographs of these samples.

The micrograph from Figure 9(a) shows the PLGA rod's surface before beginning the degradation process. It can be noted that it presents integrity and a non-porous appearance. After 6 days of degradation (Figure 9(b)), it could be observed that the aspect of the sample continued the same (integral and non-porous), with no visible beginning of degradation. After 12 days of immersion in PBS, the micrograph of the PLGA rod (Figure 9(c)) showed some small cracks on the rod's surface, most likely due to the beginning of PLGA degradation. This phenomenon becomes even clearer and more explicit when one looks at the PLGA sample micrograph, after 19 days of degradation (Figure 9(d)). It can be observed a surface with many fractures and cracks, indicating a more pronounced intensity of degradation. This behavior corroborates the one observed for the glass transition temperature, shown in Table 4, where one can observe the occurrence of the first significant reduction in T_g value, from 42°C to 31°C, within the interval of 12 and 19 days in PBS.

Through the results presented in this study, one can clearly see a sharp decrease in the values of T_g, molar mass, and pH of the medium from the 19th to the 28th day in PBS. This decrease, observed in these properties, may be due to the presence of the PLGA chains with a reduced molar mass, stemming from the in vitro degradation process. These chains can bring about a plasticizer effect within the samples, thus justifying the reported decrease. In this study, it was also observed that, when handling the samples, they become more and more brittle and fragile when reaching this period of in vitro degradation in PBS.

Table 5. Results of the rheometric analysis of the study of PLGA degradation in vitro.

Samples	Time of degradation (days)	Average molar mass ($g \cdot mol^{-1}$)
PLGA	0	1,526,000 ± 28859.8
PLGA sample 1	6	1,021,000 ± 193277.5
PLGA sample 2	12	532,400 ± 36184.3
PLGA sample 3	19	221,200 ± 41135.0
PLGA sample 4	28	74,640 ± 4528.3
PLGA sample 5	40	76,190 ± 3685.4
PLGA sample 6	55	-

4. CONCLUSION

The synthesized PLGA copolymer (82/12) in this work had a yield of nearly 90%, presented an average molar mass in the order of 10^6 $g \cdot mol^{-1}$, and showed a glass transition temperature of 45°C. The study of the PLGA copolymer in vitro degradation in PBS indicated that, after 28 days of immersion in this solution, the degradation became much more pronounced. This behavior was observed due to the reduction in the glass transition temperature values, from 45°C to 17°C; of the average molar mass, from 1.5×10^6 $g \cdot mol^{-1}$ to 7.5×10^4 $g \cdot mol^{-1}$; and of the pH of the medium, from 7.4 to 2.5. The degradation rate of PLGA significantly increased from the 19th to the 28th day in PBS. The pH of the medium has a significant influence on the copolymer degradation profile. When it diminishes, it accelerates the degradation process, resulting in smaller

PLGA polymer chains. This pH dependent degradation can be useful for drug release systems.

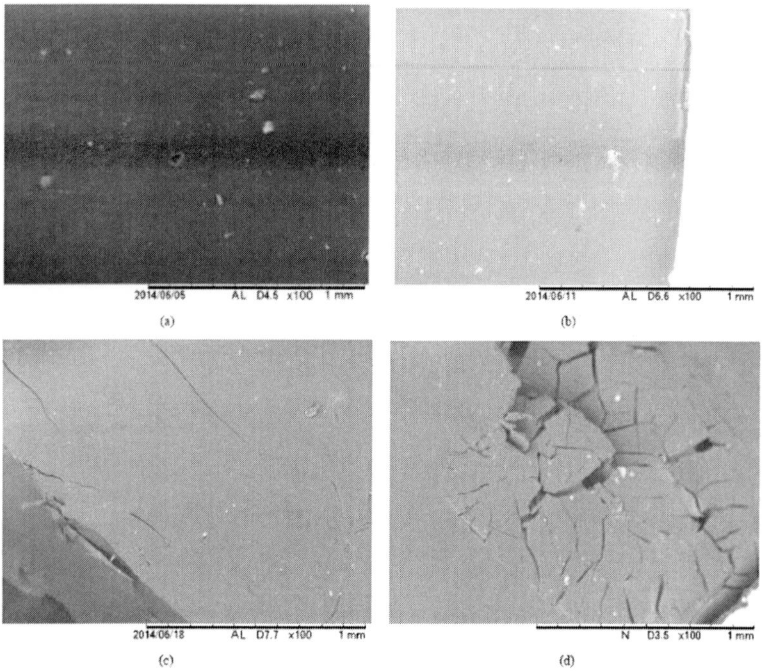

Figure 9. Micrographs of the: (a) PLGA rods before the in vitro degradation process; (b) PLGA rods, sample 1; (c) PLGA rods, sample 2; and (d) PLGA rods, sample 3.

ACKNOWLEDGEMENTS

Anamaria Teodora Coêlho Rios Silva, Barbara Camilla Oliveira Cardoso, Maria Elisa Scarpelli Ribeiro e Silva, Roberto Fernando Souza Freitas, Ricardo Geraldo Sousa The authors acknowledge the financial support of FAPEMIG, CEMIG, PETROBRAS, CAPES and CNPq.

REFERENCES

1. Ratner, B.D. and Hoffman, A.S. (1996) Biomaterials Science—An Introduction to Materials in Medicine. Elsevier, New York.
2. Langer, R. and Peppas, N.A. (2003) Advances in Biomaterials, Drug Delivery, and Bionanotechnology. Alche Journal, 49, 2990-3006.
3. Stevanović, M., Maksin, T., Petković, J, Filipić, M. and Uskoković, D. (2009) An Innovative, Quick and Convenient Labeling Method for the Investigation of Pharmacological Behavior and the Metabolism of Poly(DL-lactide-coglyco- lide) Nanospheres. Nanotechnology, 20, 1-12.
4. Hoffman, A.S. (2008) The Origins and Evolution of "Controlled" Drug Delivery Systems. Journal of Controlled Release, 132, 153-163.
5. Park, J.H., Lee, S., Kim, J.H., Park, K., Kim, K. and Kwon, I.C. (2008) Polymeric Nanomedicine for Cancer Therapy. Progress in Polymer Science, 33, 113-137.
6. Pamula, E. and Menaszak, E. (2008) In Vitro and in Vivo Degradation of Poly(L-lactide-co-glycolide) Films and Scaffolds. Journal of Materials Science: Materials in Medicine, 19, 2063-2070.
7. Duarte, M.A.T., Duek, E.A.R. and Motta, A.C. (2014) In Vitro Degradation of Poly (L-co-D,L lactic acid) Containing PCL-T. Polímeros, 24, 1-8.
8. Xiaoling, L. and Haskara, R.J. (2006) Design of Controlled Release Drug Delivery Systems. MacGraw-Hill, New York.
9. Stevanović, M. and Skoković, D. (2009) Poly(lactide-co-glycolide)-Basedmicro and Nanoparticles for the Controlled Drug Delivery of Vitamins. Current Nanoscience, 5, 1-14.
10. Stevanović, M., Radulović, A., Jordović, B. and Uskoković, D. (2008) Poly(DL-lactide-co-glycolide) Nanospheres for the Sustained Release of Folic Acid. Journal of Biomedical Nanotechnology, 4, 349-358.
11. Liang, R.C., Li, X., Shi, Y., Wang, A., Sun, K., Liu, W.H. and Li, Y.X. (2013) Effect of Water on Exenatide Acylation in Poly(lactide-co-glycolide) Microspheres. International Journal of Pharmaceutics, 454, 344-353.
12. Meng, Z.X., Zheng, W., Li, L. and Zheng, Y.F. (2011) Fabrication, Characterization and in Vitro Drug Release Behavior of Electrospun PLGA/Chitosan Nanofibrous Scaffold. Materials Chemistry and Physics, 125, 606-611.
13. Rios, M. (2005) Polymers for Controlled Release: Formulation Follows Function. Pharmaceutical Technology, 29, 42-50.
14. Mundargi, R.C., Babu, V.R., Rangaswamy, V., Patel, P. and Aminabhavi, T.M. (2008) Nano/Micro Technologies for Delivering Macromolecular Therapeutics Using Poly(D,L-lactide-co-glycolide) and Its Derivatives. Journal of Controlled Release, 125, 193-209.
15. Cohen-sela, E., Chorny, M., Koroukhov, N., Danenberg, H.D. and Golomb, G. (2009) A New Double Emulsion Solvent Diffusion Technique for Encapsulating Hydrophilic Molecules in PLGA Nanoparticles. Journal of Controlled Release, 133, 90-95.
16. Bae, S.E., Son, J.S., Park, K. and Han, D.K. (2009) Fabrication of Covered Porous PLGA Microspheres Using Hydrogen Peroxide for Controlled Drug Delivery and Regenerative Medicine. Journal of Controlled Release, 133, 37-43.
17. Jain, R.A. (2000) The Manufacturing Techniques of Various Drug Loaded Biodegradable Poly(lactídeo-co-glicolídeo) (PLGA) Devices. Biomaterials, 21, 2475-2490.
18. Soares, A.Q., Oliveira, L.F., Rabelo, D. and Souza, A.R. (2005) Polímeros Biodegradáveis: Novas Perspectivas para as Ciências Farmacêuticas. Revista Eletrônica de Farmácia, 2, 202-205.
19. Ji, W., Yang, F., Seyednejad, H., Chen, Z., Hennink, W.E., Anderson, J.M., Beucken, J.J.J.P. and Jansen, J. (2012) Biocompatibility and Degradation Characteristics of PLGA-Based Electrospun Nanofibrous Scaffolds with Nanoapatite Incorporation. Biomaterials, 33, 6604-6614.

20. Kang, S.W., Yang, H.S., Seo, S.W., Han, D.K. and Kim, B.S. (2008) Apatite-Coated Poly(lactic-coglycolic Acid) Microspheres as an Injectable Scaffold for Bone Tissue Engineering. Journal of Biomedical Materials Research Part A, 85, 747-756.
21. Qi, F., Wu, J., Yang, T., Ma, G. and Su, Z.G. (2014) Mechanistic Studies for Monodisperse Exenatide-Loaded PLGA Microspheres Prepared by Different Methods Based on SPG Membrane Emulsification. Acta Biomaterialia, 10, 4247- 4256.
22. Sun, L., Xie, Z., Zhao, Y., Wei, H.M. and Gu, Z.Z. (2013) Optical Monitoring the Degradation of PLGA Inverse Opal Film. Chinese Chemical Letters, 24, 9-12.
23. Khare, V., Kour, S., Alam, N., Dubey, R.D., Saneja, A., Koul, M., Gupta, A.P., Singh, D., Singh, S., Saxena, A.K. and Gupta, P.N. (2014) Synthesis, Characterization and Mechanistic-Insight into the Antiproliferative Potential of PLGA- Gemcitabine Conjugate. International Journal of Pharmaceutics, 470, 51-62.
24. Athanasiou, K.A., Niederauer, G.G. and Agrawal, C.M. (1996) Sterilization, Toxicity, Biocompatibility and Clinical Applications of Polylactic Acid/Polyglycolic Acid Copolymers. Biomaterials, 17, 93-102.
25. Lanao, R.P.F., Leeuwenburgh, S.C.G., Wolke, J.G.C. and Jansen, J.A. (2011) Bone Response to Fast-Degrading, Injectable Calcium Phosphate Cements Containing PLGA Microparticles. Biomaterials, 32, 8839-8847.
26. Lanao, R.P.F., Leeuwenburgh, S.C.G., Wolke, J.G.C. and Jansen, J.A. (2011) In Vitro Degradation Rate of Apatitic Calcium Phosphate Cement with Incorporated PLGA Microspheres. Acta Biomaterialia, 7, 3459-3468.
27. Li, H. and Chang, J. (2005) pH-Compensation Effect of Bioactive Inorganic Fillers on the Degradation of PLGA. Composites Science and Technology, 65, 2226-2232.
28. Merkli, A., Tabatabay, C., Gurny, R. and Heller, J. (1998) Biodegradable Polymers for the Controlled Release of Ocular Drug. Progress in Polymer Science, 23, 563-580.
29. Astete, C.E. and Sabliov, C.M. (2006) Synthesis and Characterization of PLGA Nanoparticles. Journal of Biomaterials Science, Polymer Edition, 17, 247-289.
30. Houchin, M.L. and Topp, E.M. (2008) Chemical Degradation of Peptides and Proteins in PLGA: A Review of Reactions and Mechanisms. Journal of Pharmaceutical Sciences, 97, 2395-2404.
31. Tracy, M.A., Ward, K.L., Firouzabadian, L., Wang, Y., Dong, N., Qian, R. and Zhang, Y. (1999) Factors Affecting the Degradation Rate of Poly(lactide-co-glycolide) Microspheres in Vivo and in Vitro. Biomaterials, 20, 1057-1062.
32. Tsukadaa, Y., Haraa, K., Bandoa, Y., Huangb, C.C., Kousakaa, Y., Kawashimac, Y., Morishitad, R. and Tsujimotoa, H. (2009) Particle Size Control of Poly(DL-lactide-co-glycolide) Nanospheres for Sterile Applications. International Journal of Pharmaceutics, 370, 196-201.
33. Xiao, D., Liu, Q., Wang, D., Xie, T., Guo, T., Duan, K. and Weng, J. (2014) Room-Temperature Attachment of PLGA Microspheres to Titanium Surfaces for Implant-Based Drug Release. Applied Surface Science, 309, 112-118.
34. Erbetta, C.D.C., Alves, R.J., Resende, J.M., Freitas, R.F.S. and Sousa, R.G. (2012) Synthesis and Characterization of Poly(D,L-lactide-co-glycolide) Copolymer. Journal of Biomaterials and Nanobiotechnology, 3, 208-225.
35. kasperczyk, J. (1996) Microstructural Analysis of Poly(l-lactide)-co-(glycolide) by 1H and13 C n.m.r. Spectroscopy. Polymer, 37, 201-203.
36. Maier, D., Eckstein, A., Friedrich, C. and Honerkamp, J. (1998) Evaluation of Models Combining Rheological Data with the Molecular Weight Distribution. Journal of Rheology, 42, 1153-1173.
37. Thimm, W.B., Friedrich, C., Marth, M. and Honerkamp, J. (1999) An Analytical Relationship between Relaxation Time Spectrum and Molecular Weight Distribution. Journal of Rheology, 43, 1663-1672.
38. Thimm, W.B., Friedrich, C., Marth, M. and Honerkamp, J. (1999) On the Rouse Spectrum and the Determination of the Molecular Weight Distribution from Rheological Data. Journal of Rheology, 44, 429-438.

CHAPTER 18

Synthesis and Characterization of Perfluorooctanoic Acid Anionic Surfactant Doped Nanosize Polyaniline

B. Mahalakshmi, C. Vedhi

Department of Chemistry, V.O. Chidambaram College, Thoothukudi, India

ABSTRACT

Surfactant doped polyaniline was synthesized in the aqueous solution of aniline and anionic surfactant of perfluorooctanoic acid (PFO) by chemical synthesis using potassium peroxy disulphate as an oxidant by varying the aniline to surfactant ratio. The solubility of the chemically prepared surfactant doped polyaniline (PANI) was ascertained and it showed good solubility in dimethyl sulfoxide (DMSO), dimethylformamide (DMF), acetone, acetonitrile, ethanol, aceticacid, trichloroethylene, dichloromethane, tetrahydrofuran, ethylacetate, diethylether, toluene, chloroform and sparingly soluble in n-hexane and water. The prepared polymers were characterized by fourier transform infrared spectroscopy (FTIR), UV-visible, X-ray diffraction (XRD), cyclic voltammetric (CV), EIS and scanning electron microscopy (SEM). The analysis of UV-visible and FTIR showed that aniline has been polymerized to PANI in its conducting emeraldine form. FTIR spectra showed that the peaks at 1670, 3315 and 1400 cm^{-1} corresponded to PFO. FTIR spectra showed that amine peak observed at 1593 cm^{-1} was shifted to lower wave number due to the interaction between PANI and the surfactant. SEM analysis showed that the variation in morphology of doped PANI was predominantly dependent on the concentration of the surfactant. Elemental analysis was done by energy dispersive spectroscopic (EDAX) which shows the presence of C, N, O, S and F. XRD pattern showed that the formation of nanosized (18 nm) and crystalline polymer. CV studies of the synthesized polymer exhibited good adherent behavior on electrode surface. It exhibited three oxidation peaks at approximately 0.283 V, 0.541 V and 0.989 V and two reduction peaks at 0.1421 and 0.3854 V. Electrical conductivity of PFO doped PANI was studied by impedance spectroscopic method.

Keywords: Polyaniline, Perfluorooctanoic Acid, Cyclic Voltammetry, Morphology, Electrical Conductivity

1. INTRODUCTION

Conducting polymers have become a popular basic material for advanced applications such as rechargeable batteries [1] , optical devices [2] , membranes [3] , sensors and biosensors [4] [5] , electromagnetic interference (EMI) shielding [6] , antistatic coatings [7] , etc. Among the conducting polymers family, PANI has attracted considerable attention because of its unique electrical, optical and electro-optical properties and its numerous potential applications [8] . One of the key problems related to the potential applications of polyaniline is its processability. Processability is an important requirement in conducting polymers for their possible commercial use. Since most of the conducting polymers are not processable, much of the efforts made in this field have been directed towards circumventing this problem. To solve this problem, various approaches have been tried, including addition of side groups to the polymer backbone [9] , grafting of polymers to a non-conducting polymer [10] , direct polymerization of intractable polymers into the final desired shape, making a composite of blend of conducting polymers [11] [12] and copolymerization [13] . Better processability may be achieved either by the synthesis on PANI in nano-micro scale particle, which is easier to disperse in a polymer matrix or by using an appropriate emulsifier which enhances the optical and electrical properties of PANI. Polymeric stabilizers (surfactant) affect the polymerization condition, kinetics and also the final properties of the polymer [14] [15] . Surfactants used as additives during the polymerization of aniline to effect the locus polymerization by using the emulsion [16] [17] or inverse emulsion [18] -[20] pathways, and thus to modify the molecular and supermolecular structure of the resulting PANI, and to improve the properties of the PANI with respect to conductivity, stability, solubility in organic solvents and processability. Surfactants have important implications in wetting, formation of foams, etc., in addition to aggregation of surfactants at interfaces [21] . Polymeric nanostructures are formed on surfaces due to combination of interfacial, intra- and inter-molecular forces [22] . Surfactants affect the preparation of PANI in three ways: 1) the presence of surfactant micellar and aqueous phases, thus altering the locus and the course of polymerization [18] [23] , 2) anionic surfactants may acting as counter ions for conducting-polymer polycations, and 3) the hydrophobic part of the surfactant molecules may adsorbing on the produced conducting polymers, a surfactant thus becoming a part of the resulting material.

Suspension polymerization of aniline in the presence of dodecylbenzenesulfonic acid (DBSA) with styrene-butadiene-styrene (SBS) and without SBS was carried out and results indicate that DBSA acts simultaneously as a surfactant (emulsifier) and as a dopant [24] [25] . Particle size and conductivity can be decreased by increasing the concentration of stabilizer [26] [27] . These are related to the mass of insulating stabilizer adsorbed. The size and type of the dopant (anion) affect the morphology, size and electrical conductivity of resulting polymers [28] [29] . The type and concentration of oxidant (dopant), type of protonic acids and additives affect the yield and electrical conductivity of polyaniline prepared chemically [30] . The polarity of the counter ion plays an important role in the conductivity as well as in the chemical properties. Conductivity increases which small counter ions are used [31] . The surfactants influence the physical properties (morphology, solubility) of the resultant polymer [32] . It has been reported that nanotubes of hydrophobic interiors were prepared by disassembling anionic micelles in the inner walls of positive charge bearing conjugated polymer [33] . It was also studied that supramolecular complexes involving nanoscopic amphiphilic assemblies (AAs) and polyelectrolytes have been used to prepare a variety of materials, wherein the dynamic AAs retain the structural features [34] . In this study, PANI-PFO was prepared in the aqueous solution by chemical oxidation method using KPS as an oxidant in the presence of various surfactant ratios. The effect of the anionic surfactant i.e., PFO on the chemical polymerization of aniline (An), at different surfactant ratios and potential windows is investigated. The doping of PFO changes the surface morphology of PANI. It is also found that electrical conductivity of PANI-PFO was better than that of PANI and governed by changing the concentration of surfactant during the polymerization of aniline.

2. EXPERIMENTAL SECTION

2.1. Materials

Perfluorooctanoic acid (PFO, surfactant) and potassium peroxydisulfate (KPS, oxidant) were purchased from merck AR grade and used as such without further purification. Aniline was purified by distillation in vacuum before use.

2.2. Synthesis of Polymer

Aqueous micellar dispersion was prepared by chemical oxidation method. PFO doped PANI were synthesized by keeping the concentration of the monomer (0.1 M) and oxidant (0.1 M) constant and varying the molar ratio of PFO (0.1 M to 0.1×10^{-5} M). Table 1 describes the synthetic conditions of the respective samples. Bulk polymerization was carried out by mixing various molar ratio of surfactant into 2.3 ml of 0.1 M aniline and made up to 250 ml using conductivity water. Polymerization was performed by the addition of potassium peroxydisulfate (0.1 M) in the micellar solution and stirred for 5 hrs. The colour of polymerization solution changes with polymerization step. In the initial stage, aniline/PFO/APS solution was colorless with high transparency. As the polymerization proceeded, the color of the solution changes to white through yellow, brown and finally to green, which indicates the formation of PANI emeraldine salt (ES). The precipitated particles were collected after filtering and washing with distilled water and methanol several times to remove impurities such as KPS, free PFO and unreacted aniline. These particles were then dried at room temperature.

2.3. Characterization

FT-IR spectra (Model: SHIMADZU) of the polyaniline samples were recorded in the frequency range of 400 to 4000 cm^{-1}. UV-Vis spectra of the samples in DMSO were taken using JASCO-V530 dual beam spectrophotometer in the wavelength region 200 to 1100 nm with a scanning speed of 400 nm/min. X-ray diffraction patterns of the polymers were obtained by employing XPERT-PRO diffractometer using CuK_α (k_α = 1.54060) radiation. The diffractometer was operated at 40 Kv and 30 mA. Powder X-ray diffraction pattern was recorded. The morphological study of the polymers were carried out using Scanning electron microscope (SEM Model: JEOL JSM 6360) operating at 25 kV. The electrochemical workstation (Model 650C), CH-Instrument Inc., TX, USA was employed for performing cyclic voltammmetry and impedance studies.

3. RESULTS AND DISCUSSION

3.1. UV-Vis Spectra

Figure 1 shows the UV-visible spectra of PFO doped and undoped PANI. The band observed at 278 nm corresponds to π-π^* transition. The absorption band at 373 nm was attributed to the π-π^*transition within benzoid segment (excitation of the nitrogen atom in benzoid segment) and at 452 nm was related to the polaron band-π^* transition of PANI (protonation in PANI backbone, polaron/bipolaron transition that occurred in doped PANI). A broad absorption band at 816 nm is assigned to the π-localised polaron bands of doped PANI in its emaraldine salt form (conductive emeraldine salt state of PANI) [35] -[38] . The band at around 373 nm is considerably blue shifted which indicates the interaction between PANI and the surfactant. The intensity of absorption band increases with the increasing PFO content. The shape of UV-Vis spectrum around 373 nm is changed from PANI 1 to PANI5 (Figures 1(a)-(e)). This suggests the change of doping characteristics.

3.2. FTIR Spectra

Figure 2 shows the infrared spectra of PFO doped and undoped PANI. The infrared peak at 1587 and 1496 cm^{-1} were assigned to the non-symmetric vibration mode of C=C in quinoid and benzenoid ring system in polyaniline respectively. The peaks at 1297 and 1328 cm^{-1} were corresponds to the C-N stretching vibration mode in benzenoid and quinoid ring system of polyaniline assigned to conducted protonated form [36] [37] . The absorption band at 1116 cm^{-1} was assigned to in-plane vibration of C-H bending vibration mode in N=Q=N, Q-N + H-B or B-N + H-B (where Q = quinoid and B = benzenoid). This absorption band should occur during the polymerization i.e., polar structure of the conducted protonated form. The absorption band at 882 and 816 cm^{-1} were attributed to the aromatic ring and out of plane C-H deformation vibrations for 1,4-disubstituted aromatic ring system [39] . The band at 3195 $cm^{-1}$was assigned to C-H stretching vibrations of aromatic ring. The band at 1670, 3315 and 1400 cm^{-1}were corresponds to PFO. The relative intensity of these bands increases with increasing the surfactant content in the polymer. The amine vibration peak at 1593 cm^{-1} shifted to 1587 cm^{-1}due to the interaction between PANI and the surfactant. Hence FTIR shows surfactant well doped with the polymer.

Table 1. Synthetic conditions of samples, molar concentration of aniline, PFO and KPS.

Sample	Aniline (M)	KPS (M)	PFO (M)
PANI	0.1	0.1	
PANI 1	0.1	0.1	1×10^{-1}
PANI 2	0.1	0.1	1×10^{-2}
PANI 3	0.1	0.1	1×10^{-3}
PANI 4	0.1	0.1	1×10^{-4}
PANI 5	0.1	0.1	1×10^{-5}

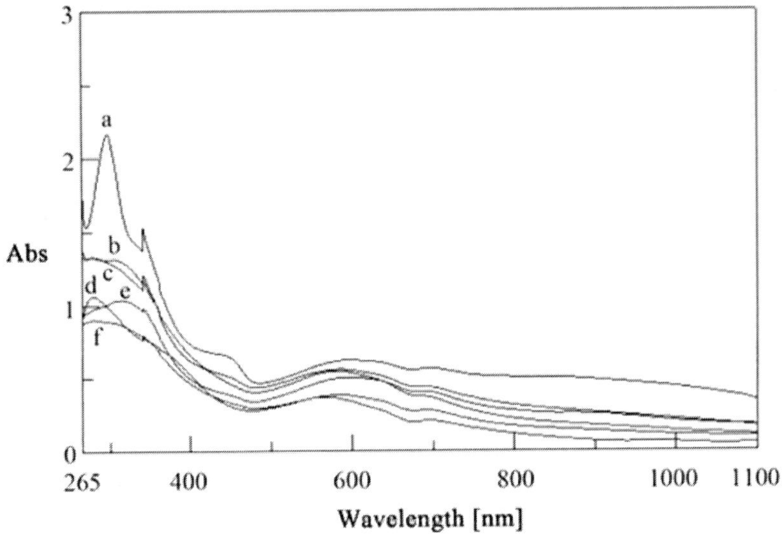

Figure 1. UV-spectra of (a) PANI 1; (b) PANI 2; (c) PANI 3; (d) PANI 4; (e) PANI 5; (f) PANI (undoped).

3.3. XRD Analysis

Figure 3 shows the X-ray diffractograms of PFO doped and undoped PANI. XRD patterns provide information in relation to the nature and structure of the samples. XRD patterns of undoped sample of PANI shows the amorphous nature. XRD patterns of doped samples show the crystallinity. The conductivity of polymers depends on various parameters such as doping level, formation of polarons and bipolarons [40] , the semi-crystalline nature of polymers arises owing to the systematic alignment of polymer chain folding or by the formation of single or multiple helices, for part of their length [41] . The pure PANI exhibited a broad hump at position $2\theta = 21^\circ$ and 25° as shown in the Figure 3(a). The PFO doped PANI exhibited sharp peak at $2\theta = 20^\circ$ and 26°, this indicated the presence of a high crystallinity and condensed structure. The intensity of the peak increases with increasing the concentration of the dopant. The nanosizes of the particles were confirmed by using Debye- Scherrer's equation. The particle size varied gradually from 18 nm in the case of PANI 1 to 77 nm in the case of PANI 5 indicating that the concentration of the surfactant was highly influenced on the size of the particles. Particle size and conductivity can be decreased by increasing the concentration of stabilizer [26] [27]. The variation in diffraction intensity with dopant concentration exhibits with the interaction of PFO in PANI network.

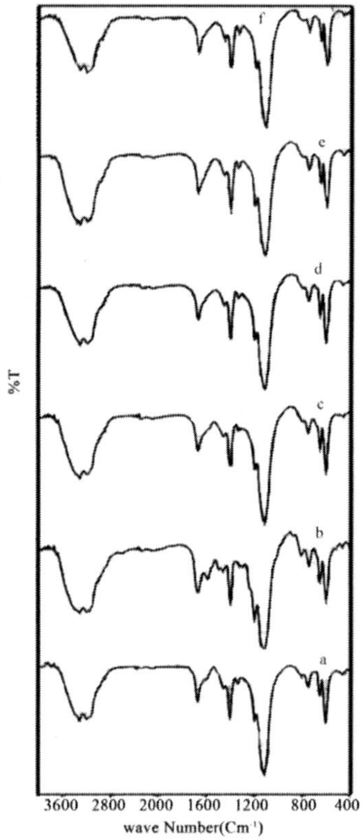

Figure 2. IR-spectra of (a) PANI (undoped); (b) PANI 1; (c) PANI 2; (d) PANI 3; (e) PANI4; (f) PANI 5.

3.4. SEM Analysis

The particle size and surface morphology were dependent on the type and concentration of surfactant, because the surfactant was adsorbed physically to the growing polymer [42] . Particle size decreases by increasing concentration of surfactant, because surfactant prevented from gross aggregation of particles. Figure 4 shows the surface morphology of doped and undoped PANI. However Figures 4(a)-(f) demonstrates the dependence of the size and homogeneity of the particles only from concentration of surfactant. Adsorption of the surface active agent on the PANI was primarily due to the hydrophobic component in the surfactant, probably via a hydrogen bonding mechanism with the aniline N-H group (graft copolymer) [27] . SEM photograph of the doped PANI show a gradual change from uneven flower like structure in PANI1 to a cage like morphology having more porous in PANI 5. The variation in morphology of the doped PANI was predominantly dependent on the concentration of the surfactant. Figure 5 Shows the EDS spectroscopy for the undoped and PFO doped PANI. It can be seen from a line that for the undoped PANI, only peaks corresponding to C, S, O and N elements were displayed. While for the PFO doped PANI, besides above peaks, peaks corresponding to C, O, S, N and F were also clearly observed, suggesting that F element originated from PFO was doped into the polymer.

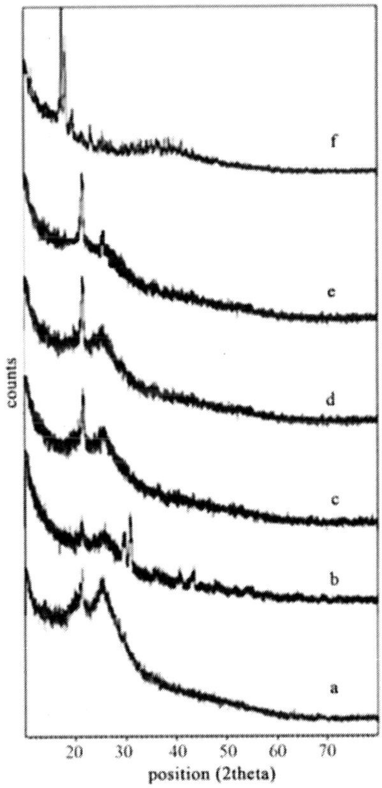

Figure 3. XRD scattering patterns of (a) PANI (undoped); (b) PANI 5; (c) PANI 4; (d) PANI 3; (e) PANI 2; (f) PANI 1.

3.5. Cyclic Voltammogram

Electrochemical behavior of undoped PANI and PFO doped PANI were shown in Figure 6. Cyclic voltammograms were recorded at room temperature by employing a three-electrode cell with platinum wire as an auxiliary electrode, as Ag/AgCl electrode as the reference electrode and coated glassy carbon (GC) as the working electrode. The CVs were obtained in 0.1 M H_2SO_4 by casting the polymer on GC working electrode and scanned between −0.6 to 1.2 V at scan rates between 25 and 500 mV/s. The redox couple peak occurred at approximately +283 mV was attributed to the transformation of PANI from the reduced leucoemeraldine (LE) state to the partially oxidized emeraldine state (ES). The redox couple peak at approximately +989 mV corresponds to transition of the PANI from LE to pernigraniline (PE) state, and was accompanied by the oxidation of aniline monomer [43] . The redox couple peak at approximately +541 mV, which was generally attributed to the redox reaction of p-benzoquinone [44] . The cathodic and anodic peak positions of PFO doped PANI shifted and increases the charge transfer with increasing the concentration of the surfactant were shown in the curve from (b-f). This increase in current was due to fast redox process at PANI-PFO matrix surface. The effect of varying scan rate ranges between 25 mV/s and 500mV/s was studied at the PFO doped PANI in 0.1 M H_2SO_4 as shown in Figure 7(a). With an increasing scan rate the peak current also increased thus indicating good adherence of the polymer onto the GC electrode surface, the peak current separation increased, and also the peak potential shifted slightly with the anodic peak to positive and the cathodic peak to negative potential directions. This is because the charging and discharging of the electro-active conducting polymer is rate determining. As the scan rate increases the peak current of PFO doped polymer increased linearly, it is indicating an adherent film on the glassy carbon electrode, this was further confirmed by a straight line graph obtained by plotting peak current vs. scan rate as presented in Figure 7(b).

3.6. Impedance Spectroscopy

Figure 8 shows the typical Nyquist plots of PFO doped and undoped PANI in 0.1 M H_2SO_4. Electrochemical impedance spectroscopy was employed to obtain equivalent circuit parameters such as the charge transfer resistance and ohmic resistance. The impedance plots show a distorted semi-circle in the high-frequency region due to porosity of electrode and a vertically linear spike in the low-frequency region. The diameter of the semi-circle gives an approximate value of the charge transfer resistance (Rct) of the PANI/electrolyte interface. The charge transfer resistances (Rct) for PANI (undoped), PANI 1, PANI 2, PANI 3, PANI 4 and PANI 5 were estimated to be around 1116, 843, 658, 437, 382 and 320 Ω respectively. The radii of the semicircle was found to increase with increase in surfactant concentration which indicates greater resistance to conduction with increase in PFO concentration which may result from more surfactant molecules bound to the electrode surface blocking the electron transfer. Thus PANI 5 with minimum PFO doping gives maximum conductivity.

Figure 4. SEM images of (a) PANI (undoped); (b) PI 3; (c) PANI 2; (d) PANI 5; (e) PANI 4; (f) PANANI 1.

4. CONCLUSION

Nano-sized PANI-PFO particles were prepared by chemical oxidation method with anionic surfactant. It had been found that the different concentration of surfactant had a considerable effect on the particle size and morphology of the resulting product. Better results were obtained for smaller concentration (PANI 5) of PFO. UV-Vis and FT-IR spectroscopy confirmed the interaction of PFO in PANI matrices. The SEM images show that the incorporation of PFO in PANI matrices changes the surface morphology of PANI. The chemical composition of polymer was further confirmed by EDAX analysis. XRD results show that the high concentration of surfactant exhibits the particle size of 18 nm and at low concentration of surfactant it exhibits the particle size of 77 nm. PFO in PANI had enhanced charge transfer and current voltage characteristics at electrode interface which was studied by cyclic voltammogram. Increases in oxidation peak current and enhanced broadening were observed due to fast redox process in PANI-PFO matrix. It shows a linear relationship between anodic peak currents versus scan rate and it indicates that the polymer has an appropriate electro activity. Electrical conductivity can be decreased by increasing the concentration of PFO. Minimum PFO doped PANI shows maximum conductivity, and it is confirmed by impedance spectroscopic method. The change in surface morphology and electrical conductivity of PANI-PFO with different concentrations of PFO show the potential for sensing application.

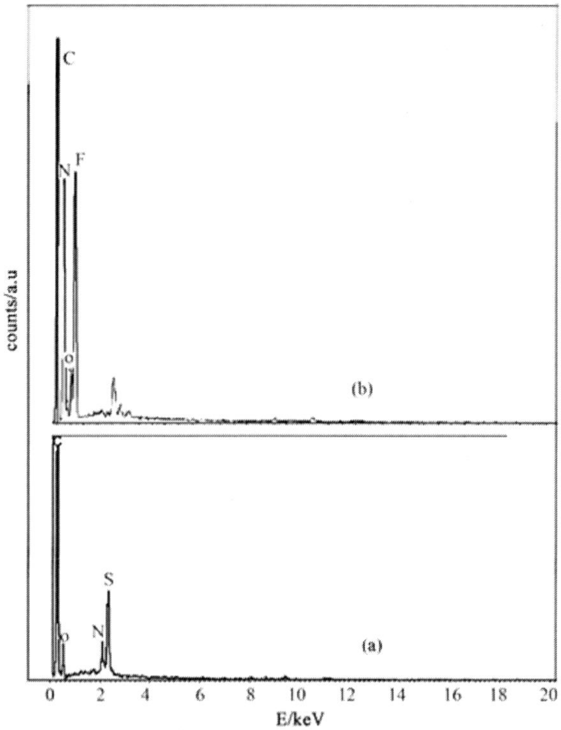

Figure 5. EDAX images of (a) PANI (undoped); (b) PANI 1.

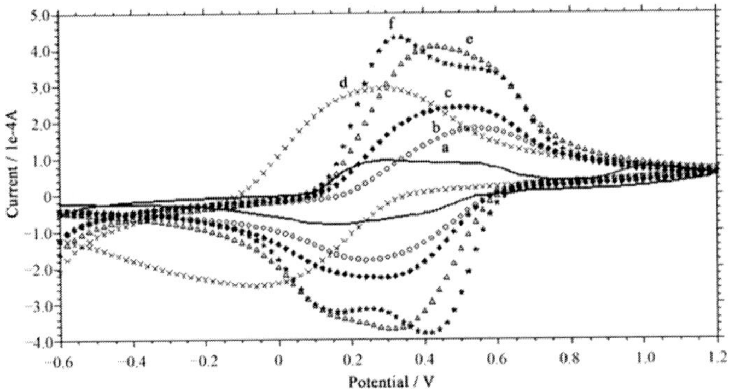

Figure 6. Cyclic voltammogram of (a) PANI (undoped); (b) PANI 5; (c) PANI 4; (d) PANI 3; (e) PANI 2; (f) PANI 1.

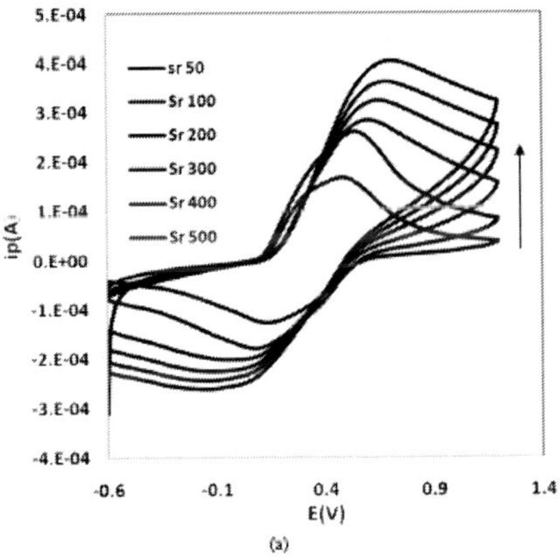

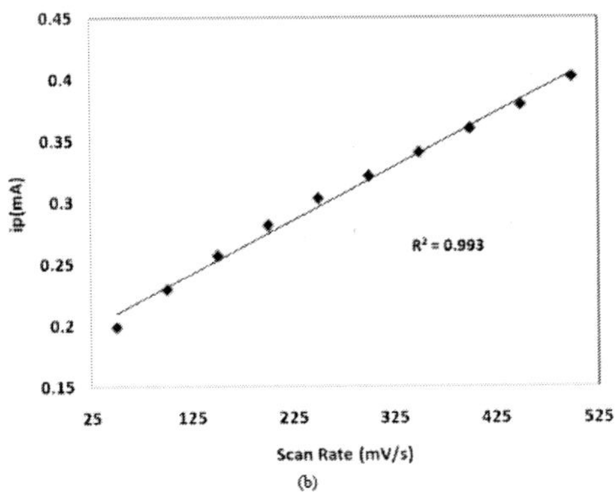

Figure 7. (a) CV taken at various scan rates for POT doped PANI; (b) A plot of anodic current against scan rate for POT doped PANI.

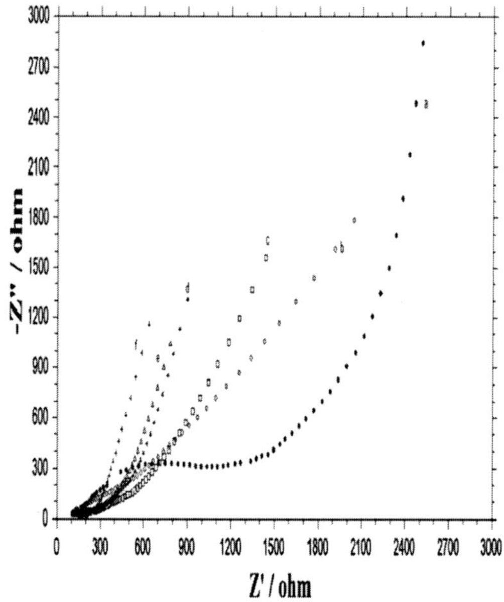

Figure 8. Nyquist plots for (a) PANI (undoped); (b) PANI 1; (c) PANI 2; (d) PANI 3; (e) PANI 4; (f) PANI 5.

ACKNOWLEDGEMENTS

B. Mahalakshmi, C. Vedhi (FAST TRACK and FIST) New Delhi, INDIA for using CHI Electrochemical workstation and Jasco UV-Vis Spectrophotometer.

REFERENCES

1. Scrosati, B. (1998) Conducting Polymers: Advanced Materials for New Design Rechargeable Lithium Batteries. Polymer International, 47, 50-55.
2. Falcao, E.H.L. and De Azevedo, W.M. (2002) Polyanilinepolyvinyl Alcohol Composite as an Optical Recording Material. Synthetic Metals, 128, 149-154.
3. Misoska, V., Ding, J., Davey, J.M., Price, W.E., Ralph, S.F. and Wallace, G.G. (2001) Polypyrrole Membranes Containing Chelating Ligands: Synthesis, Characterisation and Transport studies. Polymer, 42, 8571-8579.
4. Guernion, N., Ewen, R.J., Pihlainen, K., Ratcliffe, N.M. and Teare, G.C. (2002) The Fabrication and Characterisation of a Highly Sensitive Polypyrrole Sensor and Its Electrical Responses to Amines of Differing Basicity at High Humidities. Synthetic Metals, 126, 301-310.

5. Yamato, H., Koshiba, T., Ohwa, M., Wernet, W. and Matsumura, M. (1997) A New Method for Dispersing Palladium Microparticles in Conducting Polymer Films and Its Application to Biosensors. Synthetic Metals, 87, 231-236.
6. Dhawan, S.K., Singh, N. and Venkatachala, S. (2002) Shielding Behaviour of Conducting Polymer-Coated Fabrics in X-Band, W-Band and Radio Frequency Range. Synthetic Metals, 129, 261-267.
7. Ohtani, A., Abe, M., Ezoe, M., Doi, T., Miyata, T. and Miyke, A. (1993) Synthesis and Properties of High-Molecular- Weight Soluble Polyaniline and Its Application to the 4MB-Capacity Barium Ferrite Floppy Disks Antistatic Coating. Synthetic Metals, 57, 3696-3701.
8. Salanek, W.R., Lundstrom, I., Huang, W.S. and MacDiarmid, A.G. (1986) A Two-Dimensional Surfacestate Diagram for Polyaniline. Synthetic Metals, 13, 291-297.
9. Pandey, S.S., Annapoorni, S. and Malhotra, B.D. (1993) Synthesis and Characterization of Poly(aniline-co-o-anisidine). Macromolecules, 26, 3190-3193.
10. Andreatta, A., Heeger, A.J. and Smith, P. (1990) Electrically Conductive Poly Belend Fibers of Polyaniline and Poly- (p-phenylene Terephthalamide). Polymer Communications, 31, 275-295.
11. Nazzal, A.I. and Street, G.B. (1985) Pyrrole-Styrene Graft Copolymers. Journal of Chemical Society, Chemical Communications, 6, 375-376.
12. Aldissi, M. (1984) Polyacetylene Block Copolymers. Synthetic Metals, 13, 87-100.
13. Nalwa, H.S. (1990) Ferroelectric Polymers: Chemistry. Journal of Physics D: Applied Physics, 23, 745-748.
14. Armes, S.P. and Aldissi, M. (1989) Novel Colloidal Dispersons of Polyaniline. Journal of Chemical Society, Chemical Communications, 2, 88-89.
15. Kohut-Svelko, N., Reynaud, S. and Francois, J. (2005) Synthesis and Characterization of Polyaniline Prepared in the Presence of Nonionic Surfactants in an Aqueous Dispersion. Synthetic Metals, 150, 107-114.
16. Kinlen, P.J., Liu, J., Ding, Y., Graham, C.R. and Remsen, E.E. (1998) Emulsion Polymerization Process for Organically Soluble and Electrically Conducting Polyaniline. Macromolecules, 31, 1735-1744.
17. Palaniappan, S. (2002) Preparation of Polyaniline-Sulfate Salt by Emulsion and Aqueous-Polymerization Pathway without Using-Protonic Acid. Polymers for Advanced Technologies, 13, 54-59.
18. Xia, H. and Wang, Q. (2001) Synthesis and Characterization of Conductive Polyaniline Nanoparticles through Ultrasonic Assisted Inverse Microemulsion Polymerization. Journal of Nanoparticle Research, 3, 399-409.
19. Rao, P.S., Subrahmanya, S. and Sathyanarayana, D.N. (2002) Inverse Emulsion Polymerization: A New Route for the Synthesis of Conducting Polyaniline. Synthetic Metals, 128, 311-316.
20. Rao, P.S., Sathyanarayana, D.N. and Palaniappan, S. (2002) Polymerization of Aniline in an Organic Peroxide System by the Inverted Emulsion Process. Macromolecules, 35, 4988-4996.
21. Andrew, D.W.C., Rear, E.A.O. and Grady, B.P. (2003) Adsorbed Surfactants as Templates for the Synthesis of Morphologically Controlled Polyaniline and Polypyrrole Nanostructures on Flat Surfaces: From Spheres to Wires to Flat Films. Journal of the American Chemical Society, 125, 14793-14800.
22. Rabe, J.P. (1998) Self-Assembly of Single Macromolecules at Surfaces. Current Opinion in Colloid and Interface Science, 3, 27-31.
23. Ichinohe, D., Aria, T. and Kise, H. (1997) Synthesis of Soluble Polyaniline in Reversed Micellar Systems. Synthetic Metals, 84, 75-76.
24. Xei, H.Q., Ma, Y.M. and Guo, J.S. (1998) Conductive Polyaniline-SBS Composites from in Situ Emulsion Polymeri- zation. Polymer, 40, 261-265.
25. Osterholm, J.E., Cao, Y., Klavetter, F. and Smith, P. (1994) Emulsion Polymerization of Aniline. Polymer, 35, 2902- 2906.
26. Armes, S.P., Miller, J.F. and Vincent, B. (1987) Aqueous Dispersions of Electrically Conducting Monodisperse Polypyrrole Particles. Journal of Colloid and Interface Science, 118, 410-416.

27. Armes, S.P. and Aldissi, M. (1990) Preparation and Characterization of Colloidal Dispersions of Polypyrrole Using Poly(2-vinyl pyridine)-Based Steric Stabilizers. Polymer, 31, 569-574.
28. Hayashi, S., Takeda, S., Kaneto, K., Yoshino, K. and Matsuyama, T. (1987) Radiation Induced Effect in Conducting Polymers. Synthetic Metals, 18, 591-596.
29. Tang, M., Wen, T.Y., Du, T.B. and Chen, Y.P. (2003) Synthesis of Electrically Conductive Polypyrrole-Polystyrene Composites Using Supercritical Carbon Dioxide I. Effects of the Blending Conditions. European Polymer Journal, 39, 143-149.
30. Cao, Y., Andreatta, A., Heeger, A.J. and Smith, P. (1989) Influence of Chemical Polymerization Conditions on the Properties of Polyaniline. Polymer, 30, 2305-2311.
31. Myers, R.E. (1986) Chemical Oxidative Polymerization as a Synthetic Route to Electrically Conducting Polypyrroles. Journal of Electronic Materials, 15, 61-69.
32. Sun, B., Jones, J.J., Burford, R.P. and Skyllas-Kazacos, M. (1989) Stability and Mechanical Properties of Electrochemically Prepared Conducting Polypyrrole Films. Journal of Materials Science, 24, 4024-4029.
33. Dutta, K., Mahale, R.Y., Arulkashmir, A. and Krishnamoorthy, K. (2012) Reversible Assembly and Disassembly of Micelles by a Polymer that Switches between Hydrophilic and Hydrophobic Wettings. Langmuir, 28, 10097-10104.
34. Das, C. and Krishnamoorthy, K. (2014) Disassembly of Micelles in Nanoscopic Space to Prepare Concentric Nanotubes with Variable Hydrophobic Interiors. Chemical Communications, 50, 5905-5908.
35. Khan, R. and Dhayal, M. (2009) Chitosan/Polyaniline Hybrid Conducting Biopolymer Base Impedimetric Immunosensor to Detect Ochratoxin-A. Biosensors and Bioelectronics, 24, 1700-1705.
36. Ansari, A.A., Khan, R., Sood, K.N. and Malhotra, B.D. (2009) Polyaniline-Cerium Oxide Nanocomposite for Hydrogen Peroxide Sensor. Journal of Nanoscience and Nanotechnology, 9, 4679-4685.
37. Kaushik, A., Khan, R., Gupta, V., Malhotra, B.D. and Singh, S.P. (2009) Hybrid Cross Linked Polyaniline-WO_3 Nano- composite Thin Films for NO_x Gas Sensing. Journal of Nanoscience & Nanotechnology, 9, 1792-1796.
38. Kaushik, A., Kumar, J., Tiwari, M.K., Khan, R., Malhotra, B.D., Gupta, V. and Singh, S.P. (2008) Fabrication and Characterization of Polyaniline-ZnO Hybrid Nanocomposite Thin Films. Journal of Nanoscience & Nanotechnology, 8, 1757-1761.
39. Khan, R. (2011) Supported TritonX-100 Polyaniline Nano-Porous Electrically Active Film onto Indium-Tin-Oxide Probe for Sensors Application. Advances in Chemical Engineering and Science, 1, 140-146.
40. Fink, J. and Leising, G. (1986) Momentum-Dependent Dielectric Functions of Oriented Trans-Polyacetylene. Physical Review B, 34, 5320-5322.
41. Kazim, S., Ali, V., Zulfequar, M., Haque, M. and Husain, M. (2006) Electrical, Thermal and Spectroscopic Studies of Te Doped Polyaniline. Current Applied Physics, 7, 68-75.
42. Aldissi, M. and Armes, S.P. (1991) Colloidal Dispersion of Conducting Polymers. Progress in Organic Coatings, 19, 21-58.
43. Prasad, K.R. and Munichandraiah, N. (2001) Potentiodynamic Deposition of Polyaniline on Non-Platinum Metals and Characterization. Synthetic Metals, 123, 459-468.
44. Cooper, J.C. and Hall, E.A.H. (1992) Electrochemical Response of an Enzyme-Loaded Polyaniline Film. Biosensors & Bioelectronics, 7, 473-485.

Index